MW00838120

AIR-BORNE

Also by Carl Zimmer

Life's Edge

She Has Her Mother's Laugh

A Planet of Viruses

Evolution: Making Sense of Life

The Tangled Bank

Science Ink

Brain Cuttings

The Descent of Man: The Concise Edition

Microcosm

Soul Made Flesh

Evolution: The Triumph of an Idea

Parasite Rex

At the Water's Edge

AIR-BORNE

The Hidden History
of the Life We Breathe

CARL ZIMMER

DUTTON

DUTTON

An imprint of Penguin Random House LLC
1745 Broadway, New York, NY 10019
penguinrandomhouse.com

LIBRARY OF CONGRESS CATALOGING-IN-PUBLICATION DATA
has been applied for.

ISBN 9780593473597 (hardcover)
ISBN 9780593473610 (ebook)
ISBN 9798217046324 (export)

Printed in the United States of America
1 3 5 7 9 10 8 6 4 2

BOOK DESIGN BY LAURA K. CORLESS

The authorized representative in the EU for product safety and compliance is Penguin Random House Ireland, Morrison Chambers, 32 Nassau Street, Dublin D02 YH68, Ireland, https://eu-contact.penguin.ie.

To Charlotte,
who has delighted her parents from her first breath

Sky is omnipresent
even in darkness under the skin.

—WISŁAWA SZYMBORSKA

contents

PART 3
Afterlife

PART 4
Resurrection

PART 5
Wuhan and Beyond

prologue

THAT'S WHERE IT IS

The musicians walked onstage first: three carrying violins, one a cello. The fifth sat down at a grand piano. They were greeted with waves of sound, as the audience in the auditorium struck their hands together in applause. The applause grew stronger as dozens of singers streamed out from both wings, the men in black tuxedos, the women wearing bright scarves draped over black dresses. Older singers, some walking with canes, settled down in the two rows of chairs behind the musicians. The younger ones stepped onto the low bleachers behind the chairs.

Now all in place, they faced the audience. Out of the ninety people assembled onstage, four wore masks.

There were about a hundred seventy people in the audience on the night of May 6, 2023: a gathering of friends, families, and unconnected lovers of music from the northwestern corner of Washington State. They had traveled to McIntyre Hall in Mount Vernon for the spring performance of the Skagit Valley Chorale. One member of the choir taught fifth grade, and her tween fan army, done up in rhinestones and taffeta, buzzed overhead in the balcony. Some people in the audience wore pale blue surgical masks that fit loosely over their mouths. Others wore N95s that sealed tight. What would have seemed strange in 2019 seemed fairly normal four years later.

As the applause died down, a short woman with a gray pageboy walked to the front of the stage. She unclipped a microphone from its stand and introduced herself as Ruth Backlund. A retired high school French teacher and the president of the chorale's board, Backlund welcomed the audience.

"I have two thoughts about gratitude," she said. "I'm very grateful for Skagit County, which runs from the mountains to the sea. If you look in the woods, everything is blooming. And I am grateful for all the scientists who made the vaccines that let us be here."

The audience clapped again. Backlund introduced Yvette Burdick, the choir director. "She is the best teacher I ever had," Backlund said. Burdick strode onstage in a flowing black pantsuit and low gray pumps.

"Thank you," Burdick said with a quick bow. "We are going to start right off."

The choir began with a hymn. At Burdick's cue, the singers dropped their diaphragms. They inhaled the concert hall air deep into their lungs, into the fine alveoli at the deepest tips of their airways. The oxygen in the air seeped into their bloodstreams, while carbon dioxide outgassed. The singers then let their lungs deflate, and the altered air made its way back up through their bronchi, into their tracheas, and through their larynxes. Bands of muscle buzzed in the upwelling breeze and produced a spectrum of sound. The singers set their mouths into different shapes, to sculpt the acoustic waves as they escaped.

The sound raced across the hall, the waves jostling molecules of air and bouncing off walls. The waves ended up in our auditory canals, making our eardrums vibrate and generating electrical signals that entered our brains, where they produced the perception of sound. The physics of the air joined us in a communion as the choir shared songs about gratitude: for the Earth, for liberation from slavery, for love. "Place me like a seal over your heart, like a seal on your arm," the vibrations told us.

○ ○ ○ ○

On May 5, 2023, the day before the spring concert, the World Health Organization made a major announcement. Speaking in Switzerland at a press conference, WHO director Tedros Adhanom Ghebreyesus declared that Covid-19 was no longer a public health emergency of international concern. Three years and five months had passed since the coronavirus SARS-CoV-2 emerged in Wuhan, China. Covid-19, a disease never seen before, became the worst public health disaster of modern times, infecting the majority of people on Earth. By the time Tedros made his announcement, it had killed about 25 million of them.

Some of the first people in the world to get Covid-19 stood before us on the McIntyre Hall stage. On March 10, 2020, fifty-eight members of the Skagit Valley Chorale had become infected at a rehearsal. Before the month was out, three were in the hospital. Two of them died.

The outbreak brought horror to the choir, and also shock. They knew that some diseases can spread in droplets slathered on doorknobs, or fired at close range in coughs and sneezes. But subsequent research would reveal that the Skagit Valley Chorale outbreak was likely spread on a song. An infected singer released an invisible cloud of droplets so tiny that they resisted gravity and floated like smoke. She did not cough or sneeze to unleash the viruses: they escaped with every breath. Covid-19, in other words, was airborne.

Three years later, at the May 2023 concert, the members of the Skagit Valley Chorale were once again releasing fine droplets from their airways. The audience, listening silently, exhaled them as well. It is an inevitable part of breathing. Some of the tiny droplets drifting through the hall carried living things. Some carried harmless bacteria that feast on the traces of meals left in people's mouths long after they leave the dinner table. Some droplets carried viruses that infect bacteria that dwell in our lungs. A few harbored fungal spores.

My wife, Grace, and I sat a few rows from the stage. We hoped that no one around us was emitting a pathogen that we might inhale. While the Covid-19 emergency had just ended, the coronavirus that caused it had become a part of our lives. "This virus is here to stay," Tedros had warned the world at his press conference the day before. "It is still killing, and it's still changing. The risk remains of new variants emerging that cause new surges in cases and deaths."

If the air did indeed harbor Covid-19, we could hope that the concert hall's ventilation system would protect us. It flushed out some of the indoor air—along with the droplets and the carbon dioxide—and replaced it with fresh air from outside. I could not see this invisible traffic, but I could track it. From my coat pocket, I discreetly slipped out a white plastic box the size of a pack of cards. It displayed a number: 527.

In other words, the concentration of carbon dioxide in McIntyre Hall was 527 parts per million. Outside, the level was hovering a little lower, around 420. A puff of exhaled air coming out of a mouth has a concentration of 40,000 parts per million. Puff after puff, the choir and the audience steadily added carbon dioxide into the auditorium. If anyone was exhaling SARS-CoV-2, the viruses would accumulate around us as well. As the songs progressed—from "Jesu, Joy of Man's Desiring" to a poem by Theodore Roethke—I checked the monitor. It rose to 662, then 800. If it got much higher, I thought to myself, I would put on my mask.

○ ○ ○ ○

I did not own a carbon dioxide monitor before the Covid-19 pandemic. I did not think much about the atmosphere that we share in concert halls and kitchens and subway cars. Few people did. In the first weeks of 2020, Ruth Backlund certainly gave it no thought. Each Tuesday, she and her husband, Mark, a retired psychiatrist, rehearsed with the Skagit Valley Chorale. They would drive east from their home on Fidalgo Island onto the mainland. They would pass flat fields of tulips, daffodils, and potatoes until they reached Mount Vernon, a city of thirty-five thousand on the banks of the Skagit River. To the east, the Cascade Mountains loomed. The Backlunds made their way to the Presbyterian church on the edge of town, a building shaped like a wedge of cake turned on its side. They got out of their car and walked into Fellowship Hall, a low-slung room extending off one side of the church.

The Backlunds always looked forward to working with Burdick, who drove up from Seattle for the rehearsals. She held the singers to high standards, but she never became a dour disciplinarian. "If people are uncomfortable or unhappy, they can't sing very well," Burdick told me. "I have a tendency to be the Pied Piper. We all go off to Music Land."

When Burdick first heard about Covid-19, it seemed like a faraway disease. It certainly didn't make her rethink meeting with the choir each Tuesday. In February, when reports surfaced that Covid-19 was striking a nursing home sixty miles south of Mount Vernon, it sounded like a variation on influenza. The flu came every winter, and no one made a major change to their life when it did. The Centers for Disease Control and Prevention told the public that they could avoid getting influenza by keeping six feet away from people displaying symptoms. The droplets released in coughs and sneezes were heavy enough that they would quickly fall to the floor. If those droplets got onto a doorknob or a turnstile or some other surface, the viruses might survive long enough to be picked up by someone else. But they could be readily stopped simply by cleaning the surfaces people frequently touched.

When Covid-19 first emerged in Washington, Governor Jay Inslee applied the same public health measures to it that he might have applied to influenza. He cut down on visits that outsiders could make to nursing homes, so that they wouldn't spread SARS-CoV-2 to the residents. The rest of Washington went on with a normal winter.

Nevertheless, some members of the Skagit Valley Chorale grew uneasy. They asked about wearing masks at rehearsals, since no vaccines yet existed for Covid-19. If the singers got infected, they would have no immune response ready to fight it off. Other members of the choir recoiled at the idea of covering their mouths with masks as they sang. Hazy fears ought not rob the choir of an experience they all cherished. The choir compromised with an agreement to be prudent. "We just followed what the CDC was saying," Ruth Backlund told me.

Burdick sent out emails advising anyone who had symptoms linked to Covid-19—coughing, fever, or shortness of breath—to skip rehearsals. Anyone who might have been at greater risk of getting severely ill, whether they were getting treated for cancer or had diabetes, should consider staying home. More than a hundred people converged on Fellowship Hall for a typical rehearsal. On March 3, 2020, only seventy-eight people came.

A week later, on March 10, Skagit County health officials notified a retired nurse's aide named Susanne Jones that she was their first confirmed case of Covid-19. Jones had gotten tested after she learned that a friend with whom she had recently square-danced had died of the disease. "Suddenly my annoying allergy symptoms seemed horrifying," she later told a reporter. "And I thought of all the places I'd been."

County officials did not close down schools or stores. On the afternoon of March 10, the health department simply posted an update on its website. Howard Leibrand, Skagit County's health officer, urged people to do what they could to slow the spread of Covid-19. "The community should postpone non-essential events and gatherings of ten or more people," he said.

The 130,000 residents of Skagit County are spread thinly across nineteen hundred twenty square miles—an area larger than Rhode Island and not much smaller than Delaware. The county health department expected its recommendations to take a few days to seep across the region. When Burdick and sixty singers converged that evening on the Mount Vernon Presbyterian Church, none of them knew that someone in Skagit County had Covid-19.

Three members of the choir arrived early to set up for practice. They switched on the lights and turned on the furnace, setting the thermostat to 68 degrees. They arranged a hundred twenty chairs. Backlund recalled the choir being a little on edge that night, without the usual hugs and handshakes. "You felt like you were looking over your shoulder, but we were being really careful," she said.

At about six thirty, the singers sat down in the same seats they took every week. But that night the hall was half empty. The doors were shut. Once the temperature rose high enough, the furnace shut off, and air stopped flowing out of the ventilation system. The hall filled with singing—not with sneezes or coughs or blowing noses. No one displayed any symptoms.

For forty minutes, the choir sang unmasked. Burdick then took half the singers to the church's sanctuary to practice a number. The other half stayed behind in Fellowship Hall and rehearsed "The Shoop Shoop Song," a 1964 rhythm-and-blues hit by Betty Everett.

. . . It's in his kiss
That's where it is . . .

After forty-five minutes, Burdick's group returned to Fellowship Hall for a ten-minute break. Some singers converged around a table, grabbing oranges for a snack. Some headed to the bathroom. One woman left during the break. Carolynn Comstock heard from another singer that she had a sore throat. The rest of the singers rehearsed for another fifty minutes before finishing at nine. The singers folded up their own chairs and loaded them on a rolling rack before leaving. Comstock let her husband, Jim Owen, fold her chair for her. At sixty-two, she had retired from teaching but was helping Owen at his construction company. She had recently injured her shoulder. The singers went home. The church went dark.

The next day, Governor Inslee held a press conference to deliver an update on Covid-19. The state's cases were climbing. To check the spread, Inslee announced a ban on gatherings of more than two hundred fifty people in three counties in Washington. He also recommended that everyone across the state engage in an unfamiliar practice with an odd name: social distancing. "Individuals should try to stay six feet or at least an arm's length from each other," he announced. Beyond that distance, Inslee implied, people would be safe from infection. Burdick, Backlund, and the rest of the board decided to cancel their next rehearsal.

Three days later, Mark Backlund began to feel low. He shuffled off to take a nap. The phone rang, and Ruth picked it up. It was Comstock. She told Ruth that she couldn't come over that night because she was running a fever of 100 degrees. As the evening progressed, Ruth was overwhelmed by what she called "a weird fever feeling."

She had had her share of fevers before, but this experience felt different. "I thought, 'Oh boy, is this it?'" Ruth said. "I thought it couldn't be, because we were so careful about everything you're supposed to be careful about."

It was indeed Covid-19. Dozens of singers who had attended the March 10 practice fell ill. Before the end of the month, Nancy Hamilton and Carole Woodmansee were dead.

Ruth was both horrified and baffled. The choir had followed all the rules, but Covid-19 had managed to slip into their midst. Whoever had brought it to Fellowship Hall had not spread it to neighbors by contaminating doorknobs or by coughing droplets that fell to the floor within a few feet of others. "You couldn't all be that far apart and getting sick at the same time," Ruth said. It seemed to her that the virus must have wafted through the air.

Two weeks after the March 10 rehearsal, the World Health Organization swatted down that notion. On their Twitter account, the agency wrote:

> FACT: #COVID19 is NOT airborne.
>
> The #coronavirus is mainly transmitted through droplets generated when an infected person coughs, sneezes or speaks.
>
> To protect yourself:
> - keep 1m distance from others
> - disinfect surfaces frequently
> - wash/rub your 👐
> - avoid touching your 👁 👃 👄

The day after that tweet, the Skagit Valley Chorale outbreak became international news. A story in the *Los Angeles Times* drew the world's attention to the singing group nestled in a valley that wasn't known for much besides an annual tulip festival. CNN interviewed Burdick about the experience. Strangers sent a blast of hate mail. They said the singers had blood on their hands.

The first lesson people took from the Skagit Valley Chorale outbreak was just how easily one person could infect many others. But in the months that followed, Ruth Backlund and her fellow survivors agreed to collaborate on a scientific study that helped establish something just as important. Contrary to what WHO claimed, the study concluded Covid-19 was airborne.

○ ○ ○ ○

I am a journalist, and diseases are one of my beats. I became aware of the new virus in early January 2020, while it was still in China. By late January, a few scientists were predicting a pandemic. I started warning my friends to brace for a possible disaster. Like a paranoid doomsday prepper, I advised them to store extra toilet paper and canned food. When someone called me to plan a meeting in June, I told him meetings might not exist in June.

I was right in some ways and very wrong in others. Like the Skagit Valley Chorale, I did not concern myself with the air. If I stayed a few feet away from strangers, I'd be safe from any viruses they coughed or sneezed. The droplets they expelled would fall to the floor like ball bearings. The most worrisome risk seemed to lurk on surfaces: the skin of my hands, which I washed many times a day; the grocery bags that I disinfected with Clorox wipes.

Over the following months, I absorbed the growing consensus that Covid-19 was in fact airborne. As I recognized that floating droplets could transmit the virus from one person to another, I traded Clorox wipes for a carbon dioxide monitor. Masks became a staple. I also began to think about the air differently, as a gaseous ocean in which we all live, which infiltrates our bodies, which our own bodies transform and then return to the great transparent sea, that contains exhaled viruses that can then be inhaled. But I was also left with a question: how could such a fundamental mystery about the worst public health disaster in a century go unsolved for so long?

Once the pandemic passed its peak—after most people on Earth got infected, vaccinated, or both—I started looking for an answer. It became clear that for thousands of years the atmosphere had been an intimate, enveloping mystery. For hundreds of generations, scholars and physicians had claimed the air itself could turn dangerous. They gave bad air an assortment of names, such as miasma. Miasmas could be caused by the stars or swamps; they could spread down a street or float for hundreds of miles. When modern Western medicine took shape in the late 1800s, scientists and doctors alike tossed miasmas aside, treating them like an embarrassing relic of the Dark Ages, a concept with as much value to medicine as bleeding patients. They knew that germs spread diseases, and they knew that germs spread primarily through food, water, sex, and touch, as well as through coughs and sneezes. Germs were not airborne.

But in the 1930s, a few scientists challenged this consensus. They argued that diseases could indeed spread on currents, that germs could float for hours like smoke. They recognized that airborne pathogens posed a fundamentally different threat than the one posed by short-range coughs and sneezes. They argued that some of the worst diseases known to humanity, such as tuberculosis and influenza, spread this way. Those scientists helped create a new field: the science of airborne life. They called it aerobiology.

The aerobiologists were a motley crew. As some tracked pathogens floating inside schools and subways, others caught microbes soaring through the sky. They dazzled the world by finding spores as high as the stratosphere. The founders of aerobiology hoped their new science would unify all life of the air, whether indoors or outdoors, and make clear that the airborne diseases that afflict us are just a few species among a vast floating menagerie.

Today, aerobiologists look at the atmosphere as one of the three great habitats of life. It didn't start out that way: when the Earth formed 4.7 billion years ago, a blanket of lifeless air formed from the gases hissing out of the molten planet. Life started off aquatic—some theories point to the young ocean as its nursery, others to freshwater ponds—but it did not stay restricted to water for long. Waves sprayed droplets containing bacteria and viruses into the air. About 2 billion years ago, the ancestors of algae and other single-celled forms of life also leaped from the water and traveled for hundreds or thousands of miles.

When life spread to land, the air filled with new species. The winds scattered mats of terrestrial microbes, and then plants and fungi began releasing spores into the wind. Later, some plants evolved flowers that released pollen grains. Their airborne journeys became part of the recipe for their enormous evolutionary success. The greening land also lured animals ashore. To get their oxygen, they adapted to breathing air rather than pumping water through gills. And then some animals—insects first, then birds and bats—evolved to move through the air, with leaping legs and flapping wings.

The animals became, in turn, hosts to another kind of airborne life: pathogens that floated from one host to another. Hantaviruses, for instance, infect rodents and then escape in their urine and saliva. On the ground they can survive in dried dust. Days later, a breeze can pick up hantaviruses and carry them into the nose of another rodent visiting the same spot, causing a fresh infection.

Other pathogens turned the lungs of air-breathing animals into both a home and a launching pad. They get drawn into a host with an inhaled breath. Animals often react to a respiratory infection with an onslaught of immune cells and inflammation. This attack leaves an extra supply of mucus in the airway. To clear it out, the animals will use their lungs to deliver a powerful cough or sneeze. The contaminated mucus droplets can then strike other animals or contaminate the ground. But even regular breathing can be enough to spread some microbes. When an animal exhales air, the outgoing flow pulls droplets off the moist walls of the lungs, like a breeze passing over the ocean. Those droplets can evaporate down to droplet nuclei and float away.

Airborne diseases also took advantage of the social lives of animals. As some species evolved to live in close groups—in nests, burrows, flocks, and herds—they made it easier for a cloud of exhaled pathogens to infect a new host. It's likely that airborne diseases have fared best among animals that live together. If a solitary creature breathes out microbe-laden droplets, they may fail to reach another member of its species before they fall to the ground or get damaged by sunlight. In a herd or a den, a sick animal can release clouds of pathogens that have better odds of getting inhaled by another nearby host.

Measles, the most infectious pathogen ever found, belongs to a family of viruses that typically infect grazing mammals. Some infect seals. While seals spend much of their lives out at sea, they also haul out onto beaches where, huddling together in groups, they mate, raise their young, and breathe viruses on one another. Dolphins get their own form of measles too. While they never come ashore, they still have lungs and breathe through blowholes—a legacy of their terrestrial ancestors, which lived on land 50 million years ago. Swimming in pods, they surface together to exhale blasts of air and suck in new ones. The measles virus takes an airborne hop before the dolphins dive underwater again.

When aerobiology emerged in the 1930s, it generated great excitement, but within a few years it faltered. In World War II, the United States and other countries recruited aerobiologists to make biological weapons. And when the war ended, the aerobiologists kept on growing pathogens to wipe out cities and starve nations. A shroud of secrecy fell across much of aerobiology. Even today, the science is not entirely free of it.

In those postwar years, some aerobiologists tried to persuade public health officials to take the threat of airborne infection seriously. They largely failed. Infectious disease experts who led the fight against outbreaks and prepared for the emergence of new diseases mostly ignored the aerobiologists, even when it meant accepting some basic mistakes about the physics of air.

The Covid-19 pandemic finally rattled that consensus. In so doing, it provided an opportunity to rethink our history with the air. The Covid-19 pandemic was not a fluke. It belongs to a deep history of airborne life, one that has adapted with astonishing efficiency to our species's rapid rise—from the dawn of agriculture ten thousand years ago to the rise of cities, to the Industrial Revolution, and now to the twenty-first century's megacities and decimated wilderness. SARS-CoV-2 is only one species in an airborne habitat that we largely ignore, but would do well to understand.

○ ○ ○ ○

The Skagit Valley Chorale singers finished their concert with the Sanctus—"Heaven and earth are full of thy glory"—and then took their bows. As the audience cheered, I checked my monitor. The level of carbon dioxide had reached 903 parts per million. In our communion, we had altered the air.

After the applause ended, Grace and I filed out into the high-ceilinged lobby. We congratulated Burdick and the Backlunds, passed by the fifth graders swarming their teacher as if she were Taylor Swift, pushed open the outer doors, and walked into the night air. It was the same air we had just breathed inside McIntyre Hall, the same seamless blanket of gases. The only difference now was there was no ceiling to hold it down. I could exhale all the carbon dioxide I wanted, but my CO_2 monitor would not budge. A few miles overhead, the moisture in the atmosphere formed clouds that blocked our view of the stars. Above the clouds lay the stratosphere, a huge realm of thin gases that rises to about thirty miles over our heads, where it gives way to the mesosphere and the exosphere, the edges of the Earth's air where meteors glint as they die. Scientists now refer to the life that teems in this space as the aerobiome.

The morning after the concert, as Grace and I drove out of the Skagit

Valley, I tried to envision the aerobiome in full. Some parts of it were plain to see: a crane flapping its wings alongside the road, dragonflies darting over the meadows. But the Skagit Valley's aerobiome was mostly invisible. It was the pollen blown from the daffodils and the tulips. It was the spores of fungi rising from the tilled fields, the viruses belched by the cows. Life was rising from the soft floors of the forests that jacketed the mountains in the distance. It rose from the Pacific just beyond our western horizon, much as it had billions of years ago. Even without wings, the microbes were gliding on the wind, rising toward the stratosphere, infiltrating clouds, where they would make rain and snow and hail. Grace and I were traveling along the floor of a living ocean, one that we drew into our lungs with every breath.

PART I

To the Stratosphere

one

THE FLOATING GERMS

When Louis Pasteur died in 1895, France treated him with the kind of reverence it had once reserved for kings and saints. The streets filled with silent mourners, who watched a hearse transport his body from his home to the Pasteur Institute. The sprawling complex of buildings was nicknamed the "Rabies Palace" in honor of the vaccine Pasteur had created in 1885. Now it was prepared to receive Pasteur's body. "The entire facade is draped in mourning colors," an American correspondent wrote from Paris. "A shield three metres in diameter and bearing the initial 'P.,' surrounded by a wreath of laurel, surmounts the door." For the next four days, Parisians filed into the institute to pay their respects to Pasteur as he lay in state in the library. His body was then loaded into a mortuary car, which was pulled by six horses to the Cathedral of Notre-Dame for his funeral. The procession that followed his body included statesmen, diplomats, generals, and poets. It stretched for a mile.

The eulogies that followed painted the same picture: of a great laboratory scientist who carried out experiments that helped establish the germ theory of disease. That theory then allowed Pasteur to invent vaccines for rabies, as well as for many other diseases. Pasteur's principles made lethal infections far less likely during surgery. They made milk safe for children to drink.

It's likely that many of the mourners who listened to the memorials had an actual picture of Pasteur in their minds. In the decade before he died, portraits of the great scientist appeared everywhere. No scientist before Pasteur had enjoyed so much iconography. His portraits were printed in newspapers around the world. They hung on museum walls. The image in all the pictures was the same: a man indoors, at home among microscopes and bell jars. "Everything gets complicated away from the laboratory," Pasteur once complained to a friend.

In the most famous portrait, painted in 1885 by Albert Edelfelt, Pasteur stands in his laboratory. He is sixty-two, with a short gray beard and dressed in an ordinary suit. He leans his left elbow on a massive tome that rests on a crowded lab counter. In his right hand he holds a jar at which he stares intently. Hanging inside the jar is what looks like a red twig. It is the spinal cord of a rabbit. Pasteur had the rabbit infected with rabies, and after it died, he removed the spinal cord to let it dry. From the red twig, he later isolated weakened rabies viruses that he would use to make the world's first rabies vaccine.

But we should also try to imagine another portrait of Pasteur, one that he never posed for. It is 1860, and Pasteur—his beard still black—has traveled across France and climbed a glacier. Picture him standing on the blinding summit of ice, holding a glass globe toward the brilliant sky.

To reach the glacier, Pasteur traveled to Chamonix, a village at the base of Mont Blanc in the Alps. There he met with a guide, and the two of them headed up a trail into dark stands of pines. They were accompanied by a mule laden with baskets of long-necked glass chambers that sloshed with broth. They walked up the steep trail until they finally reached the Mer de Glace: the Sea of Ice.

The glacier resembled a frozen torrent of whirlpools. In some places the ice was bluer than the sky. The wind blew briskly over the glacier, as the vale echoed with the sound of frozen boulders crashing down the slopes. Pasteur struggled to make out the path in the glare of sunlight bouncing off the ice field.

When he reached an altitude of two thousand meters, Pasteur finally stopped. With the help of his guide, he removed one of the glass chambers from the mule's pack. He etched a groove around the sealed neck with a steel point. Pasteur then lit a spirit lamp and moved the flame under the groove to make it even thinner. He raised the glass bulb over his head, like an offering to the glacier gods. He grabbed a pair of tongs with his free hand and used them to snap off the end of the neck. The glacial air rushed inside the bulb. Pasteur then lowered it from above his head and used the spirit lamp to melt the neck closed again.

The sight of Louis Pasteur holding a glass globe of broth over his head at the top of a glacier would have baffled other travelers who might have been visiting the Mer de Glace that same day in 1860. They might have concluded that Pasteur had been driven insane by the glacier's towering indifference. It was the sort of place that inspired thoughts of madness, after all.

When the writers Mary Wollstonecraft Godwin and Percy Bysshe Shelley walked the same path forty-four years earlier in 1816, they had been overwhelmed by the glacier's scale. "The immensity of these aerial summits excited, when they suddenly burst upon the sight, a sentiment of extatic wonder, not unallied to madness," they later wrote.

The sight haunted Mary so much that she wrote it into *Frankenstein.* After the monster escapes from Victor Frankenstein, its creator visits the Mer de Glace. There he spots a grotesque form race over the ice. "I perceived, as the shape came nearer (sight tremendous and abhorred!) that it was the wretch whom I had created," Frankenstein recalls.

Pasteur was searching the glacier for monsters of his own. He was trying, for the first time in history, to capture microscopic life overhead. His quarry was, in his words, "the germs that float in the air."

○ ○ ○ ○

Pasteur's journey to the ecstatic wonders of the Mer de Glace was thousands of years in the making. From antiquity onward, people had been arguing about what was in the air.

Some philosophers in ancient Greece believed that the air reached up to Olympus, where the gods breathed a purified form, known as aether. The philosopher Anaximenes argued that air was the fundamental element of the world: everything else, from rocks to souls, was just a derivation of it. Other philosophers saw air not as the sole element, but one of four. Along with fire, earth, and water, air helped to produce all the things on Earth and in the skies.

Whatever air was, everyone agreed that it was essential to life. "Life and death are bound up with the taking in and letting out of the breath," Aristotle once observed. The physician Hippocrates and his followers believed that breaths combined with food and water to form the body's four humors. The balance of the humors was the essence of health, and it was the job of doctors to bring them out of imbalance. Changes in the air were especially good at throwing the humors into disarray, Hippocrates believed. Heat and cold, damp and dryness: they all made people sick. But the most dangerous of these threats was an invisible corruption of the air, which Hippocrates called a miasma.

The word had existed for generations. Originally it referred to a moral stain—a defilement that could be washed away only with purification. But in the fifth century BC, Hippocrates and his followers reimagined miasmas as stains on the air. They were the result of natural causes rather than crimes against the gods. Scholars created a long list of causes for miasmas. A conjunction of the planets could send a foul smell down to Earth. The ground unleashed odors. Stagnant water in marshes gave off fumes. So might a rotting corpse.

Despite their invisibility, Hippocrates was confident that miasmas were real. The way that a miasma sickened its victims was proof. "It attacks everyone, young and old, women and men, and, without distinction, those who drink wine and those who drink water, those who eat barley bread and those who eat wheat bread, those who do a lot of exercise and those who do little," Hippocrates wrote. What these victims all had in common was that they breathed the same corrupted air.

Hippocrates taught that different miasmas caused different diseases. They could even strike different species. "When the air is full of miasmas, whose properties are hostile to human nature, this is when men are ill," he wrote. "But when the air is not suitable for another type of living beings, these beings are then ill." After all, animals were made up of the same four humors that made up humans, and so they were no less vulnerable to miasmas.

The same was even true of plants. Early farmers were intimately aware of the health of their crops, although they didn't understand the causes of many of the diseases that wiped out their fields. One of the worst diseases they faced was rust, which could quickly turn a wheat field black. We know now that rust is caused by fungi, but early farmers blamed it on divine wrath. When Moses speaks to the Israelites before they enter the promised land, he warns that God will make them suffer for disobedience. One of the punishments on his list is rust. The Romans even had a deity specifically responsible for the disease. They called her Robigo. April 25 was set aside for Robigalia, a festival at which they soothed her wrath with foot races and the sacrifice of dogs. "Take your rough hands away from the harvests, and do not harm the crops," a priest would plead.

The Greek philosopher Theophrastus did not blame the gods for bad harvests. He saw humors as the cause. "For every plant, like every animal, has a certain amount of moisture and warmth which essentially belong to it," he wrote. Losing some of that moisture or warmth made crops decay. "If they fail altogether, death and withering ensue," Theophrastus warned.

The air delivered many of these failures. Scorching sunlight, cold winds, and torrential rains could all bring diseases. Theophrastus wrote that rust primarily occurred at the full moon, when sunshine was followed by dew. "Lands which are exposed to the wind and elevated are not liable to rust, or less so," he declared, "while those that lie low and are not exposed to wind are more so."

○ ○ ○ ○

When the Roman Empire declined, the work of ancient writers like Theophrastus and Hippocrates largely vanished. Their books caught flame. They became meals for worms. They rotted in abandoned libraries. What little survived did so thanks largely to monks who made copies of old texts and stored them away. A small community called Syriacs, scattered across Syria, Iraq, and Iran, translated many of the surviving books. Among their volumes were the writings of the Roman physician Galen. Syriac doctors treated Galen's works as medical textbooks, and they tried to emulate his care for patients. By the ninth century, Muslim emperors were paying scribes to translate Galen into Arabic.

From Galen, Arab doctors picked up ideas about miasmas. Those ideas helped them make sense of the ordinary diseases they saw each day, along with the terrifying ones that struck the Arab world every few generations. The most terrifying of all was the plague. It would strike suddenly, causing thousands of people to develop bulb-shaped swellings and scorching fevers, followed by swift deaths. The plague could hollow out an entire city in a matter of days.

After each plague outbreak subsided, Arab doctors would usually pin the blame on a miasma. When the plague struck Syria in 1258, they looked hundreds of miles away for its cause. Mongols had recently sacked the great city of Baghdad. A hard rain fell on the heaped corpses of hundreds of thousands of its residents, releasing a horrible smell. "This caused a severe epidemic such that it was transmitted to the air and spread to Syria," one Arab historian later wrote. "Many people died from the change in the air and the corruption of the wind."

From the libraries of Baghdad and other Arab cities, the ancient texts made their way to Europe, and miasma wafted along with them. In Italy, the odor from marshes became the sign of *mala aria*—bad air. *Malaria* was the name given to the fevers it produced. Miasmas even shaped the physical layout of medieval cities in Europe. To keep them safe from bad air, local authorities required that waste be cleaned from streets and buildings, that rivers flow briskly, and that the dead be properly buried. Pistoia, a city in central Italy, announced in 1296 that no artisan could make a stink:

> *Since it is civil and expedient for the preservation of people's health that the city of Pistoia be cleared of stenches, from which the air is corrupted and pestilential diseases arise, we establish with this law that no artisan can or must exercise his craft or carry out any work from which stench arises within the walls surrounding Pistoia.*

Despite these measures, the plague swept through Pistoia in 1348. It swept through all of Europe as well, with a spectacular fury that earned the epidemic a name of its own: the Black Death. Ancient teachings about miasmas and humors seemed useless as the Black Death raged across the continent. By some estimates, it killed off half of Europe in just a few years. "The art of Hippocrates was lost," one French doctor despaired in 1350.

○ ○ ○ ○

When European doctors published accounts of the Black Death, many of them kept up the tradition of blaming the air. Some pointed to a conjunction of Mars and Jupiter, claiming that the celestial combination drew bad vapors from the earth and sea. But a competing explanation also attracted followers, even if it wasn't sanctioned by Galen or Hippocrates. In this alternative view of medicine, diseases such as the plague were caused by contagion—a poison that grew inside the sick and then spread to the healthy.

People did not need a philosophical theory of the universe to come to believe in contagions. In the Near East, camel herders could see them at work when mange struck their animals. One camel would lose its hair and grow weak. Then the rest of the herd became sick. The mange seemed to start with the first sick camel, not with the corruption of the air itself.

The ninth-century Baghdad doctor Qusṭā ibn Lūqā recognized that such outbreaks might be the result of contagions. But that was no reason to give up miasmas altogether. Instead, ibn Lūqā struck a balance: he taught that most diseases were caused by foul vapors from fires and swamps, while others spread by contagion. "Infection is a spark that jumps from a sick body to a healthy body, so there appears in the healthy body sickness similar to what appears in the sick body," ibn Lūqā wrote.

The idea of contagion grew strong over the centuries. When the Black Death arrived, some European cities tried to prevent people from bringing the contagion inside their walls. In 1377, the Major Council of Dubrovnik set aside an island and a small town where visitors from plague-ravaged regions had to wait for thirty days before entering the city. If they did not die of the plague in isolation, they were judged free of contagion. Other cities borrowed Dubrovnik's rule. Some forced ships arriving at their ports to keep their sailors on board until it was clear they were not contagious. In Italy, authorities decided thirty days wasn't enough. They increased the isolation to forty days—a change from *trentino* to *quarantino*. Thus the quarantine was born.

The Black Death was the start of a deadly cycle. The plague would sweep back across Europe every few decades for four centuries. Struggling to make sense of the disease, some physicians married miasmas and contagions into a single explanation. People breathed in miasmas and became sick, and then their own diseased bodies produced a fresh supply of poison that wafted out of their skin and drifted away. Some physicians believed that this poison could be emitted from a sick person's eyes. A gaze could act like a magic spell, passing the plague on to healthy victims.

Plague doctors protected themselves from this contagious threat by wearing leather masks that covered their entire heads. They peered through goggles. The front of their masks formed a long beak packed with cinnamon, cloves, and opium, desiccated viper, and the ground remains of human mummies. The plague doctors believed that by breathing in the fragrance, they could block the contagion from entering their bodies.

○ ○ ○ ○

The first European scholar to think deeply about the nature of contagion was the Italian physician Girolamo Fracastoro. He drew inspiration from both ancient books and his own observations. As a physician, he saw many cases of a horrible new disease. It produced a telltale decay "born amid squalor in the body's shameful parts," he wrote. Soon it "began to eat the areas on either side and even the sexual organ." Fracastoro named the new disease syphilis, and he speculated that the contagion spread mainly through sexual intercourse.

Fracastoro tracked outbreaks among animals as well. In 1514, he offered the first known description of foot-and-mouth disease. Fracastoro watched the disease ravage herds of oxen, filling their mouths with pustules and damaging their hooves. In the sixteenth century, Fracastoro had no way of learning that foot-and-mouth disease is caused by a virus, but he recognized that it was spread by a contagion. "It was necessary at once to isolate the infected beast from the rest of the herd, otherwise all became infected," Fracastoro wrote.

A lost poem helped Fracastoro make sense of these outbreaks. In the first century BC, the Roman writer Lucretius composed *On the Nature of Things*, which offered a vision of the world profoundly at odds with those presented by Aristotle or Hippocrates. Lucretius saw the universe as made of a vast swarm of atoms. These tiny invisible particles constantly jostled into new combinations, and along the way they produced life, death, and all the variety of nature.

An Italian scholar rediscovered *On the Nature of Things* in 1417. When Fracastoro later read the poem, he was drawn to the brief passages that Lucretius wrote about diseases. The Roman poet claimed that invisible "seeds" swirled among the atoms that made up the air. "There are many seeds of things which support our life," Lucretius wrote, "and on the other hand there must be many flying about which make for disease and death."

Fracastoro seized on Lucretius's living seeds to create a theory of disease. In his 1546 book, *On Contagion and Contagious Diseases*, he wrote that disturbances among the stars released seeds that then descended to Earth. Some of them then infected animals, others plants—"so very varied are the seeds from the tainted heaven," Fracastoro claimed.

Once a seed invaded a host, it multiplied into new seeds that spread to other victims. Some kinds, like the ones that caused syphilis, could spread only through direct contact. Others were sticky enough to cling to clothes or other surfaces—Fracastoro called these sticky seeds fomites, from the Latin word for tinder. If someone stole a cloak off the corpse of a plague victim, the fomites could pass on the plague to its new owner.

Fracastoro also argued for a third kind of seed: one that traveled by air, sometimes floating over long distances before infecting another person. He knew that this invisible movement might sound like magic, but he assured his readers that these seeds had nothing to do with the occult. An airborne infection was no more magical than onions drawing tears from people across a room.

"The air is the most suitable medium, partly because it very easily receives both its own and foreign infections, and because we have to use it to live," Fracastoro wrote.

○ ○ ○ ○

Centuries later, scientists would celebrate Fracastoro for tracing the first outlines of the germ theory of disease. But they were distorting history.

Fracastoro did not have an enduring influence on his fellow Renaissance scholars. They gave serious consideration to his notion of living seeds, debated it, and then gradually abandoned it.

One reason they gave up on Fracastoro was that he made extravagant claims about the seeds of disease. To explain why syphilis caused such widespread damage in the body, he declared that its seeds were sharp, and their sharpness let them burrow deep inside their victims. Fracastoro relied on his poetic powers of imagination to see the invisible, rather than experiments.

Decades after Fracastoro's death, the invention of the microscope finally began to make invisible seeds visible. In the mid-1600s, the Dutch lens grinder Antonie van Leeuwenhoek improved the power of microscopes until he could make out miniature life-forms swimming in lake water and even in the film coating his teeth. These strange microorganisms put fresh energy into the idea that living seeds were the cause of contagious diseases. By the eighteenth century, a loud faction argued that microorganisms were responsible for plague and tuberculosis in people, for the rinderpest that killed off herds of cattle, for the mold that rotted fruit. These champions of germs—which they sometimes called insects—came to be known as contagionists.

"Mankind, Quadrupeds and Plants seem to be infected in the same manner, by unwholesome Insects, only allowing this Difference, that the same Insect which is poisonous to Man, is not so to other Animals and Plants," the British botanist Richard Bradley declared in 1721. "All Pestilential Distempers, whether in Animals or Plants, are occasion'd by poisonous Insects convey'd from Place to Place by the Air."

o o o o

Despite their sweeping confidence, the contagionists could not win over the opinion of experts. Their opponents—who were sometimes called miasmatists—kept the upper hand throughout the eighteenth century. The miasmatists did not see the need for quarantines against unwholesome insects. Instead, they tried to protect people from bad air. In Britain, they went on a crusade to clear crowded prisons of a deadly disease known as jail fever. Jail fever would eventually turn out to be typhus, caused by bacteria spread by lice. But in the 1700s, prison reformers blamed miasmas, which they tried to eliminate from the jails with ventilation.

The roof of Newgate Prison was endowed with a huge windmill that pumped breezes through the jail. When it failed to stop the outbreaks of jail fever, the miasmatists blamed bad engineering. The worst days for miasmas were those without wind—the same days that the windmill could not turn.

Other architects believed they could stop miasmas by changing the structures of the jails themselves. They created courtyards and air shafts through which the miasmas could flow out on their own, as fresh air came in to replace them. As one reformer put it, the jails became lungs of brick and stone to deliver "the wholesome breath of life in exchange for the noxious air of confined place."

Miasmatists urged that eighteenth-century cities follow the example of Pistoia by clearing filth and draining stagnant waters to keep the air free of miasmas. In Philadelphia, the physician Benjamin Rush complained that British authorities neglected that work, subjecting the city to an "unwholesome atmosphere." After the Revolutionary War, the new American government set out to destroy Philadelphia's miasmas by cleaning its streets, enclosing a creek, and restoring wasteland on the outskirts of the city. "Philadelphia, from having been formerly the most sickly, has become one of the healthiest cities in the United States," Rush boasted.

That reputation disappeared in 1793 with a devastating outbreak of yellow fever. Contagionists blamed the arrival of a ship full of French colonists fleeing the Haitian rebellion. But Rush would have none of it. He knew exactly what was to blame: bags of rotten coffee dumped on a pier. "The first cases of the yellow fever have been clearly traced to the sailors of the vessel who were first exposed to the effluvia of the coffee," Rush wrote. "Their sickness commenced with the day on which the coffee began to emit its putrid smell." The foul air then drifted down alleys and streets, spreading yellow fever along the way.

Five thousand Philadelphians died over the next few months. Rush tried to save as many of them as he could, but he did so by bleeding them and giving them mercury. His patients survived despite his medicine, not thanks to it.

Yellow fever returned to Philadelphia a number of times over the next decade, and each time Rush became more convinced that "the miasmata from our atmosphere" were to blame. Writing in 1805 to former president John Adams, Rush declared the outbreaks were bringing him victory over his contagionist rivals. "A new era has begun in the science of medicine in our city since the appearance of the yellow fever among us," Rush wrote.

These were not the declarations of a crank. When Rush died in 1813, he was hailed as the greatest American physician. "Like a primal fixed star, amid the host of heaven, he shone with a lustre wholly his own," a minister declared at a memorial for Rush. "He has done more good in this world than Franklin or Washington," John Adams wrote.

After his death, Rush inspired a new generation of anticontagionists. They held control of the medical establishment in both the United States and Europe, and they roundly condemned contagionism as bad for both health and business. Quarantines needlessly slowed down trade by forcing ships to stay isolated for weeks before unloading their goods. In 1824, the British physician Charles Maclean declared that quarantines were worse than useless: by trapping people in the very places where the air carried death, they were "little short of willful murder."

○ ○ ○ ○

In the early 1800s, contagionists remained profoundly ignorant about the microorganisms they believed spread diseases. They had yet to link any particular germ to any particular disease. Someone like Maclean could continue to mock the very idea that a germ could spread a disease. In an age dedicated to science, nothing seemed less scientific. "This unknown and incomprehensible power is endowed with the faculties of self-generation, self-annihilation, self-resuscitation, self-transportation, self-propagation, and an immense variety of other capabilities, no less wonderful, which it condescends to exercise for the destruction of mankind," Maclean sneered.

Even as Maclean attacked the contagionists, the world of microorganisms became less unknown, less incomprehensible. In 1824, when Maclean launched his withering attacks, the German naturalist Christian Gottfried Ehrenberg was in Africa on an epic series of journeys that would take him across Arabian deserts and Siberian forests. Along the way, he made the first systematic survey of microorganisms.

Back in Germany, Ehrenberg organized his tiny creatures into a new classification. He named one group of species *Bacteria.* He recognized that microorganisms were the most abundant form of life on Earth, thriving everywhere he looked. He found fossils of microorganisms in chalk and limestone—or, to be more precise, he realized that chalk and limestone were made of their long-dead cells. "Of small things worlds are built," Ehrenberg once declared.

While Maclean sneered at the idea of self-transporting germs, Ehrenberg found evidence that they could take spectacular journeys. They could even fly. Ehrenberg's first clue came from strange black sheets called paper meteorites, which were believed to have fallen from the heavens. They were so strange that museums kept them in their collections along with other oddities of nature. When Ehrenberg examined the paper meteorites, he recognized what others before him had missed: the sheets were actually dried mats of microorganisms. He speculated that they had started out growing in intertidal pools on beaches. Storm winds came along, peeling them from the coast and lofting them high in the atmosphere, where they drifted for hundreds or thousands of miles before dropping down to Earth.

Thinking back on the microorganisms he had encountered in so many places, Ehrenberg reasoned that storms might also sweep up microorganisms from deserts and ponds. If you knew what to look for, you could even find evidence for flying microorganisms in historical records. There were many accounts of blood raining down from the sky. In *The Iliad,* Zeus sends a red downpour as a sign of coming slaughter. In the fourteenth century, a shower of blood in Germany presaged the arrival of the Black Death. Ehrenberg believed that that blood rain was real, but not really blood. It was instead a red mix of dust and microorganisms raised into the sky and then washed down in storms. "How many thousand millions of tons of microscopic life forms might have been lifted and have fallen like meteorites onto the Earth since Homer's blood rain!" Ehrenberg exclaimed.

In 1844, Ehrenberg enlisted Charles Darwin in his search for flying life. Ehrenberg knew that the young British geologist had visited some of the most remote places on Earth during his voyage on the *Beagle*. He wondered if Darwin might have had any samples to spare from places like the Galápagos Islands. Darwin was happy to help, and he added a gift he thought might interest Ehrenberg: a vial of mysterious red dust.

Darwin had collected it twelve years earlier, when he was just starting on his five-year voyage on the *Beagle*. As the ship sailed across the Atlantic toward Brazil, he noticed one day that the sky was turning hazy. The cause, Darwin later wrote, was "the falling of impalpably fine dust." It dirtied the ship as it settled on every surface, and the crew of the *Beagle* complained as the dust stung their eyes. "Vessels even have run on shore owing to the obscurity of the atmosphere," Darwin wrote.

Darwin swept a smidgen of dust into a packet and stored it away in his quarters on the ship for later study. When he returned to England in 1836, he brought the dust back home along with his vast collection of fossils, rocks, and preserved animals and plants to his country estate. As Darwin developed his theory of evolution, he did not have time to even peek at the packet. Ehrenberg's request tickled his memory. He tracked down the dust and sent it to Berlin.

Darwin had suspected that the dust had come from some distant volcanic eruption, but Ehrenberg saw no evidence for that under his microscope. Instead, he was looking at life. Most of the dust was composed of microorganisms covered in shells of silica. Some of them reminded Ehrenberg of freshwater species that lived in South America.

In June 1844, Ehrenberg sent Darwin his preliminary findings. "I am truly astonished at this," Darwin wrote back.

"I will feast on them for a long time yet," Ehrenberg promised him.

Darwin persuaded other naturalists to send their supplies of Atlantic dust to Berlin. All told, Ehrenberg discovered sixty-seven species of microorganisms in the samples. He puzzled with Darwin over where they had traveled from. Ehrenberg favored South America as the source, while Darwin favored Africa. In 1846, Darwin published an account of some of the results of Ehrenberg's analysis.

Fundamentally, it didn't matter if Darwin or Ehrenberg was right. Either way, the microorganisms must have soared around a far swath of the planet, perhaps flying above the clouds. They were one more piece of evidence that Ehrenberg could marshal to show how far microorganisms could travel. In the words of one Victorian scientist, Ehrenberg's studies "proved the actual existence of an atmospheric kingdom of life."

o o o o

While Ehrenberg was examining Darwin's dust, potato farmers in Belgium were discovering a disaster in their fields. The leaves of the plants were turning black; their stalks were rotting to goo; their tubers were giving off a foul stench. Over the next few years, this strange new disease wiped out much of Europe's potato crop and caused a brutal famine. It also became a crucial piece of evidence in favor of contagious germs.

Late potato blight, as the disease came to be called, expanded its circle of destruction from Belgium across Northern Europe in a matter of months. First, farms in France and Germany started suffering huge losses. Then the blackened leaves and stalks of goo appeared across the English Channel, and then across the Irish Sea. No country suffered more from late potato blight than Ireland. For centuries, British landholders had used the country's good farmland to grow cash crops and graze cattle, which were then shipped off the island. Poor farmworkers had to survive on the food they could grow in small gardens on the edges of bogs and on rocky mountain slopes. Potatoes were one of the few crops that could reliably keep Irish families fed. By one estimate, a third of the country's population depended on potatoes alone for almost all their nutrition in the 1840s. When late potato blight came to Ireland, it took away their food, and a million Irish died. Millions more fled overseas.

Late potato blight threw naturalists into a fierce scientific debate about its cause. When microscopists inspected rotting potatoes, they discovered something alive: a new kind of fungus that some researchers called a water-mould. But many experts refused to believe that the fungus caused the disease. One correspondent to *Gardeners' Chronicle* blamed "destruction by absorption of miasm floating in the air." Others speculated that stagnant pools of toxic air were to blame; others suggested electricity in the atmosphere.

But if an atmospheric disturbance made the potatoes rot, the miasmatists did not have a good explanation for why the same water-mould appeared so often on the dying plants. According to one theory, the tissues of the dying plants reorganized themselves and produced the microorganisms, in a process known as spontaneous generation. The idea of spontaneous generation was centuries old by then; some scholars thought that flies and even mice arose from decaying bodies. By the 1840s, naturalists had long given up the idea that animals could form from stray molecules in a corpse, but many still believed that dying tissues could turn into bacteria or fungi. Miles Berkeley, a British vicar who was among the first naturalists to identify the water-mould, rejected spontaneous generation as its cause. He found it not just on rotting potatoes, but healthy ones. "The decay is the consequence of the presence of the mould, and not the mould of the decay," Berkeley said.

The potato famine ended, but the debate over late potato blight did not. Berkeley was joined in his research by Anton de Bary, one of Ehrenberg's former students. De Bary grew water-moulds on potatoes in his laboratory, documenting every bizarre stage in their life cycle. He sprinkled spores on healthy potato leaves and observed them push microscopic tentacles into the plants' cells. He watched them develop into meshes of filaments and then produce miniature treelike structures filled with new spores. The spores were the only cause of late potato blight; De Bary never saw the potatoes decay without them. De Bary's work led him to join Berkeley in rejecting spontaneous generation. Only moulds made moulds.

The contagionists had a question of their own to answer, though: if late potato blight was caused by a microorganism, how had it spread across Europe in a matter of weeks with the speed of a storm? They returned to the air for their answer. Every infected plant produced vast numbers of spores. The rain washed some of them into the soil, where they could infect other plants. But the wind also carried spores away to ravage distant farms.

Berkeley believed that many other species of fungi traveled through the sky as well. Their spores, he wrote, "are wafted about in the air, where they may remain for a greater or less period, till, obeying the natural laws of gravity, they descend in some distant regions. The trade winds, for instance, carry spores of Fungi mixed with their dust, which must have travelled thousands of miles before they are deposited."

Here was a new image of how the air delivered diseases: not with miasmas caused by foul gases, or with seeds from the stars. Instead, the air was a skyway along which invisible organisms flew like wingless birds, searching for a place on land where they could grow.

o o o o

Louis Pasteur got pulled into the debate over spontaneous generation almost by accident, thanks to some spoiled beet juice. Within a few years, he had become so obsessed with the question that it lured him to the top of the Mer de Glace, where he waved a globe of broth over his head.

In 1856, Pasteur was a young chemist teaching at the University of Lille, far from the hub of French scientific life in Paris. He worked on esoteric questions about the chemistry of wine that brought him little public attention. One day the father of a student asked for a favor. Louis-Dominique Bigo-Tilloy owned a distillery where he used yeast to ferment beet juice into alcohol. Something had gone wrong, and the juice was turning sour instead. Aware of Pasteur's work on wine, Bigo-Tilloy asked if he could find the cause of the trouble.

Pasteur accepted the challenge. He looked at it as a chance to learn about how fermentation happens. He taught himself how to use a microscope and peered at slides of spoiled and unspoiled juice. Pasteur spotted a difference between them that was both simple and profound. In the unspoiled juice, he saw yeast—chains of bead-shaped fungi that turned sugar into alcohol. In the spoiled juice, he saw something else: swarms of dark rods.

Further experiments led Pasteur to realize that he was looking at two kinds of microorganisms. He called the dark rods lactic yeast because they turned sugar into lactic acid. (They would later turn out to be bacteria, not yeast.) With his discovery, Pasteur could give Bigo-Tilloy a simple solution for his problem. If his beet juice turned sour, he should boil it. Once Bigo-Tilloy had killed the lactic yeast, he could reseed the juice with his alcohol-producing yeast, and he would be back in business.

The discovery benefited Pasteur as well, drawing praise from Paris's leading scientists. In 1857 he was appointed the administrator and director of scientific studies at the École Normale Supérieure. In later years, the French would treat Pasteur like a secular saint, but at the École he was an academic terror. His dictatorial leadership led seventy-three students to resign. He proposed that the university stop admitting women to science classes. "Their presence is always a nuisance," Pasteur said. In 2022, on the occasion of his two hundredth birthday, a pair of scientists at the Pasteur Institute wrote an essay to mark the event. They called the institute's namesake "unfair, arrogant, haughty, contemptuous, dogmatic, taciturn, individualist, authoritarian, careerist, flatterer, greedy, and ruthless with his opponents."

Pasteur often tried to erase the achievements of rival scientists even as he promoted his own. He presented himself as the first scientist to plumb the nature of fermentation, despite the fact that another scientist, Antoine Béchamp, had reached many of the same conclusions a few years earlier. "The most brazen plagiarist of the nineteenth century and of all centuries: it is Pasteur," Béchamp later grumbled.

○ ○ ○ ○

At the École, Pasteur continued to direct his ruthless brilliance to lactic yeast. "Where do they come from, these mysterious agents, so feeble in appearance yet so powerful in reality?" Pasteur asked. "This is the problem that has led me to study so-called spontaneous generation."

Pasteur considered the possibility that beet juice spontaneously generated lactic yeast, and then cast it aside. A far simpler explanation was that the yeast spread through the air. The microorganisms might float on currents, randomly landing in vats. Pasteur made a passing mention of his opinion in a paper, and he soon received a scolding letter from France's leading proponent of spontaneous generation.

Félix-Archimède Pouchet was twenty-two years Pasteur's senior. He had come late in life to the spontaneous generation debate, having first made a name for himself by building a natural history museum in Rouen, eighty miles from Paris. At the museum, Pouchet carried out experiments on everything from turtle lungs to insect guts. The late potato blight of 1845 seems to have first drawn Pouchet's attention to spontaneous generation. He came down firmly in favor of it, rejecting Berkeley's claim that the blight was caused by an airborne water-mould. In the 1850s, Pouchet built an elaborate theory in which all microorganisms formed from lifeless molecules. "Spontaneous generation is the production of a new organized being that lacks parents and all of whose primordial elements have been drawn from ambient matter," he declared.

In his letter to Pasteur, Pouchet haughtily assured the young chemist that he must be wrong about beet juice. It was the fermenting juice that produced the microorganisms, not the other way around. To show Pasteur the errors of his ways, Pouchet described experiments of his own that he believed provided clear proof of spontaneous generation. When Pouchet placed sterile hay in a sealed flask of germ-free water, the hay produced germs.

Pasteur was not cowed. He informed Pouchet that his experiment had a fatal flaw in its design. He accidentally gave air a chance to enter the supposedly sterile flask. Airborne germs had landed on the hay, Pasteur guessed, and later multiplied. Yet Pasteur also recognized that his own work did not make a compelling case against spontaneous generation. If he wanted to confidently reject it, he would have to run some new experiments.

The private spat soon turned into a public spectacle. France's Academy of Sciences recognized the importance of the spontaneous generation debate and launched a contest with a prize of twenty-five hundred francs for the best study casting new light on the question. Pouchet and Pasteur both signed up.

o o o o

As the experiments unfolded, it was hard for the public to decide which view of life was stranger. Spontaneous generation had the whiff of blasphemy since it did not require divine intervention for the creation of living things. But the idea of the atmosphere teeming with germs—invisible seeds that could turn juice to wine and potatoes to slime—also strained the nineteenth-century mind. In 1860, a French journalist informed Pasteur that he was going to lose the contest. "The world into which you wish to take us is really too fantastic," he said.

To take people into that world, Pasteur would have to catch germs from the air. He worked with glassblowers to invent a series of devices for the hunt. One device—a linked set of thin tubes—let Pasteur draw air through a piece of sterile cotton by sucking at one end like a straw. He could then wash germs out of the cotton and look at them under a microscope.

For another set of experiments, Pasteur poured sterile broth into long-necked flasks. Germs drifted down through the necks and grew in the broth, turning the flasks cloudy. If Pasteur bent the necks down in a swanlike curve, the broth remained clear. The germs in the air could not make their way up through the necks, Pasteur argued.

When Pouchet heard about Pasteur's experiments, he sneered. Did Pasteur really believe that every germ in fermenting and decaying organic matter came from the air? If that were true, Pouchet argued, every cubic millimeter of air must be packed with more germs than all the people on Earth. "The air in which we live would almost have the density of iron," he said.

Pouchet had a point, and Pasteur changed his hypothesis in response. Now he held that germs were not everywhere in the air at all times. They drifted in clouds and were more common in some places than others. To prove his updated hypothesis, Pasteur left his lab to capture germs in his flasks. In August 1860, he opened eleven straight-necked flasks in the courtyard of the Paris Observatory, a seventeenth-century building near the center of the city. Afterward, they all turned cloudy with multiplying germs. Pasteur then went down into the observatory's cellar, where germ-laden dust presumably fell to the floor without any breezes to kick it back up or deliver fresh germs from elsewhere. He opened ten flasks. Only one came to life.

○ ○ ○ ○

Pasteur then left Paris on his hunt. In the summer of 1860, he brought dozens of portable flasks to Arbois, the town in eastern France where he had spent much of his youth. Pasteur wandered the country roads, opening his flasks. He held them over his head in order not to contaminate them with the germs on his clothes. Eight out of twenty flasks he opened in Arbois ended up with microorganisms growing inside. He traveled to a small mountain nearby for another survey. At an altitude of five hundred eighty meters, only five out of twenty flasks sprang to life.

It made sense to Pasteur that he had a harder time finding germs on the mountain. The farther he got from human settlements, the sparser airborne life became. To put that idea to an extreme test, Pasteur imagined taking a hot-air balloon thousands of feet into the air. He expected that as he climbed toward the clouds, he would capture no germs at all. But the more Pasteur thought about taking a balloon ride, the less practical it seemed. He decided instead to travel to the Alps in order to climb the Mer de Glace.

His first foray to the glacier ended in failure. When he reached the ice and tried to seal the flasks, the glare of the sunlight made it impossible for him to see the flame of his lamp. Pasteur fumbled as he tried to hold it under the neck of his flask. He started to worry that he might be contaminating the liquid with germs he carried on his skin or his tools. Pasteur gave up and trudged to a mountain lodge—more of a hut than a lodge, really—to spend the night.

He left the flasks open as he slept. Only in the morning did he remember to reseal them. In little time, all thirteen flasks were rife with microorganisms. That did not surprise Pasteur. He reasoned that the travelers who visited the lodge had brought clouds of germs from around the world.

On the second day, Pasteur left the lodge and tried again. He modified his lamp so that the flame would burn bright enough for him to see it. When he climbed back up the glacier, the experiment worked flawlessly. Pasteur opened twenty flasks and then sealed them up for the trip back to Paris. Only one of them turned cloudy with germs. The other nineteen remained sterile.

Pasteur returned to Paris with his seventy-three flasks, and in November 1860, he brought them to the Academy of Sciences to make his case. He entered the domed auditorium, walked up to the table where the prize committee sat, and laid out the flasks before them. The judges peered at the clear and cloudy broths as Pasteur described his journey. The collection of fermented and sterile liquids, he announced, "gives us the indubitable proof that the inhabited places contain a relatively considerable number of fertile germs."

Pouchet refused to accept the evidence, but he withdrew from the contest anyway. No one is quite sure why, but it's possible that Pouchet suspected the judges were biased against spontaneous generation. Or perhaps he decided he was at a disadvantage as an aging outsider with a bright young light of Parisian science as a foe.

Still, even after Pasteur was awarded the prize in 1862, Pouchet continued to spar with him. He climbed mountains with flasks of his own and claimed his own results favored spontaneous generation. Pasteur brushed off the expeditions as sloppy work. The rivalry remained so intense that the Academy of Sciences set up a new commission to evaluate their new work. Pouchet delayed the proceedings over and over again, demanding more time to run additional experiments. As the deliberations at the academy dragged on, Pasteur decided it was time to seize public opinion. He would put on a spectacle.

○ ○ ○ ○

The Sorbonne had recently launched a series of public lectures called scientific soirees. Pasteur secured the evening of April 7, 1864, for a performance. That night, the university's amphitheater filled with Paris's elite, who had followed the battles between Pasteur and Pouchet for four years now.

Pasteur rose to speak to the crowd. He was surrounded by lab equipment and a lamp to project images on a screen. "You will not leave here tonight without being convinced that the spontaneous generation of microscopic creatures is a chimera," he promised.

He started by attacking Pouchet, pointing out flaws in his rival's experiments. Pasteur then declared that the air in the amphitheater was rife with germs. "We can't see them now, for the same reason that, in broad daylight, we can't see the stars," he said.

At Pasteur's command, the lights went out, save for a cone of light that revealed floating motes of dust. Pasteur asked his audience to picture a rain of dust falling on every surface in the amphitheater. That dust, he said, was alive. To demonstrate this strange fact, Pasteur used one of his glass pumps to suck air onto a sterile piece of cotton. After soaking the cotton in water, he put a drop on a glass slide under a microscope. The projector threw the image of the slide on the screen for Pasteur's audience to see. Alongside soot and bits of plaster, they could make out squirming corpuscles. "These, gentlemen, are the germs of microscopic beings," Pasteur said.

Next, Pasteur turned to a swan-necked flask filled with clear water. "I have deprived it of air-borne germs, or of life itself—for life is the germ, and the germ is life," he said. "The doctrine of spontaneous generation will never recover from the mortal blow inflicted by this experiment."

By destroying spontaneous generation, Pasteur suggested, he was not just performing science but also protecting the Church. "What a victory would be won by materialism, gentlemen, if it could cite in its support the demonstrable fact that matter organizes itself, brings itself to life—matter, in which all the known forces of nature may already be said to reside!" he said.

To end his lecture, Pasteur recounted his travels in search of microorganisms. He talked about capturing germs in the cellar of the Paris Observatory, in the countryside, and, finally, on the Mer de Glace. Germs were everywhere in the air, kicked up in dust, taking flights of unknown distances, and then settling back to the ground, where they worked their magic of fermentation. Germs broke down "everything on the surface of this globe which once had life, in the general economy of creation," Pasteur said. "This role is immense, marvelous, positively moving."

The lecture ended with a standing ovation. And within a few months, Pouchet dropped out of the academy's review. Perhaps he decided that a second outflanking was enough.

Pasteur's hunt for floating germs had now elevated him to the highest ranks of French science. And yet he quickly abandoned it to run other experiments. Those experiments would, over the next two decades, help establish the germ theory of disease. And they would take place entirely in the comfort of Pasteur's lab. Nevertheless, his germ-collecting forays represent an important scientific milestone. Earlier scientists, like Darwin and Ehrenberg, found microorganisms on the ground or at sea. They could only guess that the germs had flown to where they found them. But Pasteur had, for the first time, systematically grabbed them in flight.

two

THE SANITARIANS

As Pasteur turned out the lights at the Sorbonne in 1864, he offered his audience a bold claim. He drew their attention to the germs that floated through the amphitheater. Some airborne microorganisms, he said, caused humanity's most feared diseases.

"They sometimes carry sickness or death," Pasteur said, "in the form of typhus, cholera, yellow fever, and so many other kinds of scourges."

At the time, most doctors believed that such diseases were indeed airborne, but not as Pasteur imagined. The prevailing notion was that the atmosphere itself became corrupted, so that a gulp of the miasmatic air was enough to throw a healthy body out of balance, to spark a disease that could kill. Pasteur was talking about something fundamentally different: he believed cholera and other diseases were caused by contagious germs. As microorganisms multiplied inside the human body, they caused the symptoms of diseases. Their progeny then floated to their next victim.

But Pasteur had no proof that floating germs caused any diseases. He had simply let his imagination run wild. His speculations did not impress many doctors. One of the few doctors who quickly recognized the importance of floating germs was the British surgeon Joseph Lister.

Lister was a master at repairing broken bones and amputating limbs, but he was frustrated by how many of his patients died after his surgery. If their skin was torn by a protruding bone, it would often turn red, become inflamed, and fill with pus. Deadly gangrene frequently followed. Up to a half of amputations in Britain ended in death. Surgeons disagreed about what caused wounds to turn putrid: some blamed miasmas; others, oxygen in the air.

When a colleague drew Lister's attention to Pasteur's work in 1865, he was dazzled by its "simplicity and perfect conclusiveness." And he was particularly taken by Pasteur's suggestion that the decomposition of dead bodies was the result of germs falling out of the air. Lister wondered if germs were also falling into the open wounds of his patients, triggering their flesh to ferment like tainted beet juice and turning their wounds putrid. To satisfy himself that Pasteur was right, Lister replicated his experiments. Instead of yeasty water, he used boiled urine, but he got the same results.

Lister showed off his foul flasks of urine to his fellow surgeons, but they could not understand why he was obsessed with microbiology. Their skepticism did not slow him down, and soon Lister came up with a new way to do surgery. He would protect his patients by applying "some material capable of destroying the life of the floating particles," as he wrote later.

Thinking about what might be capable of the job, Lister picked out carbolic acid. He had heard it could rid sewage of its stink, which might mean that it killed germs. Lister had an opportunity to try out carbolic acid in 1865, when he treated an eleven-year-old boy named James Greenlees. James had been run over by a cart that broke a bone in his leg and left it sticking out of the skin.

Lister set the bone and covered the open wound with lint soaked in carbolic acid. Then he covered the lint with a sheet of tin, over which he put a bandage. Every day thereafter, Lister would check on James, remove the bandage, and paint a fresh coat of carbolic acid on the lint. The wound healed perfectly, and the boy survived.

Over the next year Lister treated eleven more compound fractures in the same manner, and the procedure was overwhelmingly effective at stopping infections. When Lister published his results, Pasteur praised him for carrying out such careful experiments even as he worked as a surgeon and ran a large hospital. "I do not think that another example of such a prodigy could be found in my country," Pasteur said.

Lister saved countless lives by demonstrating how to safeguard patients from infection. But he did not confirm Pasteur's greater vision of airborne microorganisms causing diseases. Nothing in Lister's work demonstrated that cholera was caused by some particular germ, for example. Medical experts on cholera in the 1860s continued to reject contagion as its cause. At the time, the world's leading authority on the disease was Max von Pettenkofer, a professor of hygiene at the University of Munich. "No living epidemiologist has dedicated more time or greater talent to the study of cholera than von Pettenkofer of Munich," the *Lancet* wrote. His name, *Nature* declared, "was known throughout the civilised world." And Pettenkofer assured the world that cholera was caused by inhaling a poisonous gas that seeped from the ground. He called it *"das Cholera-Miasma."*

○ ○ ○ ○

When Max von Pettenkofer was born in 1818, cholera had only just emerged as a terrifying new disease, in the same league as older scourges like smallpox or the plague. Some people called it "King Cholera." Others just called it "the monster."

Cholera was exceptionally humiliating as it killed its victims. It struck suddenly, sometimes causing people to collapse in the streets. They vomited, suffered a torture of cramps, and sometimes went blue in the face. Victims of cholera often expelled a form of diarrhea called rice water stool, causing them to dehydrate so quickly that about half of them died in a matter of days.

In the late 1700s, colonial physicians working in coastal cities in India first reported thousands of people succumbing to cholera, but the disease only gained the world's attention in 1817, when a ferocious epidemic spread out of India, west to Africa, east to Indonesia, and then north to China. Year after year, the epidemic widened, finding thousands of new victims, until it abruptly vanished in 1823.

Doctors were left to debate what had just happened. Some thought a contagion had spread cholera from one person to the next, but Charles Maclean—Britain's champion of miasmas—declared it was nothing of the sort. In an 1824 report, he pointed out many ways in which cholera failed to behave like a contagion. Cholera did not seem to spread in chains of transmission; instead, it often flared up simultaneously in many towns at once. Its abrupt disappearance in 1823 also undermined the contagion theory, Maclean argued, because a microorganism that could spread so easily should not vanish and leave so many vulnerable people untouched. "It can scarcely be necessary to observe how utterly incompatible these facts are with the existence of a specific virus, as the cause of this disease," Maclean wrote.

Then in 1827, cholera roared back. It struck the Ganges River delta and then spread to Europe for the first time, before reaching the Americas. With cholera now raging in their own cities, European doctors debated its cause even more intensely. Believing cholera was contagious, German soldiers set up military cordons to stop people from bringing the germ from Russia. They halted trains to inspect passengers for disease. They even fumigated letters and money. Others who favored miasmas believed the air must be cleaned and called for garbage to be cleared from the streets. But no measure seemed to stop cholera. City after city burst into epidemic flames.

John Snow was an eighteen-year-old doctor's apprentice working in the north of England when the second cholera pandemic arrived there in 1831. The doctor he worked for treated miners who dug coal deep underground. Snow could not understand how miasmas could have caused their cholera if they were supposedly being produced far overhead, in the air over sewers, garbage pits, and swamps.

Some doctors shared Snow's skepticism, but the consensus of miasma held firm. The Westminster Medical Society held a debate on the cause of cholera that lasted for six months. In April 1832, they took a vote. "The evidence brought forward to prove the said malady a contagious disease has completely failed," the chairman announced.

○ ○ ○ ○

Pettenkofer was eighteen when cholera first arrived in his home city of Munich in 1836. It suddenly killed 918 residents before it vanished—not just from Munich, but from all of Europe. At the time, Pettenkofer was studying to become a pharmacist, following the example of his uncle, who served as the pharmacist to the royal court of Bavaria. But Pettenkofer abandoned his studies when he discovered poetry and plays, running off at age twenty-one to join a theater troupe. He gave himself the stage name Tenkof—chopping off the front and back ends of his name—and toured Bavaria for a year. But his acting career failed to launch. Tenkof went back to Pettenkofer and returned to the University of Munich to finish his studies.

Instead of becoming a pharmacist, however, he proved a chemical prodigy. He invented a new method to detect arsenic while he was still a university student, which Munich authorities used to test a corpse for poison. At age twenty-nine, he was appointed as a professor at the university, and he went on to create a string of inventions, including a new formula for cement, a copper-based amalgam for filling cavities, and a meat extract for making soup. Pettenkofer concocted a means to extract gas from burning wood, and the streets of many German cities came to be illuminated with lamps burning his fuel.

In 1848, the third cholera pandemic arrived in Europe. As it killed people in Berlin, Paris, and London, Bavaria's ministry of the interior established a commission to investigate its cause. The commission included two doctors, along with a pair of physicists to study the electricity and magnetism of the air. Pettenkofer was invited as well, despite having no expertise in diseases. He would look for clues to cholera in its chemistry.

Since Bavaria was still free of the disease in 1849, Pettenkofer could study only medical reports sent from places already under attack. He looked over analyses of blood and urine for a chemical signature of cholera. He could find nothing. After years of searching, Pettenkofer got so frustrated that he decided chemistry was useless for finding the cause of the disease. He turned his attention instead to the statistics of the epidemic. Who had gotten sick and who hadn't might tell him something about cholera that their blood and urine could not. And in Pettenkofer's research, he discovered John Snow.

By the time cholera returned to England in 1848, Snow had become a prosperous doctor in London, specializing in the new practice of anesthesia. London's authorities continued to act on the belief that cholera was caused by miasmas, but Snow was more skeptical than ever that an "inhaled poison," as he called it, was to blame. As he analyzed the timing of outbreaks in London, their locations, and other clues, he grew increasingly convinced that cholera was caused by a microorganism that spread from one victim to the next through contaminated water.

Snow found that the disease ravaged the houses on one side of a street, while sparing the other. The residents who lived on the cholera-stricken side dumped their dirty water out their windows into a channel running down the street. Snow reasoned that a contagion in the waste seeped into the well that supplied the residents with their drinking water. In another study, Snow compared cholera among people who received their water from two companies—one taking in clean water, the other taking in water polluted by sewage. The vast majority of people who fell ill with cholera drank the contaminated water.

In August 1853, a yellow flag was hoisted in Snow's own neighborhood in Soho to warn of a new outbreak. Snow drew a map of the cases. He could see that seventy-three of the eighty-three deaths in Soho occurred in houses served by the same water pump in Broad Street. Snow persuaded local health officials to take the handle off the pump, after which the outbreak ended.

Pettenkofer was impressed by Snow's work and tried to replicate it in Germany. But he did not find the same patterns. There was no difference in infections between the people who got their water from different companies. If cholera was contagious, Pettenkofer reasoned, then doctors and nurses who treated sick patients should have often become sick themselves. But they were at no greater risk than other Germans. "I have disposed, once and for all, of the spread by drinking water," Pettenkofer declared.

In 1854, cholera returned at last to Munich. At the peak of the outbreak, Pettenkofer took a walk around the city that would haunt him for years. "It looked as if a breath of pestilence had been breathed over the city from whose poison people began to die in all parts of the town," Pettenkofer later recalled. Over the next few weeks, that breath of pestilence killed seven thousand people.

As Pettenkofer analyzed the deaths across Munich, he developed a theory of his own—one that could account for all the findings that had baffled him. Snow had been right to look to geography, Pettenkofer decided, but he had not read his maps correctly. The difference between neighborhoods that were spared and those that were ravaged had to do with the ground on which they stood. Houses built on porous, loamy soil suffered terribly. But nearby houses that stood on compact soil escaped punishment.

From observations like these, Pettenkofer developed his cholera-miasma theory. The cycle of the disease started with some microorganism that scientists had yet to discover. It entered people's bodies and multiplied, but it did not make people sick with cholera. Those healthy carriers released the microorganism in their feces or urine. If it reached airy soil, it could continue to thrive, fermenting the soil in much the same way that yeast ferments beer. The fermentation released a poisonous gas, Pettenkofer believed.

The poison wafted from the ground, crept into houses, and found its way into people's lungs. As victims breathed in more and more of the gas, it quietly wreaked damage on them until they suddenly collapsed. When the fermentation finished underfoot, the gas stopped seeping from the Earth, and the epidemic vanished.

Pettenkofer created his theory of underground fermentations to explain cholera, but in time he became convinced that it could also account for other diseases, such as malaria and typhoid. Miasmas emanating from the soil were particularly dangerous to humans, he believed, because they were so difficult to avoid. People could avoid the sick and tainted food, but they had no choice but to breathe thousands of times a day. "We live in the air, like fish in water," Pettenkofer observed.

○ ○ ○ ○

Two decades after Tenkof's last performance, Pettenkofer deployed his theatrical skills again. This time, he used them to promote his cholera-miasma theory. His public lectures were so compelling that journals in other countries printed them verbatim. Pettenkofer promised that cities could prevent cholera by preventing miasmas from forming out of filth and stagnant water. "Cleanliness acts as a deterrent to cholera," he declared.

Munich took Pettenkofer's advice to heart. The city installed pipes to bring fresh water from the mountains, giving them enough pressure to make toilets flush properly. Waste was no longer dumped on Munich's open ground; a sewage system moved it clear of the city and poured it into the nearby river.

Pettenkofer also believed in clearing the air inside houses. "It is possible by increasing the rapidity of the change of the air in our houses to prevent it becoming too much contaminated," he said. Indoor pollution took the form of dirt, dust, and poisonous carbon monoxide from charcoal. Fires for cooking and heat could create updrafts that pulled up noxious air full of cholera-miasmas and other disease-causing gases from the ground. Ventilating buildings would flush out the bad air and bring in the good.

Pettenkofer discovered that he could measure the ventilation of air by gauging the carbon dioxide inside a building. In unventilated rooms, people would raise the level of the gas as they exhaled. Pettenkofer found dangerously high levels in houses, schools, and theaters around Munich. Measuring carbon dioxide became a standard way to judge the ventilation of indoor spaces. Many experts agreed on 1,000 parts per million of carbon dioxide as a safe level. That figure is known today as the Pettenkofer number.

Pettenkofer was far from alone in fighting for fresh air. In the mid-1800s, an international movement of physicians, nurses, and reformers came together, calling themselves sanitarians. They looked on in horror at the rising death toll in the fast-growing cities of Europe and the United States, and blamed the carnage on the squalid, crowded housing into which poor migrants from the countryside were forced. Feces and urine, mixed with manure from horses and pigs, ended up flowing down the streets. The filth corrupted air, which then made people sick.

To fight these diseases, the sanitarians did not search for cures. Instead, they sought to prevent sickness by keeping the environment—especially the air—free of corruption. They enlisted journalists to the crusade. New York newspapers warned of the dangers posed by marshy, garbage-strewn neighborhoods—"a great laboratory of fever-producing gases," as one described it.

Sanitarians set out to keep hospitals clean as well. The leader of that effort was Florence Nightingale, who became committed to clearing the air when she served in British military hospitals in the Crimean War. As she battled outbreaks among the wounded soldiers, she scoffed at the notion that some contagion spread disease among their ranks. "Suffice it to say, that in the ordinary sense of the word, there is no proof, such as would be admitted in any scientific inquiry, that there is any such thing as 'contagion,'" she once declared.

Instead, Nightingale believed that hospital air became corrupted—from the filth on the floor, from urine in chamber pots, from noxious moisture released by sick patients. Shutting up a hospital ward made it more dangerous, because the healthy patients were forced to breathe the stagnant air. Nightingale tried keeping the air clean with regular changes of bedding and by washing down hospital walls. After the war, she returned to England and campaigned to make the same changes to civilian hospitals as well. She even changed the architecture of hospitals, designing wards with wings branching off a main corridor, each wing sporting windows through which foul air could escape and fresh air could take its place.

Sanitarians also changed the architecture of houses and apartment buildings to purify the air. In 1869, the New York City Board of Health laid down new rules for ventilation, forcing forty-six thousand windows to be cut into dark interior rooms across the city. New buildings had to be built with light shafts to allow fresh air to expunge the miasmas.

The sanitarian movement made Pettenkofer its hero. In the early 1800s, Munich had suffered hundreds of deaths every year from typhoid fever. But after the city followed Pettenkofer's advice, the rate fell to close to zero. The University of Munich built Pettenkofer an entire institute where he trained new hygienic scientists who came from as far away as Japan. Meanwhile, his cholera-miasma theory became an international consensus. At cholera conferences, Pettenkofer's name was on everyone's lips, while John Snow was barely mentioned.

As his fame grew, Pettenkofer sneered at anyone who still held on to the "drinking-water faith." But even as Pettenkofer enjoyed the peak of his celebrity, the followers of the drinking-water faith were engineering his downfall. And with his downfall, the age of miasmas would come to an end.

○ ○ ○ ○

Louis Pasteur shared John Snow's conviction that a contagion caused cholera, but they disagreed on how that contagion spread. Snow thought it was in the water, while Pasteur believed it was in the air. In 1865, when an outbreak of cholera struck Paris, Pasteur set out to find it in a Paris hospital. He pumped air from the cholera wards through cotton filters. He checked the currents rising up into the attic. He collected dust from the floors and bedding. Pasteur found many kinds of spores, but no germ on which he could blame cholera.

As cholera tormented Paris, another epidemic was raging in the countryside. This one attacked silkworms, on which France's huge silk industry depended. Silkworm breeders raised the insects in such great numbers that their munching on mulberry leaves sounded like rain striking leaves in a storm. But now a disease called pébrine was covering the insects in spots and leaving them too weak to eat. The breeding rooms went silent, and France's silk industry faced doom.

The government dispatched Pasteur to find the cause of pébrine. Pasteur knew that in the 1830s an Italian lawyer named Agostino Bassi had investigated an outbreak of another silkworm disease called white muscardine. Bassi had blamed it on a fungus. Pasteur followed Bassi's example and searched for a microorganism in the silkworms sick with pébrine. He found a single-celled fungus inside them, and he was able to spread pébrine by infecting other worms with it.

The fungus did not spread through the air, however: Pasteur found that sick caterpillars released it in their droppings and contaminated the leaves they ate. Just as Pasteur had had practical advice for fermenting beet juice a decade earlier, he now had practical advice for silkworm breeders: they should inspect their insects to be sure they were free of the fungus. Thanks to his guidelines, pébrine stopped killing worms, and the silk industry was saved.

For Pasteur, pébrine also revealed something deeper about life. The fungi in sick silkworms were not spontaneously generated by pébrine; they were its cause. In this work, Pasteur was taking some preliminary steps toward what would later become known as the germ theory of disease, a theory that would explain cholera, tuberculosis, and many other human diseases. But the scientist who would provide the first clear-cut evidence of a germ causing a human disease would not be Pasteur. It would be a young doctor who had never published a single study of germs before.

○ ○ ○ ○

Robert Koch earned his medical degree at the University of Göttingen in 1866, while Pasteur was working on his silkworms. The young doctor struggled to start a medical practice, ending up as the district medical officer for Wöllstein, a small city surrounded by farms and forests in what is now Poland.

Koch and his family occupied four rooms on the second floor of a house in Wöllstein. Many of his patients would show up at night after having spent their days working their fields. They often paid for his exams with wild game. Eventually Koch's wife hung a curtain to divide the clinic room in half, and in the half by the windows, Koch amused himself in his free time with a microscope. He would look at algae and other single-celled organisms he collected in the forests and ponds around Wöllstein. In 1873, Koch obtained blood of a sheep that had died of anthrax. And when he looked at the blood under his microscope, his hobby turned into a project that would ultimately win him the Nobel Prize.

Anthrax was a menace for Wöllstein farmers. Cows and sheep would develop fevers, stumble, bleed out of their noses, and die. Farmers who handled their diseased carcasses or hides sometimes developed pustules on their skin, but usually recovered. Certain pastures seemed to be hot spots for anthrax, and outbreaks seemed to track the weather. When scientists inspected the blood of animals killed by anthrax, they noticed peculiar microscopic rods. But they could not determine if the rods were alive or were some kind of lifeless crystal.

Koch looked at the rods for himself. He noticed that if he put a flame under a slide, the warmth triggered the rods to swell and grow shiny. "They looked like a pile of glass threads or like a climbing plant with long parallel vines," Koch later recalled. He suspected that he was looking not at crystals but living things. After growing for fifteen hours, the rods produced small grains along their length. Koch recognized that they were making spores.

As Koch peered through his microscope, he thought back to the lectures he had heard at Göttingen from a professor named Jacob Henle. Henle was a contagionist, but he recognized that the theory lacked compelling proof. That proof would come in the form of links between germs and diseases. Microbiologists would need to isolate a particular kind of germ from hosts that suffered the same disease. They would then have to cause the same disease in a new victim by transferring the germ.

Koch suspected that the rods and spores he was observing were microorganisms that caused anthrax. The rods that grew inside infected animals produced spores. When their hosts died, the spores spilled out and spread into the soil. They were hardy enough to survive in the ground, where they could wait to infect healthy animals that passed through the contaminated fields to graze.

Koch wanted to prove that cycle, but he needed to learn a lot more about microbiology before he dared try. He left Wöllstein in 1875 to take a grand tour of Germany's scientific centers. A stop at Pettenkofer's institute in Munich allowed him to learn more about the great epidemiologist's theory about cholera brewing in the soil. The conditions of the soil might help cause anthrax too by controlling which pastures became deadly with spores.

On his return to Wöllstein, Koch was ready to carry out the kind of experiment Henle had envisioned. Koch got spleens from sheep and cows that had died of anthrax. He poked the organs with slivers of wood. Using the slivers as primitive syringes, he delivered the tissues into the bloodstreams of mice. The mice soon developed the symptoms of anthrax and died.

Their deaths demonstrated that anthrax could be transmitted by blood. But they didn't eliminate the possibility that something other than anthrax bacteria in the blood of sheep and cows might have been the cause. Koch figured out the right ingredients that anthrax bacteria needed to grow in a flask. He injected some of his homegrown bacteria into another batch of mice, and they died from anthrax as well.

By completing the cycle of death and life, Koch sealed his case. His simple experiments had demonstrated for the first time how one particular microorganism acted as a contagion, causing a particular disease as it moved from host to host. That achievement suddenly thrust Koch into the top ranks of German science. "The man has made a magnificent discovery, which, for simplicity and the precision of the methods employed, is all the more deserving of admiration," one microbiologist declared.

○ ○ ○ ○

Koch left Wöllstein for Berlin, where he was given a post at the Imperial Health Office. There he used the same method he had developed for anthrax to study tuberculosis.

Tuberculosis did not crash down on humanity in devastating waves like cholera. Instead, it slowly destroyed lives year in and year out. The disease started with coughs that left victims exhausted, and over years their faces took on a hollowed-out pallor. According to one estimate in Koch's day, one in seven humans across the planet died of TB. Doctors could not agree on what caused it, though. It was miasma according to some, a hereditary curse according to others.

Koch reasoned that tuberculosis might be caused by bacteria. He discovered a distinctive microorganism in the sputum of TB patients in 1882, which he learned how to grow in his lab. When he injected the bacteria into rabbits, their lungs developed TB's distinctive scars and became rich with the same bacteria.

Here, Koch concluded, was the cause of tuberculosis, a species of microbe that would come to be known as *Mycobacterium tuberculosis*. It was a staggering discovery, although it left one fundamental question about TB unanswered: Koch did not know how the bacteria spread. After all, tuberculosis did not become one of the world's worst killers one injection at a time.

Koch's success at identifying the causes of both anthrax and tuberculosis set the standard for linking germs to diseases—a set of methods that came to be known as Koch's postulates. Researchers who followed them quickly uncovered the causes of many of the worst diseases known to medicine. In the process, they assembled overwhelming proof that the germ theory of disease was correct.

○ ○ ○ ○

In 1883, cholera made a devastating return. It first flared in Damietta, an Egyptian port city on the Mediterranean, and then swept up the Nile. As it approached Cairo, forty thousand residents fled the city. The British soldiers who had recently taken control of Egypt used cordons to keep infected people from traveling to untouched towns. They boiled their own water before drinking it, a measure of which John Snow would have approved. Despite their efforts, the epidemic grew, and soon it was killing thousands of Egyptians each week.

The cholera epidemic was not just a human catastrophe but a geopolitical crisis. Much of the world's trade moved through Egypt's Suez Canal, over which Britain now had profitable control. France and Germany called for ships bound for Europe to stop when they reached the canal and discharge their crews, in case they were sick with cholera. Only after safely quarantining for days would the crews be allowed back on their ships. Quarantines threatened to slow the trade through the canal to a molasses-like flow.

Britain dispatched the eminent physician William Hunter to Egypt with a team of ten assistants to determine the cause of the cholera outbreak. Hunter seems to have made up his mind before he left. He later wrote that he was among those "who do not believe in the contagiousness of cholera, and who, therefore, do not look for a specific entity or germ." He looked for a cause instead in the physical conditions of the country. Filth, Hunter said, was "a powerful factor." So was the odd weather that British doctors in Egypt described to him: "a very peculiar condition of the atmosphere was observed—a yellowness of the air, somewhat of the nature of a fog."

In his report for the British government, Hunter declared that "the peculiar atmospheric state noticed would probably act as an exciting cause in the immediate production of the epidemic outbreak." He provided the British government with a convenient conclusion: Egypt's own air had produced the epidemic rather than a germ transported by Britain from India. No trade-jamming quarantines were required.

But Hunter's team was not the only one to come to Egypt. The Egyptian government also invited France to send a delegation. Louis

Pasteur dispatched four of his disciples, who searched for a microorganism that might be killing Egyptians. They had to abandon the hunt when one of the researchers died of cholera. Germany sent a team as well. Pettenkofer was not asked to take part in the expedition, despite his reputation as the greatest cholera expert in the world. Instead, Robert Koch was appointed to lead the mission.

Soon after Koch's team arrived in Alexandria, they identified a distinctive type of bacteria in the guts of cholera victims. Each of the microorganisms had a little twist to its cell, leading Koch and his colleagues to call it the comma bacillus. Other scientists had observed the same microorganism years earlier, but now Koch built a more compelling case that it caused cholera. His team found the comma bacillus in every cholera victim they autopsied, and never found it in people who had died of other causes.

As compelling as those results were, Koch had yet to follow all his postulates—to grow the comma bacillus and inject it into animals to cause cholera anew. But he ran out of time in Egypt. After having killed fifty-eight thousand people, Egypt's cholera epidemic abruptly ended.

Koch decided it was not yet time to go home. He traveled on to India, where small cholera outbreaks had become a regular part of life. There he found more evidence for the comma bacillus. His team found it in the feces of patients with cholera and in the bowels of people it killed. If Indians recovered from cholera, Koch found, the comma bacillus disappeared from their stool. The fact that the same bacteria was growing inside cholera victims in both Egypt and India made the link stronger. It grew stronger still when Koch's team learned of a new outbreak of cholera in an Indian village. They tested the water in a storage tank used for drinking, cooking, and washing clothes. Koch's team isolated the comma bacillus once more.

For all this evidence, Koch never completely fulfilled his postulates. When he grew comma bacilli and injected them into mice, the animals did not get sick. Still, Koch decided he had found enough proof. He announced that the comma bacillus was the cause of cholera, and Germany agreed. Koch came home a hero, a German Pasteur.

On his return, Koch visited Pettenkofer. He was no longer an obscure frontier doctor; now he met Pettenkofer as a scientific equal, even a rival. The two men had fundamentally different theories for how cholera spread. Koch believed the comma bacillus traveled in contaminated water and multiplied in people who drank it. The microorganism caused the deadly symptoms of cholera, and its progeny escaped in diarrhea to infect new hosts. Pettenkofer continued to believe that microorganisms released into the soil produced a miasma that was then inhaled by people, leading to their deaths.

The meeting did not lead Pettenkofer to change his cholera-miasma theory. He simply absorbed Koch's work into it.

"Koch's discovery of the comma bacillus alters nothing, and, as is well known, was not unexpected by me," Pettenkofer declared. After all, he had long argued that some microorganism was essential to cholera—to produce deadly fumes, rather than deadly infections. Koch's failure to cause cholera in mice by injecting comma bacillus into them seemed only to strengthen his theory.

To promote his updated theory, Pettenkofer let loose a torrent of lectures and essays. They were crammed with details on geology and weather. He recalled trips to Italy and Malta, where sudden storms were followed by explosions of cholera. "If a weather storm can create a 'cholera storm,' then the cholera must be existent in the soil," he said in 1884.

○ ○ ○ ○

In 1892, cholera returned once more to Europe. It did not come by storm, but by rail. The wave of death began in Afghanistan and spread west along train lines to Baku, Moscow, and Saint Petersburg before finally arriving in Western Europe. Hamburg, a city in northern Germany, got hit especially hard.

Over the course of the nineteenth century, Hamburg had grown to a population of 800,000. People were drawn there by its booming shipping industry, which delivered wheat to Britain and brought British goods back to Europe. Immigrants streamed into Hamburg for passage across the Atlantic to the United States. The city's leaders—elderly businessmen rather than full-time politicians—were quick to spend money on projects that would bring more business to Hamburg, such as a new port and shipbuilding yards. They were less eager to spend money on things that would mostly benefit the growing ranks of Hamburg's poor. Immigrants waiting for ships squeezed into a foul neighborhood where they had to relieve themselves in latrines and chamber pots. The city's reservoirs fell into such disrepair that people regularly found eels swimming in their drinking water.

Just across the Elbe River, the town of Altona had no eel trouble. In 1859, it set up a filtering facility to pump river water through sand. After the cholera outbreak in 1873, Hamburg's leaders grudgingly agreed to filter their own water, but they spent years bickering about how to pay for it. The delay did not cause an uproar, because most of Hamburg's doctors were miasmatists. The chief medical officer of the city, Johann Kraus, was a well-known supporter of Pettenkofer. One Hamburg doctor published a pamphlet on cholera in which he blamed miasmas produced by melting Arctic icebergs.

On August 14, 1892, a sewer inspector in Altona began vomiting. Soon people across the river got sick too. Hamburg's leaders failed to respond for days, while Altona's doctors quickly sent specimens to Koch in Berlin to analyze. In them, Koch found the comma bacillus.

After Koch notified the German government, it dispatched him to Hamburg to set things straight. The city was in chaos by the time he arrived. "I felt as if I was walking across a battlefield," Koch wrote on August 25.

Koch determined that the comma bacillus had contaminated the Elbe River and spread from there to Hamburg's reservoirs. As the bacteria flowed through faucets, it sickened people by the thousands. Koch forced the city's leaders to treat cholera as a germ-driven disease rather than as a miasma. Wagons delivered clean water to neighborhoods across Hamburg. Disinfection crews moved through the houses of the sick. They doused bedding and clothes in Lister's carbolic acid to kill comma bacilli that might be clinging to them.

Once Koch's orders were put into action, he headed back to Berlin, leaving his protégé Georg Gaffky as the city's new hygienic advisor. The outbreak ebbed under Koch's rules, but it took a brutal toll on Hamburg: 8,606 people died in less than three months—1.34 percent of the city's entire population. Across the Elbe in Altona, where people drank sand-filtered water, only .2 percent died.

The two communities, drinking from the same river, seemed to Koch like a vast experiment. And that experiment, in which Hamburg had a death rate almost seven times higher than Altona, refuted the idea that cholera was caused by corrupted air. It was clear to Koch that a water-borne germ was to blame. To prevent other German cities from suffering the way Hamburg had, Koch demanded a national law for managing epidemics based on the germ theory of disease. On September 26, 1892, an expert commission met to work out its details. Along with Koch, its members included Pettenkofer.

When the commission met, Pettenkofer refused to give up his theory. He adamantly blamed the disaster in Hamburg on some sort of drastic change to the city's soil. For days, Pettenkofer and Koch battled. Pettenkofer argued against quarantines and water purification. Koch insinuated that Pettenkofer, now seventy-three, had fallen behind the times. After they finished sparring, the commission voted in favor of Koch's new measure. The drinking-water faith had now become gospel.

○ ○ ○ ○

Pettenkofer returned to Munich in defeat, but he was not yet ready to surrender. He came up with an experiment to prove he was right. Tenkof would put on a final one-man show.

In a letter to Koch, Pettenkofer asked for some comma bacilli isolated from a cholera patient. He did not indicate what he would do with them. Gaffky supplied a sample from the outbreak in Hamburg, and two of Pettenkofer's colleagues used it to grow a cholera broth.

On the morning of October 7, 1892, Pettenkofer prepared for his performance by eating a light breakfast. He waited two hours for the acid in his stomach to subside. To neutralize any leftover gastric juices, he drank a mix of water and baking soda. Pettenkofer—now white bearded and toothless—went to a lecture room at his Institute of Hygiene. He raised the vial of comma bacilli in front of his acolytes and announced he would drink it.

They begged him not to. They would drink it for him. Pettenkofer refused and tipped back the vial. He swallowed the bacteria and waited to see if he would die.

Pettenkofer survived to tell his story. In fact, he told it with dramatic flair a few weeks later to an audience at the Munich Medical Society. He described how he refused to hand over the vial to his students because an old man ought to risk his own life for his convictions. "Even if I had deceived myself and the experiment endangered my life, I should face death calmly, for it would not be as a thoughtless and cowardly suicide," Pettenkofer declared. "I should die in the cause of science, like a soldier on the field of honor."

After he drank the comma bacilli, Pettenkofer's bowels became noisy. He experienced some mild diarrhea. He provided his stool to his followers, and in it they found swarms of comma bacilli. But Pettenkofer never suffered agonizing cramps, rice water stool, or any of cholera's other deadly hallmarks. The morning after his experiment, Pettenkofer woke up at his regular hour, dressed, and strolled to the institute. He conducted a regular day's work and then strolled back home. After a few days, the comma bacilli vanished from his stool.

"I could not but conclude that, although comma bacilli may cause diarrhea, they cannot cause cholera," Pettenkofer concluded.

Afterward, he sent off a thank-you note to his rival: "Herr Doktor Pettenkofer has now drunk the entire contents and is happy to inform Herr Doktor Professor Koch that he remains in good health."

○ ○ ○ ○

The applause Pettenkofer hoped for did not come. His experiment could not destroy the drinking-water faith. Researchers determined that the comma bacillus—which would later gain the Latin name *Vibrio cholerae*—often failed to cause severe symptoms. People who carried it without feeling ill unwittingly helped the bacteria spread.

Years later, Pettenkofer's biographer asked Georg Gaffky about the old man's cholera experiment. When Pettenkofer asked for comma bacilli, Gaffky recalled that "we could guess what he was going to do." Gaffky had selected a patient who suffered only a mild case of cholera to take a sample from, rather than a corpse. He did not tell Pettenkofer that. When Pettenkofer failed to develop cholera, he believed he had proven his cholera-miasma theory. He did not realize he was the victim of mercy.

With each passing year, the germ theory kept shuffling medicine into a new configuration. Researchers used Koch's postulates to show that bacteria were not the only form of life to spread diseases. A fungus produced ringworm. Malaria was the result of a single-celled protozoan called *Plasmodium* carried by mosquitoes. Mosquitoes also carried the cause of yellow fever, an exquisitely small type of pathogen called a virus. Other viruses would turn out to cause some of the worst diseases humans faced, including smallpox, influenza, and rabies.

Once scientists identified germs, they could trace the routes they traveled from one host to another. Ringworm spread by ordinary skin contact. Syphilis was caused by a kind of bacteria called *Treponema* transmitted only by sex. Contaminated water delivered *Vibrio* and *Salmonella*. The rabies virus required the service of live animals, which bit their victims. Jail fever proved to be caused not by prison air, but by lice-borne bacteria, and its old name changed to typhus. Rats carried fleas infected with *Yersinia pestis*, the bacteria that caused plague.

By the end of the 1800s, miasmas were fading away, looking like an obsolete illusion. And while Louis Pasteur had offered an alternative vision of the air carrying invisible floating germs that spread diseases such as cholera, typhus, and yellow fever, he was wrong on all three of those counts. Now microbiologists were left to wonder if floating germs caused any diseases at all.

Robert Koch suspected that tuberculosis might spread by air. When people sick with TB coughed up bacteria-rich phlegm, it dried out and turned to infectious dust, he speculated. "Thus, it is probable that tubercle bacilli are usually inhaled on dust particles," Koch wrote in 1882.

Following Koch's lead, doctors urged housewives to keep their homes clean to protect their families. Spitting was no longer just offensive. Now it spread a deadly disease. Sanitarians urged that spittoons be placed in homes, offices, and factories. New York City went further, making spitting in public a crime.

Yet Koch's claim that tuberculous was spread by dust was just speculation. Almost a decade would pass before his close friend Carl Flügge launched a series of experiments to track the movement of invisible germs through the air.

Flügge ran an experiment to see if guinea pigs, which were known to be very susceptible to TB, could get infected by inhaling dust. At the Institute of Hygiene in Breslau, which Flügge led, a team of scientists collected sputum from tuberculosis patients and spread it on boards and cloth. They then blew puffs of air to lift up the dried bacteria. Guinea pigs inhaling the dust rarely got sick.

That experiment left Flügge wondering if Koch was wrong. Perhaps TB spread without the help of dust. He refitted some of his labs for a much bigger experiment. Healthy volunteers stepped into an empty room with Petri dishes scattered across the floor. They washed their mouths with a bacteria-laced solution that tasted like rotten herring. The bacteria, called *Serratia marcescens*, grew naturally in soil and on food; as far as microbiologists could tell, it was harmless to people. After enduring a *Serratia* mouthwash, the volunteers coughed, sneezed, sang, whistled, and talked.

Later, Flügge's assistants collected the dishes and let them incubate. After a few days, bright red colonies of *Serratia marcescens* began growing. Flügge's experiment showed that people could send bacteria into the air, and the bacteria could land in Petri dishes several yards away.

In another series of experiments, the volunteers expelled their *Serratia*-infused breath into rooms without Petri dishes. Only after the volunteers left the room did Flügge's scientists set out the dishes. They could wait as long as an hour and still catch *Serratia* from the air. The experiment proved that bacteria could float for a long time after being expelled from a mouth—long enough to be inhaled by others. Flügge then switched from healthy volunteers to ones suffering with tuberculosis. He had them speak and cough. The results were similar: he could capture *Mycobacterium tuberculosis* yards away.

"The dissemination of germs through the air has been underestimated," Flügge concluded.

Flügge was now convinced that some germs spread through the air, even if others used food, water, or physical contact to reach a new host. He believed that talking could release tiny drops that could ride air currents for hours. And he suspected that tuberculosis was not the only disease that spread through the air. Respiratory infections—such as influenza, diphtheria, and whooping cough—might also take the same route. Flügge even speculated that smallpox and plague—which were not considered respiratory diseases at the time—might spread by air. Germs that floated outdoors were mostly harmless, Flügge believed, because the constant circulation of fresh air diluted droplets to near oblivion. But indoors, floating in unventilated air, they posed a threat.

It occurred to Flügge that one of the riskiest indoor places might be an operating room. A surgeon chatting to his colleagues would exhale drops, some of which might carry germs. Even if they lived harmlessly in the mouth, they might turn vicious once they fell into a patient's opened body.

Flügge persuaded the Breslau surgeon Johannes Mikulicz of the invisible dangers of the air. They agreed that surgeons should wear a mask big enough to cover the mouth, nose, and beard—the standard facial fashion for the men who performed surgeries in German hospitals. From a piece of gauze, Mikulicz fashioned a "mouth bandage," as he dubbed it, and persuaded his colleagues to wear it as well.

To test the masks, Flügge had volunteers rinse their mouths with a rotten-herring *Serratia* cocktail, put on his mouth bandages, and then count to five hundred fifty. A single layer of gauze was not enough to stop the bacteria from escaping from their mouths and landing in Petri dishes. But two layers provided a strong barrier.

"Now we are breathing through it as easily as a lady who wears a veil in the street," Mikulicz said in 1897, speaking on behalf of all the bearded men performing surgery in Breslau. Before long, both men and women working in operating rooms around the world were putting on masks as well.

o o o o

Beyond the operating room, however, Carl Flügge had much less influence. Demonstrating that people could release bacteria-infused droplets from their mouths was not proof that real outbreaks were caused by airborne germs. To many young sanitarians, Flügge's theory about airborne germs seemed barely more plausible than the old ideas about miasmas.

The American sanitarian Charles Chapin emerged as the leading skeptic. Chapin had been trained as a medical student in the 1870s that miasmas caused a long list of diseases. "Air was the chief vehicle of infection," he later wrote. "Nay, it was infection itself."

As soon as Chapin discovered Koch's work, he threw aside his miasma lessons and started investigating how germs spread diseases. When he was appointed as the superintendent of health for Providence, Rhode Island, he took the then radical step of building a microbiology lab where he could isolate germs that sickened the city. Chapin also studied the epidemiology of Providence—the patterns of disease that emerged across space and time—to understand how the germs spread.

Chapin found plenty of evidence that diseases could be transmitted by tainted milk, contaminated food, and the bites of infected mosquitoes. But he could also see how easy it had been for his predecessors to blame miasmas instead for outbreaks. "The most natural explanation is that the infection is carried by air," he said.

It was a natural explanation, but a wrong one. "A bundle comes home from the store," Chapin once told an audience. "The clerk puts the string in her mouth when doing it up. As soon as it arrives home, it is unwrapped, and the contaminated string is given to the baby to put in its mouth. No wonder it gets diphtheria in some mysterious way." Everywhere Chapin looked, he saw germ-coated surfaces that could spread disease. "Many a lifelong infection has come from a drinking glass, a roller towel, a dentist's instrument, or the hands of a nurse," he warned.

Chapin believed that the germ theory of disease made miasmas unnecessary, much as Copernicus had made epicycles unnecessary to explain the movements of the planets and the sun. He didn't see anything more compelling about Flügge's proposal that airborne germs could spread diseases over long distances. "The breath of a patient is not the carrier of disease," Chapin maintained. "In fact it is entirely free from disease germs."

In ruling out airborne germs, Chapin made a subtle distinction. While he did not believe that people's breath contained germs, he recognized that the droplets sprayed out in coughs and sneezes might carry them. But he figured those spray-borne germs could fly only a short distance before falling to the ground, pulled down by gravity. "They have not wings," Chapin declared. "They get into the air in a mechanical way such as by loud talking, sneezing, and coughing, all of which may be a real source of danger to a doctor, nurse or anyone who comes within a few feet of a sick person."

○ ○ ○ ○

Chapin's new paradigm helped to wipe out Max von Pettenkofer's reputation. In 1894, Pettenkofer retired from his institute, left Munich, and moved into a house on Lake Starnberg. He spent much of his time rowing across the water. Friends who visited Pettenkofer at the lake thought that he was growing weary with life. In 1901, a throat infection brought him agonizing pain and plunged him into deep depression. On February 10, at age eighty-two, Pettenkofer put a revolver to his right temple and pulled the trigger.

The first wave of Pettenkofer's obituaries offered polite praise. "The death of Professor Pettenkofer under tragic circumstances removes from among us one of the founders of public hygiene," the *Lancet* announced shortly after his death.

But as more time passed, the memories of Pettenkofer grew harsher. In 1935, the British epidemiologist Major Greenwood declared that, outside of Munich, "Pettenkofer and his teaching are as obsolete as Galen." The *British Medical Journal* was even more ruthless: "To men of this generation, Pettenkofer is the shadow of a name."

Chapin replaced Pettenkofer at the summit of public health. In 1910, he published *The Sources and Modes of Infection*, a book that would become a classic of the field. In it, Chapin recounted how medicine had moved from superstition to science, brushing quickly past Pettenkofer and his discredited cholera-miasma. "It will be a great relief to most persons to be freed from the specter of infected air," Chapin assured his readers, "a specter which has pursued the race from the time of Hippocrates."

In 1930, Chapin went to the Statler Hotel in Boston to be feted by the American Public Health Association. Surgeon General Hugh Cumming gave Chapin, then seventy-four, the association's highest award. "I believe that perhaps we owe the entire formulation of the modern viewpoint on communicable diseases to Dr. Chapin," Cumming declared.

That modern viewpoint would endure for generations. But even in the 1930s, a small group of scientists were starting to question Chapin's paradigm. They came to believe that the specter of infected air was real: that some pathogens could travel long distances to reach humans, as well as animals and plants. Much of their work would go neglected for decades. But nearly a century after Chapin earned his award, their work came back to light. And Chapin's views about airborne infection would become obsolete in turn.

three

A WATERMELON DOCTOR

In July 1933, New York reporters buzzed around a hangar near Flushing Bay. Charles Lindbergh, the most famous man on Earth, had left his guarded mansion in New Jersey to come to Queens and prepare his Lockheed Sirius airplane for a new flight. In 1931, Charles and his wife, Anne, had flown the plane west to Japan and China, and now, two years later, they were about to take it on an even more dangerous journey to the east, soaring over Greenland to Europe. "They have flown so much together that the fact in itself means little," one reporter wrote. "But it sets them against their natural scene, the clouds and the blue sky, storm and fog and sleet, the most romantic adventurous couple in American life."

The reporters knew a lot about the upcoming flight. They knew it would be the first journey that the Lindberghs would take after the kidnapping and death of their first child. And they knew that the Lindberghs were traveling in order to scout flight routes for the Pan Am Corporation. Six years had passed since Charles Lindbergh had taken the first solo flight across the Atlantic, but commercial planes were not yet flying over the ocean. The Lindberghs would search for a series of safe landing spots along the way from the United States to Europe where planes could refuel.

The reporters in Flushing Bay chronicled the preparations for the flight. Engineers scrambling over the Lindberghs' red-winged monoplane installed a new engine to let it reach speeds over two hundred miles an hour. They fitted it with a new radio for Anne to operate. They swapped the wheels for a pair of aluminum pontoons so that the plane could make water landings. Supplies were loaded into the plane—a rubber boat with a twelve-foot mast on which Charles and Anne could float at sea, and a snow sledge in case they crashed in the Greenland mountains. A month's emergency rations were loaded on board, as well as a rifle and a machete in case the food ran out.

Yet the reporters failed to notice the eleventh-hour arrival of a man who flew in from Washington to Glenn Curtiss Airport. Fred Meier—a dapper, charming forty-year-old government scientist—brought a last batch of equipment to stow aboard the Lockheed Sirius. Meier's delivery included dozens of aluminum tubes, each the size of a roll of quarters and individually wrapped in sterilized paper. He also handed over a strange pair of slender metal rods held together by a series of brackets that looked like a pool cue for a machine-age game of billiards.

On July 9, thousands of onlookers—possibly including Meier—thronged to watch the Lindberghs start their journey. Charles was dressed in gray trousers, a blue cotton shirt, and a tie. Anne wore a khaki flying suit and a white helmet, with a sweater tied across her shoulders. They climbed into the plane, which then slid down a ramp into the bay. With Movietone cameras filming onshore, the Lindberghs taxied half a mile out and back as they warmed up their engine. Then Charles opened the throttle and the plane kicked up spray as it gained speed. He lifted the plane into the air and dipped its wings in a farewell salute. "We circle and circle over the river, the barges, the dump island—wheeling and turning to avoid camera planes," Anne wrote in her diary. She and Charles raced ahead and disappeared over the hills of Westchester.

None of the newsreels and articles about their takeoff mentioned Meier's strange delivery. That was by design. Months before the flight, he and the Lindberghs had agreed to keep their project a secret. Meier himself was starting to gain a little fame for exploring the air over cities, prairies, and forests in search of life. One newspaper described him as "the government's ace-spore hunter." Now the Lindberghs were going to secretly take that hunt over the mountains of Greenland and, for the first time, to the middle of an ocean. They were going to help Meier learn just how far, and how high, life could reach.

○ ○ ○ ○

More than seven decades had passed since Pasteur had caught floating germs on the Mer de Glace. He had inspired other scientists to improve on his equipment, creating new contraptions that would come to be known as aeroscopes. Some scientists followed up on Pasteur's idea that airborne germs caused diseases, but they failed to catch any clear candidates inside hospitals. In the 1870s, a medical student named Pierre Miquel was so dazzled by the aeroscopes deployed in his Paris hospital that he abandoned his career as a doctor to search for airborne life outside.

Miquel found a place to carry out his work in the Parc Montsouris, a stretch of greenery at the southern edge of Paris. On the park grounds stood an enormous replica of a Tunisian palace, complete with circular windows, high arches, and Arab-style domes. In one corner of the palace, the builders had put a weather station where meteorologists measured the temperature and atmospheric chemists analyzed the molecules in rain. Miquel joined their ranks as the station's chief of micrographic services.

In his new job, Miquel invented an aeroscope far more sophisticated than the flasks Pasteur had used in 1860. It steadily drew in a volume of air that Miquel could set precisely. A sticky glass slide inside the aeroscope could trap tiny airborne objects. The aeroscope then pushed out the old air and took in another gulp, collecting more debris on the slide.

Miquel installed aeroscopes on an exterior wall of the palace, where they gulped twenty liters of air every hour, day and night. He regularly collected slides and inspected them for pollen grains, fungal spores, and bacteria. Because Miquel knew exactly how much air he sampled each hour, he could estimate the density of airborne life. He lumped together all the microorganisms he found, referring to them as spores. Every cubic meter of air in the park, he concluded, contained thirty thousand spores.

Sometimes Miquel would leave the Parc Montsouris with an aeroscope, taking it to a hospital or a laboratory. In those indoor spaces, he found just a few thousand spores in a cubic meter of air. He climbed down into the Paris sewers and found just a few dozen in the undisturbed atmosphere. But it was Miquel's work at the palace that became his most important effort, because he made measurements there with relentless regularity. He tracked the rise and fall of spores through day and night, and he followed them across the seasons.

It was, Miquel confessed, a lonely job. "There are too few scientists devoted to this laborious and delicate research," he complained. "I understand that the courage can weaken in front of the tiredness imposed by this task."

In his oceans of data, Miquel glimpsed faint patterns. Rain seemed rife with spores, but spores became scarce immediately after a storm. Bacteria were more common in the winter than in the summer, although their peaks changed from one year to the next. Miquel found that wind blowing south over the city could deliver more than twice as many bacteria to him as the winds coming north from the countryside. Miquel used these figures to estimate that 5 trillion bacteria rose into the atmosphere over Paris each day.

Looking at thousands upon thousands of gulps of air taken over the course of decades, Miquel struggled to find the laws of spores. In his line of work, Miquel said, "an inference made yesterday is often destroyed by the facts of the next day." The facts became simply too much for him to handle. Miquel once tried to sort through all the fungi he was catching, only to give up. "The micrographer who has leisure could make some nice studies of this subject," Miquel said.

Although Miquel published two books about his work, he was largely forgotten after his death in 1922. He left behind a legacy of confused respect for the life of the air. For all his labors, he never made sense of it. "It is necessary to travel, so to speak, in the dark, probing at every moment a terrain strewn with obstacles, on which many distinguished minds have already made a mistake," Miquel said. He considered his years of painstaking work only a start. "I have succeeded in clearing the edges of a still unexplored path," he wrote.

○ ○ ○ ○

Miquel gathered airborne life close to the ground. Just how high it could reach, he could not say. The first reports from high altitudes came in 1873, when an English doctor named Charles Harrison Blackley recounted his search for the cause of his hay fever by flying kites.

For much of his life, Blackley suffered for days on end from coughs, a stuffy nose, a scratchy throat, and difficulty breathing. The symptoms would suddenly vanish, only to return weeks or months later. In 1857, Blackley got a clue about his misery while taking a walk. He started at the ocean and strolled inland past flowering fields of hay. "When I had got within the distance of six or eight miles from the seashore, I felt that my old enemy was coming on again," he later recalled. Five days later Blackley returned to the same place. The hay had been cut, and the wind was now blowing from the sea. This time he barely suffered a brief spell of hay fever.

The two starkly different walks left Blackley convinced that airborne hay pollen was to blame for his illness. His suspicions sharpened when he picked up a vase at his house one day. It held a bunch of grasses his children had picked. "A small cloud of pollen was detached and came in close proximity to my face," he later recalled. "I commenced sneezing violently in the course of two or three minutes."

By then botanists had discovered some fundamental facts about pollen—that most species of plants produced this living dust, that the grains were carried to other plants by insects or the wind, that a pollen grain could mysteriously trigger a flower to produce seeds. But no one had given much thought to the journey that pollen grains made through the air or to what happened if people inhaled them. To test his hunch, Blackley rubbed pollen into his eyes and breathed it into his airway. His hay fever immediately grew worse.

Blackley decided to launch a systematic survey of the pollen in the air. He borrowed one of Pasteur's tricks for capturing floating germs, pumping air through cotton to trap pollen grains that he could count under a microscope. Blackley found that the pollen count was high in the summer months when his hay fever was at its worst.

In 1864, Blackley came across Charles Darwin's report that dust laced with microorganisms floated out to the middle of the Atlantic. It left him wondering if pollen could also travel long distances. Perhaps his hay fever was caused not just by pollen from a nearby vase or a field of hay, but by grains that flew in from other countries. These long-distance travelers might soar in the upper altitudes. To catch them, Blackley built a kite from tissue paper, waterproofed with linseed oil and tree resin. It measured six feet long and three feet wide. To its tail Blackley tied brass cases, each holding a sticky glass slide. He launched the kite into the wind. It soared as high as fifteen hundred feet.

When Blackley pulled the kite down to Earth, he expected to find only a scant amount of pollen in his traps. He found a shocking abundance. There was more pollen at fifteen hundred feet than closer to the ground. Blackley managed to catch pollen even when he flew his kite in the high winds blowing in from the plantless ocean. He concluded that pollen and perhaps germs as well could be carried long distances across the sea at high altitudes.

When Blackley published his book about his research, Charles Darwin sent him some unexpected fan mail. "The power of pollen in exciting the skin & mucous membrane seems to me an astonishing fact," Darwin wrote. "I have seen an account of *buckets-full* of coniferous pollen having been swept off the deck of a ship off the American coast."

Blackley was grateful for Darwin's praise. But it did not help him win over doctors. Some critics could not understand how an atmosphere filled with pollen could spare most people while tormenting a few. "Hay-fever is essentially a neurosis," the American doctor George Miller Beard declared.

Microbiologists had their own objections. Symptoms like itchy eyes and scratchy throats could be caused only by germs spread from person to person, they believed, not by stray pollen. With enough research, they believed, the hay fever germ would be revealed. None of those skeptical microbiologists recognized that the body's defenses against germs sometimes reacted badly against pollen grains—or against cat dander or peanuts. Blackley died in 1900, six years before a word was coined for the phenomenon he had discovered: *allergy*.

o o o o

As Blackley stood on solid ground and flew his kites, a few explorers were rising above the clouds. Hot-air balloons had been invented in the late 1700s, and within a few decades physicists were using them to take readings of the upper atmosphere. When the British scientist James Glaisher rose thirty-seven thousand feet in 1862, he felt as if he was drifting through an empty ocean. "Absolute *silence* reigns supreme in all its sad majesty," he wrote. "Our voices have no echo. We are surrounded by vast desert."

Blackley never rode in an air balloon. Neither did Pasteur or Miquel or any other early aerobiologist. It was not until 1892 that a Swiss doctor named Hector Cristiani soared almost two miles into the air to search for life.

It happened almost by chance. The famed balloonist Eduard Spelterini descended one day out of the sky over Geneva in a magnificent craft made of yellow silk. Spelterini offered to take the city's scientists up in his balloon, the *Urania*. Cristiani could not resist. He was the opposite of Miquel: where Miquel was monomaniacal, Cristiani was catholic in his scientific tastes. At the University of Geneva, he studied tumors that grew in muscles, cut thyroid glands out of snakes, and inspected bacteria from a patient's infected penis. When Cristiani got the chance to ride aboard the *Urania*, he rushed to have ten aeroscopes built for the flight, rugged enough to withstand a bumpy, windy voyage.

On the cool, crisp afternoon of September 11, 1892, Cristiani loaded his equipment into Spelterini's gondola. As *Urania* rose to three thousand meters, Cristiani opened his aeroscopes one at a time and then opened more of them as the balloon descended again. When Cristiani inspected the slides, he discovered fungal spores and bacteria from as high as a thousand meters. But he was cautious about drawing lessons from a single journey. It was possible that some of the microbes he had caught were shed by the *Urania* itself, rather than floating in the high atmosphere.

Over the next two decades, only a few other balloonists captured microbes from the air. In the meantime, a new means for reaching the sky emerged. The German engineer Otto Lilienthal built himself a glider from willow rods. He used it to fly on updrafts that blew over hills near Berlin. In 1895, he became the first person ever photographed airborne. The image shows Lilienthal frozen in flight above a barren field. After two thousand successful liftoffs, Lilienthal's glider failed him in 1896, and he plunged fifty feet to the ground. "Sacrifices must be made," he reportedly declared, and then died.

The news of Lilienthal's death caught the attention of Orville and Wilbur Wright, two Ohio brothers who ran a print shop and repaired bicycles. They began building flying machines, and in 1903, on a hill near the coast of North Carolina, they launched the Wright Flyer. By World War I, pilots were using planes to drop bombs and fire machine guns on troops below. After the war, American military pilots looked for peacetime work. Some delivered airmail. Others entertained farm towns with barnstorming shows. The Department of Agriculture turned army planes into crop dusters to spray newly developed pesticides on farm fields.

Around 1920, a plant pathologist at the University of Minnesota came up with a scheme to use airplanes for research. The question he wanted to answer, as he later described it, was simple: "What goes on up in the air?"

○ ○ ○ ○

Elvin Stakman grew up in Minnesota wheat country, and he paid his way through college by working summers on threshing crews. In the summer of 1904, he came face-to-face with a national disaster: American wheat farmers lost up to half their harvest that year to rust—the same fungus that had figured as divine punishment in the Bible and inspired Romans to slaughter dogs to the god Robigo.

At the dawn of the twentieth century, rust could still cause biblical destruction. Unlucky farmers would walk out one morning to discover their fields were turning black. The wheat was speckled with spores that grew into pustules that produced more spores, which traveled to neighboring stalks. "The terrible devastation over thousands of square miles can scarcely be realized by any one who has not seen it himself," Stakman later said.

After finishing college and working for a few years as a schoolteacher, Stakman went back to the University of Minnesota to get a PhD. In graduate school, he studied *Puccinia graminis*, the species that caused wheat rust in the United States. It did not travel directly from one wheat stalk to another. It first had to switch to another plant known as barberry to complete its life cycle. Barberry was a common wild bush in Europe and Asia, and English settlers had brought it to the colonies, to enjoy its deep green foliage and metallic-red berries. Birds ate the berries, flew for miles, and released them in their droppings. The rust spread west with the barberry, ready to devastate the new wheat fields that farmers planted.

Outbreaks of rust in the nineteenth century prompted wheat experts to search for strains that could resist the spores. One Ukrainian breed proved especially hardy, and by the early twentieth century, it was planted across most American wheat farms. But this solution turned into its own disaster. As the rust multiplied and mated, it quickly produced strains of its own, some of which could attack the new breed of wheat. The new outbreaks were even worse, because now the rust could spread across miles of monoculture fields.

In graduate school, Stakman investigated the absurdly complicated life cycle of the fungus. Once rust ravaged a stalk of wheat, it produced spores that could drop to the ground and survive the winter. In the spring, the spores roused back to life, growing into tiny stalks that produced a second type of spore that had to land on a barberry plant in order to develop. On the barberry leaves, the spores produced pustules, which then produced a third type of spore that was picked up by insects, which carried them to other barberry plants. When the spores from two plants encountered each other this way, they could have a kind of fungal sex, mixing their genes together and then growing into a yellow patch on the underside of a barberry leaf. And that patch produced a fourth type of spore, which then traveled to another wheat plant, to make a fifth spore that started the cycle all over again.

Stakman earned his PhD in 1913. A month later, at age twenty-eight, he became the head of the university's plant pathology department. The science had changed dramatically since the days when Miles Berkeley, a vicar, made hugely important discoveries about late potato blight in his spare time. It had become a professional field, and a matter of national security. When World War I broke out in 1914, the United States used its huge harvest of wheat to keep Allied soldiers well fed. Germany, meanwhile, was hit by a new wave of late potato blight that wiped out two-thirds of its potato crop in 1916, leading to the deaths of 700,000 civilians. "This epidemic of late blight did about as much to win the war for the Allies as the bullets did," Stakman later said.

The United States government wanted to maintain its advantage in food. Rust threatened that edge in 1916, when it caused an outbreak that swept across much of the American Midwest, destroying two-thirds of the country's wheat crop. Even that gigantic loss did not bring the United States anywhere close to a national famine, thanks to the vast supply of wheat that farmers were growing. But the government now treated rust as a threat to its military strategy.

Stakman advised the government that the best way to fight rust was to attack the barberry bushes it depended on. "We don't know whether all rust comes from them but we know there's a lot of rust comes from them," he later recalled saying at the time. "So it seemed sensible to eradicate barberries."

A number of plant pathologists agreed with him, and together they paid a visit to the secretary of agriculture. At the meeting, they persuaded him to systematically destroy barberry. As soon as the secretary agreed to the proposal, the plant pathologist sitting next to Stakman punched him in the stomach, declaring, "Fine, and you put this man in charge." With the wind knocked out of him, Stakman accepted the job.

Stakman lined up support from farmers, flour-milling companies, and railroad operators. He lobbied states to ban the planting of barberries and started a propaganda campaign to persuade ordinary Americans to kill every barberry plant they spotted—not just on farms, but even in city yards. Every dead plant was a small victory for the war effort. A poster for Nebraska's Barberry Eradication Day in 1918 declared, "This barberry bush is a kin of the Kaiser."

In November 1918, Germany surrendered, and the wartime demand for wheat began to ebb. But the battle against rust rolled on. "Thousands of bushes have already been destroyed, but thousands remain as a standing menace to our grain crops," Stakman declared in a 1918 report. "Destroy the barberry and protect the grain." The government no longer linked the fungus to Germany; now they paired it with peacetime enemies such as anarchists and even the devil himself.

As airplanes became more common in American skies, Stakman saw a chance to learn more about his enemy. He knew that a single infected barberry plant could produce billions of rust spores that were light enough to be carried by breezes to neighboring plants. But how far could they travel? "If there was a virulent race down in Louisiana, we had to know in self-protection whether we could expect that that would be carried to North Dakota, Minnesota, and up into Canada by the air," Stakman later said.

Stakman became the first scientist to fly in airplanes to survey the life in the air. He arranged for pilots from the United States Army Air Service to take him and his team on dozens of flights. They soared over the Great Plains: from Waco to San Antonio; from Sidney, Nebraska, to Pine Bluffs, Wyoming. On some flights, Stakman held out a glass slide smeared with Vaseline for five minutes at a time. He sometimes collected spores in a wooden box fixed to the lower wing of a biplane. By pulling on a wire running to the cockpit, Stakman could open and shut doors in the box, exposing slides to the wind.

As high as sixteen thousand feet, Stakman caught spores. On some flights he caught thousands. The rust from a barberry bush could threaten more than just neighboring fields, Stakman realized. It could destroy farms hundreds of miles away.

o o o o

Stakman spent the early 1920s surveying the air before turning his attention back to Earth. He began breeding new varieties of wheat, hoping that some could resist the newly evolved strains of rust. But he held out hope that someone would continue his barnstorming research, and in 1931, Fred Meier revived the air campaign.

Eight years Stakman's junior, Meier had grown up in Massachusetts and gone to Harvard to study cryptogamic botany—a subject that includes fungi, algae, and lichens. In the summer of 1915, he traveled from Harvard to Washington, DC, to work as an assistant at the Department of Agriculture. He spent much of his time looking down the tube of a microscope at molds and other fungi that ruined crops. One afternoon, Meier left the lab to walk to the Potomac and spend the afternoon fishing. Along the way, he passed a freight yard where piles of watermelons shipped from the South were rotting on the tracks. It was possible that one of them had become infected with some disease while it still grew in a farm field. After the melons had been loaded into a boxcar, the disease ruined the whole batch. Meier picked up an armful of rotten melons and staggered back to the lab.

When Meier inspected the fruit, he could see that the meat had turned a slimy black. Some of the melons had rotted only around the stem, while others were ruined from rind to core. It looked like the work of some kind of fungus. To find it, Meier sterilized a razor blade in a flame and sliced out a thin portion of a watermelon, just where the rot was advancing into the ripe red meat. He picked off bits with a platinum needle and moved them into Petri dishes. After two days, the Petri dishes had filled with gray fungal threads. The threads produced ball-shaped fruiting bodies, inside which spores grew.

For centuries, contagionists and miasmatists had argued over the causes of the diseases that wiped out crops. But thanks to the work of nineteenth-century naturalists like Miles Berkeley and Anton de Bary, plant pathologists embraced the germ theory of disease. When Meier discovered spores on his rotten watermelons, he imitated Robert Koch to prove that the fungus was responsible for the disease.

First, he injected threads into three spots on a healthy watermelon. The fruit began to rot at all three places. For a new experiment, Meier bought sixteen watermelons at a wharf in Washington. He inoculated eight of them with the gray threads and left the other eight intact. For thirty-six hours they sat by an open window. All of the watermelons injected with the fungus began to decay. "The 8 checks remained perfectly sound throughout the course of the experiment," Meier noted with satisfaction. Still in college, Meier had discovered a new disease. He named it watermelon stem rot.

"This piece of research led to my being inextricably mixed up with the watermelon industry," Meier later recalled. He shuttled between Harvard and Georgia, where he studied the fungus on watermelon farms. When the United States prepared to enter World War I, Meier tried to join the army, but was rejected on the grounds of physical disability. Even out of uniform, he wanted "to be of as much service as possible," as he later put it. He jumped at the chance to join the Department of Agriculture when he was offered a full-time job, abandoning his work toward a PhD at Harvard.

"In March 1918, I left the University to take charge of a laboratory at Miami, Florida," Meier later wrote, "and since then have given all my time to the government." He called himself a watermelon doctor.

∘ ∘ ∘ ∘

The *Miami News* heralded Meier's arrival in Florida: "Man from Department of Agriculture Here to Learn What Ails Melons." He had come to continue his research into watermelon stem rot. Meier soon discovered that the "weakling parasite," as he called it, was not strong enough to invade healthy fruit. It could gain entry only to a damaged one—even a nick on a stem was enough.

To fend off the fungus, Meier concluded, farmers and train operators needed to handle their fruit carefully, so as not to damage the watermelons and make them vulnerable to attack. To fight the fungus in farm fields, he concocted a spray of copper sulphate and other compounds. Thanks to Meier's efforts, watermelon farmers saw a drastic drop in their losses in the early 1920s.

As Meier traveled from one diseased melon field to another, he tried to imagine how the stem rot traveled between them. He discovered that watermelons were not the only host on which it could grow. The stumps of trees surrounding melon fields could also foster the fungus, producing spores that could waft into the air. "Since the spores are wind borne, they are abundant in the air and are likely to find lodgment on the cut stems," Meier observed. How far the spores could travel, he could not say. But he did wonder.

Meier left Florida for New York to run a plant disease laboratory, and then headed to department headquarters in Washington. As he rose through the ranks in the 1920s, the mission of the Department of Agriculture was changing dramatically. The government cut off price supports for farmers after the end of World War I, causing their income to crash and the value of their farmland to fall. The department tried to keep farmers from falling on hard times by helping them make their farms more like factories. Its scientists and economists came up with new strategies to increase production. They encouraged farmers to go into debt to buy tractors and other new machines, along with chemical fertilizers to spur the growth of crops.

For his own part, Meier helped farmers by reducing crop losses from diseases. He crisscrossed the United States, inspecting fields of potatoes, beets, and honeydews. He would later declare that he had traveled more than anyone in the entire Department of Agriculture. Meier had an easy charm that allowed him to move comfortably among farmers, food inspectors, and melon distributors. He offered advice on fighting smuts, molds, and blights. He taught fruit inspectors how to recognize signs of disease. He urged farmers to protect their seeds from infection and to blanket their farms with even stronger fungicides when the old ones failed. He would go on the radio to talk about the uncertain future of the coming harvest, about how the weather might or might not bring outbreaks of apple scab or rose mildew.

In 1930, Meier, then thirty-seven, took charge of the government's campaign to fight rust. Over the previous twelve years, the United States had destroyed 18 million barberry bushes, and losses from rust had dropped from 57 million bushels of wheat to 9 million. "Excellent progress has been made," Meier declared as he took over the fight. Meier's task was to search for the hundreds of thousands of individual bushes still lurking across the country, each a potential source of a new wave of devastation. Meier enlisted 4-H clubs and Boy Scouts to look for them. "To an increasing degree, we are led to locations of bushes by reports from school children," Meier said.

For all his efforts, though, rust and other pathogens still managed to ravage farms. "Too often have farmers disinfected their seed and planted it in disease-free soil only to have their hopes for a good crop blasted by the mysterious appearance of a rapidly spreading disease," Meier said. He wanted to understand that mysterious appearance. When he walked through a rust-stricken wheat field, he could see a cloud of what looked like red dust—which was, in fact, a vast number of fungal spores rising into the air. The spores were flying. And before long, Meier was flying too.

○ ○ ○ ○

Stakman persuaded Meier to go beyond rooting out barberry bushes and learn more about how wheat-killing rust spores moved through the air. But it wasn't as easy for Meier to fly off into the clouds as it had been for Stakman a decade earlier. Now the country was in the midst of the Great Depression, and the Department of Agriculture was busy with aiding farmers devastated by the Dust Bowl and foreclosures. Meier's superiors gave him permission to look for life in the sky, but only in his off hours.

Meier tacked Petri dishes onto sticks, creating instruments that looked vaguely like squash racquets. On vacations and weekends Meier charmed his way onto planes flown by the army, navy, and coast guard. To stay warm in the open cockpits, he wore a thick leather coat, leather mittens, and a leather helmet that completely covered his head, except for his goggled eyes. He looked like a barnstorming plague doctor.

Meier even persuaded authorities to let him travel on the USS *Los Angeles*, a 656-foot-long airship. "The crew of the *Los Angeles* and some of the airplane pilots were somewhat skeptical as to the success of the hunt," he later recalled, "for immediately after exposure the plates looked fully as clean as before." But it took only two or three days in an incubator for those seemingly clean plates to grow lush blotches. Meier pricked the blotches with a sterile needle, which he dipped into test tubes that quickly filled with dozens of species of fungi.

James Glaisher may have thought the atmosphere was a lifeless desert, but Meier was discovering an invisible zoo. The diversity of life in the clouds gave Meier hope of making some major discovery that could allow him to finish the Harvard PhD he had abandoned twelve years before. Some of the fungi he found were known to attack potatoes. Others grew on lettuce and celery. Meier found a species that ferments cheese. On a flight over Washington, DC, he set a record for finding life at high altitudes, trapping the spores of a tree-killing fungus at eighteen thousand feet.

"Judging from our success in germinating the spores collected on recent flights, many of the crop-destroying fungi are none the worse for wear, after a period of air travel," Meier said. Tracking spores at high altitudes would be essential to controlling plant diseases, Meier concluded, thanks to "the certainty of their returning to earth to start new generations of crop destroyers many miles from the diseased fields."

Meier was finding a remarkable amount of life in the air despite the fact that he could take only short hops into the sky. He had no idea how much higher or farther spores could travel. If a living spore could rise to eighteen thousand feet over the capital, where else could it go? Could high-altitude winds carry it over the Atlantic? Would it fall into the water and perish, or might it travel all the way to another continent?

Other plant pathologists thought the questions Meier was asking weren't worth answering. He later recalled how another plant pathologist, Herbert Whetzel, "ridiculed the idea of there being any value in such studies." The ridicule got under Meier's skin. When he found out that the Lindberghs were planning a trip across the Atlantic, he saw a chance to show the skeptics they were wrong.

○ ○ ○ ○

Seven years earlier, Charles Lindbergh had been a college dropout flying mail from St. Louis to Chicago. But once he made the first nonstop solo flight across the Atlantic in 1927, he became a celebrity on par with kings. He also became rich, thanks in large part to the businessman Juan Trippe, who enlisted Lindbergh to lobby for his new company, Pan American Airways. Together, Trippe and Lindbergh secured lucrative contracts to deliver mail from the United States to Latin America and the Caribbean. Lindbergh also advised Trippe on the design of new passenger planes that would ultimately fly across the Atlantic and the Pacific. These flying boats, as they came to be known, had to carry so much fuel that they would have to land on water instead of tarmacs.

Lindbergh used his celebrity travels for a "girl-meeting project," as he later put it. The project ended in Mexico when he met Anne Morrow, the daughter of the American ambassador. A quiet college student who wrote poetry, she dismissed Lindbergh at first as a "baseball-player type." But Anne eventually fell in love with him, and he with her. After they married in 1929, Charles taught Anne to fly, and they traveled from Los Angeles to New York in fourteen hours and twenty-three minutes, setting a new record. Anne made the trip seven months pregnant.

Reporters gawked at her pregnancy, disgusting Charles so much that he mostly avoided the press for the rest of his life. But his silence made the public hunger even more for news of the famous couple. Lindbergh honed his skill at evading reporters. On November 28, 1930, he slipped out of his home in New Jersey and drove unrecognized into New York. He made the trip in order to talk to a Nobel Prize–winning surgeon about an idea that had obsessed him for years: how to fix broken hearts.

Lindbergh had always fashioned himself a scientist. As a boy growing up on a Minnesota farm, he wondered what distinguished living animals from dead ones. "I was spellbound by the body of a dead horse I encountered while walking through a woods adjacent to our farm," he later wrote. "I had seen dead things before, but never of such magnitude. The difference between life and death was so apparent in that rotting hulk, and yet it was not understandable! What stopped life from living?" When Charles and Anne settled in New Jersey, he bought a library of books about medicine and biology and began building a lab. But he had no idea what to do in it.

Soon afterward, Anne's sister suffered a bout of rheumatic fever that damaged a valve in her heart. Lindbergh wondered why no one had invented a mechanical heart to replace defective originals. When he asked Anne's obstetrician, the doctor suggested Lindbergh talk to his former boss, Alexis Carrel. Carrel, a professor at the Rockefeller Institute for Medical Research, had won the Nobel Prize in 1912 for keeping tissue alive outside the body. He was still carrying out experiments that blurred the line between life and death. When the two men met in New York, Lindbergh immediately fell under Carrel's guru-like spell. "In Carrel," Lindbergh would later write, "spiritual and material values were met and blended as in no other man I knew."

Inspecting Carrel's lab equipment, Lindbergh quickly pointed out how to make it work better. Carrel made him his unofficial technician, and Lindbergh threw himself into designing a better chamber for preserving living tissues. He proved a natural bioengineer. He also enjoyed Carrel's early morning lectures about human nature. Carrel, like many scientists at the time, was a eugenicist who believed that the white race was naturally superior to all others. But he put his own peculiar spin on this ideology: he hoped to perfect the art of organ transplantation so that the elites of Western civilization could live longer and exert their superiority over the world. Lindbergh lapped it all up.

Lindbergh left Carrel's company for another flight with Anne on July 27, 1931. Their son, Charles Junior, was only thirteen months old at the time. Charles and Anne boarded their Lockheed Sirius and flew across the North Pacific to scout stopovers for Pan Am's flying boats to Asia. The world followed every leg of their journey, which took them to Alaska, Japan, and China. Only a few months later, in March 1932, the Lindberghs made a horrific, unplanned return to the limelight. Charles Junior disappeared from his nursery, and his lifeless body was found weeks later. The catastrophe left Anne profoundly depressed for months. Charles sought peace in Carrel's laboratory, building pumps he imagined could conquer death.

In August 1932, Anne gave birth to a second child, Jon. Charles at the time was "attempting to build a somewhat normal life," as he later wrote. For him, that meant researching potential routes that flying boats could take across the Atlantic. With few islands to serve as stopovers on the way to Europe, Charles turned his gaze up toward the Arctic. "You could evade the oceans if you were willing to fly far enough north," he later said. Charles announced to Anne they would be taking another journey the following summer over Greenland and Iceland. Anne was reluctant to leave Jon, but Charles insisted he needed her on the voyage.

○ ○ ○ ○

It's not clear how Fred Meier learned of the Arctic flight, or how he managed to pitch the Lindberghs his idea for a survey of the air. By 1933, Charles was getting tens of thousands of letters each year. A mechanical engineer asked Lindbergh to visit his house, which he had built entirely of newspapers. An American steel company wanted help getting a contract in Portugal to build a bridge. The governor of Minnesota invited Lindbergh to come to the Swedish Midsummer Festival. A woman in Los Angeles sent him the antisemitic tract *The Protocols of the Elders of Zion*, darkly promising it would help him understand what was happening to the United States. However Meier managed to get through the crush of requests, Lindbergh agreed to his plan. But Lindbergh demanded that the government stay mum about the survey. He would tolerate no publicity.

The two men met to discuss how Lindbergh would carry out the survey. Meier's Petri dish squash racket would not work on a high-speed flight through bitterly cold Arctic air. Lindbergh tapped the creativity he had used to help Carrel search for immortality. An invention came to him: "an extremely simple device which permits opening aluminum cylinders containing sterile glass slides, after they have been thrust into the air stream of a plane in flight," as Lindbergh later described it.

"I made a rough sketch of this in a few minutes," he recalled. "It was done so hurriedly and was so simple that I would not consider placing it under the heading of an invention."

He and Meier dubbed the device a sky hook "because of the connection of that word with the early days of aviation," Lindbergh later said. The name probably referred to a metal rod that military planes used to latch themselves onto dirigibles. Their new creation, Lindbergh hoped, would "hook organisms out of the sky."

○ ○ ○ ○

The Department of Agriculture still barred Meier from working on his aerial surveys during official hours, but that didn't stop his superiors from enjoying the proximity to fame he provided them. Henry Wallace, the secretary of agriculture, sent a personal letter of thanks to Lindbergh for his cooperation. The spread of diseases through the air remained such a great mystery that the survey would be of great significance, Wallace said. "While it is realized that the collection of these samples will be incidental and perhaps cause some inconvenience on your trip, we believe the effort worthwhile for the findings will be of practical interest to our plant pathologists and may prove valuable from the meteorological point of view," he wrote to Lindbergh.

In his off-hours, Meier had the American Instrument Company build a sky hook according to Lindbergh's plans. Along with the four-foot aluminum rod, they created cartridges to hold Vaseline-covered slides. Meier loaded the slides into the cartridges in a special room at the Department of Agriculture normally used to grow cultures of microbes. The air in the room was kept still to prevent stirring up germs that might settle on Petri dishes. After anchoring each slide in a cartridge, he slid an aluminum sleeve over it, locked it shut, and rubbed it down with alcohol. Once the sky hook was ready, Meier rushed to New York to deliver it to the Lindberghs before they departed on July 9.

After taking off from Flushing Bay, they flew to Maine. There they landed at the Morrow summer home, where Jon was staying with Anne's mother and an armed guard was walking the property to prevent another kidnapping. Anne played hide and-seek with Jon, crouching behind chairs in the library as the baby crawled to find her. The thought of leaving Jon tortured her: "That dreadful ache; can't say anything," she wrote. Charles had little to say about their children on the voyage. When a relative sent Anne photographs of Jon, he only said, "Very bad for you to look at these."

On July 11, the Lindberghs left Maine for Canada. Anne took over the controls of the plane and Charles put the sky hook to use for the first time. He slipped a slide cartridge onto the top of the aluminum rod, opened the canopy of his cockpit, and put the base of the sky hook in a bracket on the fuselage. The top of the sky hook extended two feet above the edge of the cockpit. Pulling down on a handle, Charles drew down the cartridge's aluminum sleeve and exposed the slide within. He let the winds blow over it for an hour before pushing the sleeve back up and removing the cartridge.

Every day or two, Charles exposed another slide as he and Anne made their way north. On July 22 they laid eyes on the southern coast of Greenland. "The mountains towered up magnificently to meet us, a great wall against the sky," Anne later wrote. After hugging the western coast of Greenland for a few days, the Lindberghs set out over the ice cap covering most of the island. "That smooth white dome seen beyond the rim of mountains was so unreal that I had nothing in my mind to compare to it," Anne recalled. As they flew on, the mountains shrank behind them to black spots on the horizon. "Ahead—nothing but dazzling white," Anne wrote, "a white that had no depth or solidity but looked like clouds or fog under the glaring sun."

Charles exposed a slide over the ice cap, as he and Anne flew as high as twelve thousand five hundred feet. The outside temperature had fallen to 10 degrees Fahrenheit. The cold caused the rubber cork at the base of the cartridge to fall out as Lindbergh was handling it in the cockpit. For a fraction of a second, the sleeve slid open, exposing the lower part of the slide to the cockpit air before Charles plugged it shut again with cotton.

"This may be sufficient to cause an erroneous positive result," he jotted in his notebook.

No one knew that the Lindberghs were performing an experiment over Greenland. Virtually no news came from the famous couple at all. On August 10, a radio operator in Copenhagen thought he caught a message that they had crashed shortly after takeoff. In a matter of hours, rumors raced around the planet that Charles and Anne were entombed in Arctic ice. A few newspapers even got the story into their evening editions. "He had a brief but magnificent career," a newspaper in Paris declared that night. When morning arrived in Greenland, word came from Julianehaab: "Both Lindberghs safe."

The sky hook continued to open and shut as the Lindberghs completed their journey over Greenland. They operated it as they flew to Iceland and crossed the Norwegian Sea. The survey ended on August 26, when Charles and Anne landed at the entrance of Copenhagen Harbor.

The Lindberghs went on with their journey, stopping in cities across Europe, before heading south to Africa, crossing the Atlantic to Brazil, and turning north to the United States. Moviegoers watched the couple land in Miami on a British Pathé newsreel entitled "Lindy & Anne Home Again!" After twenty-nine thousand miles, the couple grinned at the reporters as they hurried into a waiting car. The newsreel announcer said nothing about the secret collection of twenty-six slides that Charles and Anne had shipped back to Meier in Washington.

When the slides arrived, Meier was eager to get them under his microscope. But his attention was pulled in other directions. The Department of Agriculture needed him to help administer the Civilian Conservation Corps, a program that was hiring millions of laborers left unemployed by the Depression to plant trees and build terraces. And Meier now had a new voyage to prepare for. He had been invited to search for life at altitudes far higher than the Lindberghs could ever reach. A balloon would soon be rising to the stratosphere.

four

ETHEREAL SPACE

Fred Meier's invitation to the roof of the world came from Chicago. The city was preparing to host the 1933 World's Fair, otherwise known as A Century of Progress International Exposition. For the occasion, the organizers commissioned the construction of a balloon that could hold half a million cubic feet of hydrogen gas. The craft, also named *A Century of Progress*, would lift two pilots in a metallic sphere eleven miles into the sky, higher than any human had traveled before. The balloon would take scientific instruments as well as people. Its payload would include a set of glass tubes, loaded with spores supplied by Meier.

Only three decades had passed since scientists had even realized that the stratosphere existed. When unmanned balloons rose through the air, their thermometers registered colder and colder temperatures until they reached about seven miles. And then the chilling stopped. At first the finding didn't make sense, because scientists thought that the atmosphere was a uniform mix of gases that gradually thinned out at higher altitudes. In 1902, the French meteorologist Léon Teisserenc de Bort recognized what those readings were telling scientists: the atmosphere had layers. The lowest level, which he called the troposphere, turned out to be home to Earth's clouds and snow, to the wind and rain that make the weather, to the dense gas that provides enough lift for planes to fly. Above the troposphere Teisserenc de Bort recognized a new region. The stratosphere, wispy and dry, seemed like a place where life could not survive. Not only was the air pressure perilously low, but cosmic rays and ultraviolet radiation might be intense enough to kill anything that ventured too high.

The Swiss physicist Auguste Piccard first got the idea of traveling to the stratosphere inside a metal sphere. In 1931, he and his assistant climbed into a gondola of his own design and rose over Germany until they reached an altitude of 51,775 feet. "Here our ascent was of a fairy-like beauty," Piccard later recalled. "We saw the sun rising above the horizon long before the earth beneath us had been touched by the sun's rays." After visiting the stratosphere for the first time, the two balloonists landed their craft on a glacier in Austria. They spent a cold night in the gondola, had a breakfast of glacier ice and arrowroot flour, and trekked to the nearest village.

The organizers of the Chicago World's Fair asked Piccard to design an even bigger balloon, hoping it would set a new record. Meier was invited to prepare an experiment to see if life could survive in the stratosphere. From his flights on planes and dirigibles, he knew that spores could survive eighteen thousand feet above sea level. But he could not say how much higher life could rise before being killed by cosmic rays or ultraviolet radiation. Meier decided to send the hardiest spores in his collection aboard *A Century of Progress*. He picked out rust and other types of fungi he had captured on his own airborne journeys and packed them into cotton-stuffed glass tubes.

Just after midnight on August 5, 1933, thousands of spectators converged on Soldier Field to watch *A Century of Progress* prepare for flight. Bathed in spotlights, the balloon inflated like a gargantuan jellyfish extending its bell. After three hours, it was ready for launch. Its pilot, Lieutenant Commander Thomas Settle, gave the order for the ground crew to let go of the ropes. As the balloon rose, it lifted the gondola off the ground, with Meier's tubes of spores lashed to the rigging. But after a few minutes, the balloon began to leak and sink. It collapsed in a nearby rail yard.

That was not the end of the mission, though. *A Century of Progress* was taken to Akron, and after three months of repairs, it was ready for another flight. This time Settle traveled with Major Chester Fordney. On November 17, 1933, the balloon carried the two men smoothly into the

Ohio sky. The ascent was so uneventful that Settle and Fordney chatted on their radio with people in Akron, Chicago, and New York. Meier's tubes of spores rose with them. Settle and Fordney became the first Americans to reach the stratosphere, rising to an altitude of 61,237 feet before they began releasing hydrogen from their balloon to start their descent. They had beaten Piccard's 1931 record, but they would later learn they had been bested. Weeks earlier in September, a Soviet balloon had reached 62,304 feet.

Settle and Fordney drifted east as they descended, finally landing in a New Jersey swamp. They spent the night sleeping on the canvas, and in the morning they waded for five miles until they found a house where they could make a call for a rescue. The tubes of spores survived the flight and were delivered to Meier in Washington. For the first time ever, a scientist now had the chance to see what direct exposure to the stratosphere did to living things.

o o o o

Meier now had two sets of samples to inspect: the tubes that had traveled to the stratosphere on *A Century of Progress*, and the slides that Lindbergh had exposed to the Arctic air. But his superiors did not want him spending any official hours on either. So Meier packed up the tubes and slides and brought them to his house in Chevy Chase, Maryland. "The problem was settled by setting up a laboratory at my home in which work is being done at night and in early morning hours," Meier explained to Lindbergh.

To inspect the spores from *A Century of Progress*, Meier provided them with food. All seven types of fungi germinated and began to grow tendrils. Life, it seemed, could survive the harsh conditions near the edge of space. Still, Meier could not rule out the possibility that the spores had been protected from the full brutality of the stratosphere. The cotton in the tubes might have shielded them from some of the ultraviolet rays streaming down from the sun. The spores might also have enjoyed some protection from the gondola against which they rested, rather than floating out on their own in the near void.

In spite of those caveats, the journey brought Meier fame. "It was the first time in the history of science that living spores had been sent to such a height above the earth and brought back for scientific analysis," the *Boston Globe* announced. "Life in the form of fungus spores or molds can survive 11 miles above earth where human beings would die if unprotected."

Meier then turned to Lindbergh's haul. He ran each sky hook cartridge through a flame to sterilize the exterior so that he wouldn't contaminate the slides when he extracted them. Many slides were speckled with volcanic ash and glass. Along with those geological traces, Meier also found remnants of life: filaments of algae, spicules of sponges, wings of insects, and grains of pollen. The slides were also loaded with spores of fungi: *Macrosporium, Cladosporium, Leptosphaeria, Mycosphaerella, Trichothecium, Helicosporium, Uromyces, Camarosporium, Venturia.* In some cases, Meier could see that the fungi were still alive, pushing out tendrils into the Vaseline.

"Your slide exposures were very successful," he wrote to Lindbergh in December 1933. "It seems certain that they will contribute materially to our knowledge concerning the spread of both harmless and plant disease causing fungi by means of spore movement in upper air currents. The pollen counts obtained should be of interest to botanists and to medical men."

To Meier, Lindbergh's trove of spores was proof that microscopic life could fly like the Lockheed Sirius from one continent to another. "Definitive evidence has been obtained that air currents play a part in the dissemination of pollen and spores of fungi between northern lands," Meier later said.

Meier went so far as to invite Lindbergh to Maryland to work with him. "Could you find time to spend a night at my home in order that we might put in some time at the microscope and at making notes on condition of terrain at different points where exposures were made?" he asked. "You would be entirely safe from interruption from anyone except those whom you would wish to see or talk to on the telephone."

There's no evidence that Lindbergh responded to Meier, even to turn him down. But Meier's letters to Lindbergh hint that he was worried that his collaboration with Lindbergh was slipping away. He sent the sky hook to Lindbergh, as if he were offering a gift to a modern god. "We have made photographs and drawings which will enable us to make another, so that you may now put it with the museum collection if you care to do so," Meier wrote.

° ° ° °

In 1934, Meier looked for other ways to continue finding life in the air. He took flights of his own over Texas farms. "A tremendous amount of rust inoculum was being liberated into the air," he later recalled. He recruited some of his young barberry plant warriors to track the flight of the fungus. Children in Nebraska set slides on the ground and then retrieved them after rainstorms. When they sent the slides to Meier, he found that the rain had washed rust spores down on them as well.

His surveys supported Elvin Stakman's hunch that winds were blowing rust spores in vast numbers from the south to the north. If the government could make regular sweeps of the air, they might be able to pinpoint the fungal hot spots in the Great Plains. Farms in those places would do well to try planting resistant varieties of wheat.

"Adjoining countries may have a mutual interest in the problem," Meier added. Canadian wheat growers might get influxes of rust flying in from the United States. Rust outbreaks in Texas and Oklahoma might have been launched from Mexico. Meier believed that the United States should offer help to Mexico to plant resistant wheat, not just for their own benefit, but to protect American farms as well. The life of the air might become a matter of international diplomacy.

Rare flights would not be enough to allow scientists to understand the highways in the sky on which rust traveled. Meier saw a chance to piggyback sustained surveys on the fast-developing science of weather forecasting. Until the 1930s, the US Weather Bureau took measurements with giant box-shaped kites flown at the end of two miles of piano wire. Now the bureau was switching to a small fleet of biplanes that could fly weather instruments as high as three miles.

Meier persuaded the Weather Bureau to put a new sky hook on its planes, and the Department of Agriculture agreed to provide him with the lab space, table lockers, and glassware he needed for his work. But his superiors also made it clear "that Mr. F. C. Meier shall serve as leader of this project, confining his work to unofficial hours." His research remained a private sideline, even as he was about to make some of his most spectacular discoveries.

○ ○ ○ ○

In early 1934, an army captain named Albert Stevens convinced the National Geographic Society to sponsor the construction of a balloon bigger than any yet flown. Stevens would pilot the craft, called *Explorer*, with Captain Orvil Anderson and Major William Kepner. Their goal would be to reach the staggering altitude of seventy-five thousand feet. As Stevens prepared for the voyage, he gave Meier another chance to send spores up to a record-setting height. Meier designed a new container for the *Explorer* voyage, one that would expose his fungi to the full harshness of the stratosphere. He packed eight kinds of spores in hollow beads of quartz, through which ultraviolet radiation and cosmic rays could pass.

The launch site for *Explorer* was a valley in South Dakota that came to be known as the Stratobowl. Construction crews filled it with supplies, including fifteen hundred tanks of hydrogen, and then began assembling the craft. On July 9, 1934, the wife of the governor of South Dakota christened *Explorer*, but instead of champagne she poured liquid air on the gondola from a long-necked flask. When it hit the metal shell, it burst into vapor. In the days that followed, the crew fitted the gondola with cameras and cosmic ray detectors. They brought a spectrograph on board to measure the color of the sky, and they adorned the instrument with a necklace of Meier's tubes.

Once *Explorer* was ready, the crew had to wait for the right weather. On July 27, a high-pressure system arrived, and soldiers spread out the balloon on a two-hundred-foot-wide bed of sawdust. Into the night, they held the balloon steady as it inflated with hydrogen. "The weather was ideal," Stevens later wrote. "Hardly a movement could be detected among the acres of cloth." It took three hours to lash the gondola to the inflated balloon. Anderson and Stevens slipped inside the metal globe, while Kepner stood on top of it to direct the launch, caged in by the ropes. *Explorer* gently rose and safely cleared the rim of the Stratobowl.

Kepner continued standing atop the gondola as the balloon rose to fifteen thousand feet. The crew vented some gas to stop their ascent, and Stevens climbed out to join Kepner on its roof. The two men now had to lower the spectroscope, festooned with Meier's tubes, until it dangled well below *Explorer.* "I wonder if any of my readers have ever had the problem of lowering 125 pounds of swaying bulk a distance of 500 feet on a quarter-inch rope!" Stevens later wrote in *National Geographic.* It took half an hour for Kepner and Stevens to finish the job.

The two explorers joined Anderson back inside the gondola, and they dropped ballast to resume their ascent. When they entered the stratosphere, they paused again at forty thousand feet to turn on a Geiger counter that would detect cosmic rays. The clicks of the device sounded like chickens pecking grain from a pan. The crew resumed their ascent, and then chatted with radio stations below. The sky grew dark overhead.

At sixty thousand feet, the routine of the flight was suddenly interrupted by the clatter of a small rope that had fallen onto the gondola roof. When Stevens peered through a porthole, he spotted a rip in the lower surface of the balloon.

The crew was able to manage the crisis for a while. Despite the rip, the balloon was still holding on to some hydrogen, and so they remained afloat. But as the hydrogen heated up in the sun, it expanded and swelled the canvas. The rip lengthened, letting out more gas. Stevens realized that his goal of reaching seventy-five thousand feet was now out of reach. Instead, he had to make sure he and his crew didn't fall to their deaths. Before they started their return to Earth, they performed one last piece of research, pumping stratospheric air into five flasks. Then the descent began. *Explorer* had reached 60,613 feet—a few hundred feet shy of the *Century of Progress.*

Over the Geiger counter chicken scratch, Stevens heard the whoosh of fresh rips. The crew adjusted the air pressure and checked their oxygen tank to make sure they had enough of the gas to breathe. In just over an hour, they dropped more than 7.5 miles. At twenty thousand feet, Kepner and Anderson opened the hatches, letting in fresh air.

"We all climbed out on top and took a good look at the balloon," Stevens later recalled. They could now see that the balloon was flapping in the wind as it descended, each wave of the fabric tearing it open even more. Suddenly the lower section of the balloon fell away. Only a parachute-like dome remained, and now the *Explorer* began falling fast.

The first order of business was to cut the line on which the spectrograph and Meier's flasks were hanging. As Stevens watched the rope plummet from the gondola, he could only hope that the payload's parachute would open, so that it could survive the fall. The crew then shed weight in the hopes of a soft landing. They dumped the last of the liquid oxygen, fastened the tanks to parachutes, and hurled them overboard. They poured out the lead shot they used for ballast. But they were still falling too fast. It was now necessary to abandon ship.

Stevens and his crew put on their parachutes, but they did not immediately jump. They worried that they would land so far from the gondola that they would struggle to find it after it crashed. "At 6,000 feet, we again talked the matter over and decided we had better leave," Stevens wrote.

Anderson jumped first. While the other two men got ready, they heard an explosion. A little cloud of hydrogen was still trapped under what remained of the balloon, and it now ignited. There was nothing left to slow the gondola's fall. Kepner quickly bailed out, but Stevens struggled for so long that by the time his parachute deployed overhead, the remnants of the *Explorer* balloon fell on top of it. Fortunately, his parachute worked itself free, and he gently descended to a Nebraska cornfield.

It took Stevens just a few minutes to wrap up his parachute and meet Anderson and Kepner at the gondola. Hundreds of curious locals and souvenir hunters were already there to greet them. Stevens went to a nearby farmhouse to shed his two suits of long underwear, which he put on the clothesline. After he made a few telephone calls and sent some telegrams, he stepped back outside. Someone had stolen his underwear.

Stevens headed back to inspect the gondola. The impact had crushed it like an eggshell, and he had to borrow an ax to hack his way in. "Inside, the instruments of which we had been so proud were a heart-breaking mass of wreckage," Stevens wrote. Meier's spores, on the other hand, had avoided the gondola's ugly fate. The payload had landed "as gently as if it had been handed to a skilled workman," Stevens later wrote.

The chaos of the crash delayed the delivery of the quartz beads back to Meier. When he inspected the spores, he found that five of the eight types of fungi were still alive. It was possible that the other three had died in the stratosphere. But it was also possible that their delayed journey back to Washington had been too much for them to survive. "Results were somewhat inconclusive," Meier later wrote.

Explorer's sponsors did not let that ambiguity stop them from using Meier's experiment to burnish their story. Rather than photograph Meier at his home, where he had actually investigated the stratospheric spores, a photographer came to the Department of Agriculture. Captain Stevens came as well, in his military uniform. Stevens sat in front of a microscope and peered at a slide of rust spores that had traveled with him to the stratosphere. One of Meier's superiors showed up too and posed standing nearby. Behind the two men stood Meier, jacketless. He looked anxious as he clutched a set of tubes. In October 1934, the photograph appeared in Stevens's account of the voyage in *National Geographic*. The caption reads, wrongly, "All came back to earth alive."

○ ○ ○ ○

Meier was fast becoming a fixture in the press. Two months after his debut in *National Geographic*, in December 1934, he stunned reporters by letting them know he was about to publish a paper chronicling Lindbergh's spore-trapping efforts the previous summer.

The sky hook had remained a secret for well over a year. For all the attention the press lavished on the Lindberghs, no one had discovered that they had carried out biological research over the Arctic. Now Meier unveiled the expedition in the *Scientific Monthly*. He did not offer a dry, sober catalog. Instead, he recounted "an unusual botanical collecting trip," accompanying his story with Anne's photographs and Charles's handwritten notes.

Charles Lindbergh didn't say a public word about Meier's report, but newspapers celebrated the achievement as another triumph for him. The Associated Press released a report that appeared on the front page of newspapers on Christmas Eve. "A vast unseen world of unbelievably tiny forms, some deadly to crops, some harmless, that floats high over earth on the wings of the wind, is spotlighted by the announcement that Col. Charles A. Lindbergh has been invading it for science," the AP announced.

To explain the significance of the achievement, the *New York Times* reached back more than seventy years to the Mer de Glace. "In the brave days when the relation of bacteria to disease was still both an alarming and a reassuring discovery, the great PASTEUR thought of collecting micro-organisms from a hot-air balloon and thus throwing some light on the manner in which epidemics spread," a reporter wrote. But Pasteur could only manage to climb a glacier. Now Charles Lindbergh (with Anne unmentioned) was going far beyond Pasteur's dream.

"With Colonel Lindbergh's proof that these organisms can cross wide expanses of water," the Associated Press wrote, "it is reasonable to suspect that currents of the upper air carry them all over earth, says Fred C. Meier."

The discovery was both a marvel and a worry. "From a wheat field stricken with black stem rust, for example, or from barberry bushes, a host to the disease, tiny spores can be easily wafted upward by rising air currents until they reach the steadily blowing, long-distance winds of the upper sky lanes," the AP reported. "They may not be washed to earth again by rains until they have been blown a thousand miles from their starting point, and then farmers wonder whence came the sudden outbreak of rust in the fields."

And the fungal spores might not be alone high over the Arctic, the *Times* speculated. "There can be little doubt that pathologists will be in a better position to explain how plant and possible animal diseases spread with a rapidity that sometimes baffles the public health authorities."

○ ○ ○ ○

With high-altitude life making headlines, Meier looked for his next opportunity to advance his unofficial research. In February 1935, he pitched Lindbergh a new expedition—or, rather, a series of them. They would fly together in Pan Am's growing fleet of flying boats. Lindbergh had flown the first flying boat, the *American Clipper*, on its maiden voyage from Florida to Colombia in 1931. Meier now proposed that they travel in a flying boat to South America. They could later fly across the Pacific in one of the flying boats Pan Am was building for that journey.

The surveys would give Meier an unprecedented view of life across the atmosphere. "Perhaps this work is the best means I will ever have of completing a Doctor's thesis for Harvard," he confessed to Lindbergh. "If not carried forward soon, however, some one will do it for us."

As proof, Meier anxiously shared a letter with Lindbergh. It had come from Herbert Whetzel, the plant pathologist who had ridiculed Meier two years earlier for going on wild-goose chases. Meier's success had made Whetzel change his tune. He congratulated Meier and informed him of his own plans to take a plane up above Cornell University in New York, where he was a professor. Meier knew that Whetzel traveled regularly to the Caribbean to study plant diseases; soon, he might be flying there as well.

To stay a step ahead of the competition, Meier proposed that Lindbergh meet him in Miami as early as the following month. They would board a flying boat with a radio operator and fly south to Belém in Brazil. Lindbergh would pilot the plane, Meier would take a few samples of the tropical air with Petri dishes, and the radio operator would deploy the sky hook.

"If in the course of your busy days you find time to pave the way for an interview with Pan American airways re collections in the south, I will be glad to come to New York," Meier said. But March came and went without a reply from Lindbergh. Finally, on April 22, he sent Meier a telegram. A Pan Am executive would meet with Meier in New York two days later. "Have told him about your work and believe company will cooperate in any way possible," Lindbergh wrote.

Meier came away from the meeting with a new plan that he presented to the Department of Agriculture. He would take a two-week voyage without Lindbergh. Pan Am would let Meier hop flights from Miami to Port-au-Prince in Haiti, to San Juan in Puerto Rico, and onward through the Caribbean and Mexico. Meier argued that the information he would collect would be valuable to the department—perhaps he might gather more evidence as to whether rust traveled from Mexico to the United States, for example—and so he proposed that he move temporarily to the Bureau of Plant Industry, which was dedicated to basic research on plant diseases.

The chief of the bureau gave Meier's proposal his blessing. "As the Bureau considers this work worth following up, it will be appreciated if the Extension Service would loan Mr. Meier's services to the Bureau of Plant Industry for the duration of the trip," Frederick Richey wrote. But Meier's director at the Extension Service insisted he go on two weeks of official leave for the trip; what's more, he would allow Meier to go only "at such time when he can get away without detriment to his work." Meier's new plan came to nothing.

Lindbergh did not leave a record of why he wouldn't accompany Meier on the trip, but 1935 was certainly not a good year to approach him about a high-profile expedition. In January of that year, a carpenter named Bruno Hauptmann was put on trial for the death of Charles Lindbergh Jr. After he was found guilty the following month, he was sentenced to death. The trial made the press hungrier than ever for news. Lindbergh decided he would have to leave the United States altogether, taking Anne and Jon to England to live on a remote country estate.

But by the end of 1935, Meier enjoyed a spectacular consolation prize. He was invited to create an experiment to send on a third trip to the stratosphere, the most ambitious yet.

○ ○ ○ ○

Not long after Captain Stevens parachuted safely from the wreck of *Explorer*, he got to work on *Explorer II*. The goal remained the same: to reach an altitude of seventy-five thousand feet. To avoid any more disasters, Stevens's team made the balloon stronger and designed it to use helium instead of hydrogen. Aboard the new gondola, Stevens would study the same scientific questions as before, but with better equipment.

For the new trip, Meier constructed quartz tubes in which he would send more fungi. This time, he filled them with seven types of fungi, six of which he had personally caught on his flights. He selected black mold he found ninety-seven hundred feet above Texas plains, and blights floating forty-four hundred feet above Brunswick, Georgia. Those spores were veterans of the air, he reasoned. If they had survived at altitudes of several thousand feet, they might be more likely to survive a trip to the stratosphere.

On the *Explorer II* mission, Meier wanted to do more than test the mettle of his spores. He hoped, as he later put it, to carry out the "first attempt to explore the stratosphere for living things." Perhaps life already existed in the stratosphere, or at least paid it regular visits.

To find out, Meier created an ingenious stratospheric sky hook that would be dropped from *Explorer II*, collecting spores as it fell. In April 1935, Meier tested the device aboard the Goodyear airship *Enterprise*. As it sailed over Washington, DC, he tossed the sky hook out of the gondola. Its parachutes opened flawlessly.

In July, Stevens returned to the Stratobowl with another small army to launch *Explorer II*. The balloon inflated without a hitch, but it then ripped open and collapsed on top of the men rigging it to the gondola. It took months to repair the balloon, and Stevens did not get back to the Stratobowl for a second attempt until November. "Camp life, even in heated tents, loses some of its charm in zero weather," he complained.

On November 11, Stevens and Orvil Anderson finally took off, only for a stiff downdraft to push the balloon toward the rim of the Stratobowl. Stevens and Anderson dumped hundreds of pounds of ballast to escape, shocking the thousands of onlookers with a rain of lead shot. Having lightened their load, the pilots got past the rim, and the rest of the ascent went smoothly.

Explorer II passed seventy-two thousand feet, far above the previous record. When it reached its flight ceiling, Anderson released some helium to let them hang in the void. For an hour and a half, they drifted through the dark blue sky, the gondola humming and clicking with experiments. Meier's spores sat in their quartz tubes, exposed to the full brunt of the stratosphere. "We were now floating in the nearest approach to a natural vacuum in which man has ever placed himself," Stevens later wrote. "We were temporarily almost divorced from Mother Earth."

Anderson opened the valves again to begin their descent. Stevens pulled a release cord, dropping Meier's sky hook from a metal arm that extended from the gondola. "We watched in vain for its parachute to open," Stevens recalled. At seventy-two thousand feet, the air was too thin to spread the folds of the parachute's fabric. The bag vanished from sight, leaving Stevens to wonder if it would crash to the ground. But after the sky hook fell a couple thousand feet, the air finally became dense enough to unfurl the chute.

Its upward jerk yanked the bag open, releasing Meier's aluminum cylinder. A second parachute then deployed, slowing the cylinder's fall. The force of the parachutes pulled sections of the cylinder apart, revealing baffles through which stratospheric air could flow. It swirled past a glass tube covered in a sticky coat of glycerin, which could catch any spores or bacteria roaring by. As the stratospheric sky hook fell toward Earth, the air pressure increased. At thirty-five thousand feet, it became strong enough to push a piston up into the cylinder, sealing it closed.

While Meier's cylinder parachuted to Earth, Stevens and Anderson were preparing for their own landing. They put on leather football helmets borrowed from a local high school. The balloon dropped through the clouds, toward South Dakota farm fields. This time, the descent was so flawless that the gondola reached just a few feet above the ground when Stevens and Anderson pulled the rip cord, opening the top of the balloon. Gravity took over, pulling *Explorer II* the last yard down to Earth. The air inside the gondola was "momentarily filled with flying clothing, empty cans, ballast sacks, small cameras, tools, and other objects," Stevens wrote. Stevens and Anderson were unharmed, and everything aboard *Explorer II* survived intact, including Meier's quartz tubes, still lashed to the gondola.

His stratospheric sky hook was also recovered. Within a few days Meier was busy working on both experiments, and at the end of November 1935, newspapers began reporting the results.

"Plant Diseases Live 13 Miles Up" the *New York Times* declared. Five out of the seven types of fungi Meier had sent up in the quartz tubes had still been alive when he got hold of them. "Tubes containing the spores were exposed to cold lower than 65 degrees below zero, Fahrenheit, extremely low atmospheric pressure which would kill a man, ultraviolet rays from the sun which do not penetrate to the earth's surface and are capable of killing some lower forms of life, ozone and extreme dryness," the newspaper reported.

Meier next turned to his stratospheric sky hook. He opened the aluminum cylinder, coated the glass tube with a nutrition-loaded paste, and waited to see if anything grew. Over the course of a few days, stratospheric colonies appeared on the glass.

Five of the colonies contained bacteria. While they looked indistinguishable from one another, Meier found that they fed on different kinds of food and thrived at different temperatures, suggesting they were different species. The other five colonies in the sky hook contained fungi—common molds and rots, all of which Meier had previously collected from airplanes at lower altitudes. Now he shattered his own record, demonstrating that life existed at least thirty-five thousand feet above the Earth.

Meier could only speculate how they had managed to soar above the troposphere, rising above the weather. "Perhaps we see from the results of this experiment an additional explanation of the spread of organisms which causes diseases of plants and animals," he said, "as well as a basis for understanding why identical species of micro-organisms are constantly found in widely separated parts of the world."

The discovery provoked speculations in the press about life beyond Earth. "Scientists have debated whether spores—infinitesimally small organisms—could be swept through the universe from one planet to another," the United Press wrote. Now Meier had shown that spores survived in an imaginably harsh environment. "If they can live under these conditions, experts logically asked if they can not live under the only slightly more severe conditions of ethereal space."

o o o o

In 1936, Meier continued to fly. On a foray over the Caribbean, he caught spores on the trade winds. When he encountered a rainstorm, he dove through it to sample life at different altitudes. At ten thousand feet, he plucked bacteria and fungi from the air, but when he descended to two hundred feet, the Caribbean air seemed to become practically sterile. It was as if the rain had washed it free of contamination.

In that same year, Meier also started making bigger plans. He felt ready at last to propose making his unofficial pastime his official job. By then he had collected five years' worth of airborne life. And even if Lindbergh was slipping away, Meier had managed to build a huge network of collaborators. Amelia Earhart agreed to bolt a sky hook to her Lockheed Electra for her upcoming flight around the world. Other pilots agreed to sample air on their voyages. Meier's network also included scientists who, like him, were using their free time to study the life of the air. Bernard Proctor, an MIT microbiologist who mostly studied bacteria that contaminated food, also inspected microbes collected on daily flights of a Boston weather plane. In California, Claude ZoBell specialized in bacteria that lived in the ocean. But he also found time to put Petri dishes on the roof of a seaside building to catch microbes blowing in from the sea.

Meier decided it was time for them to create a new science: a living meteorology. They would systematically collect data, which they would analyze to discover fundamental rules. Meier envisioned them exploring "the extent to which the air, out-of-doors and indoors, functions in dissemination of human, animal and plant diseases and the degree to which objects carried by air currents can be used to indicate origin and movement of air masses."

Meier brought his plan to the National Research Council, an organization of government advisors that guided the course of science in the United States. And in March 1937, the council agreed to the creation of a committee to act on his idea.

In that same month, Earhart was supposed to start her journey. But her landing gear collapsed as she tried taking off from a runway in Hawaii, and she had to wait till May for it to be repaired. Leaving from California on her second attempt, she and her navigator, Fred Noonan, made their way to South America, then to Africa, and onward to India. On a call to her husband from Java, Earhart mentioned she was collecting samples for Meier and taking notes about her work. Earhart and Noonan left New Guinea on July 2, 1937. They were planning to travel twenty-five hundred miles east and land on Howland Island. But they failed to reach it and were never heard from again.

Earhart's sky hook and her collection of slides disappeared with her. In newspaper interviews, Meier mourned both the lost aviators and the lost opportunity for science. "Miss Earhart, in this phase of her research program, was utilizing the airplane to advance knowledge in a field opened by Louis Pasteur in classical experiments he reported in 1860," he said.

Meier turned back to his plans, redoubling his efforts. He finally persuaded the Department of Agriculture to pay him to study the life of the air as his official job, and then he persuaded a group of prominent doctors and scientists, including his old mentor Elvin Stakman, to join his committee. On November 12, 1937, four months after Earhart's disappearance, the committee gathered at the headquarters of the National Research Council in Washington.

One of the council members, the biologist Robert Coker, opened the meeting. "I don't want to take your time or embarrass the Chairman by going into detail as to why he is Chairman," he said. Coker was referring to Meier, who was sitting nearby. "Everyone knows he is the logical and proper man to lead this movement."

Coker talked instead about this movement. "I believe we are starting here something which may go far," he said. "The undertakings of this group may have a very great importance to the health of man, animals, and plants and to other branches of science."

One by one, the committee members talked about what they might do. As the ideas piled up, the researchers worried that they might end up in chaos. "This is of such far-reaching importance that a central organization of some kind is absolutely essential," Stakman said. "I hope it will be possible to have some one individual devote all his time to the problem." That person would naturally be Meier.

For his own part, Stakman hoped the new effort would include bigger surveys in the sky for rust and other threats to farms. Virulent strains of rust were spreading across much of the United States at a speed that could be explained only by high-flying spores.

Earl McKinley, the dean of George Washington University School of Medicine, told the committee about his own modest attempts at catching airborne germs. "It was simply a matter of holding a plate out from an open plane," he said. McKinley believed that his fellow medical experts needed to take the air as seriously as plant pathologists did.

"It may be that atmosphere may be much more important than we think," he said. "The interesting thing about some of the most explosive diseases we have is the observation that these diseases frequently break out almost simultaneously in many different places apparently without enough time for transportation from one place to another by actual carriers."

There were two researchers who could have spoken with supreme authority on that question, but they had not come to the meeting. William Firth Wells and his wife, Mildred Weeks Wells, had just opened the Laboratories for the Study of Air-borne Infection at the University of Pennsylvania two months earlier. There they were carrying out experiments on the spread of viruses and bacteria. While the Wellses were absent, they were the subject of much talk at the Washington meeting. Esmond Long, another professor from the University of Pennsylvania, described their work in detail.

"Long ago it was felt necessary to purify the water supply, but up to the present time we have done relatively little to purify air," he said. "The Doctors Wells have already made a big start through apparatus for sterilizing air."

Meier adjourned the meeting at four fifteen, telling his fellow committee members they had achieved all he had hoped for. "It would appear that today we have launched a new branch of science—aerobiology," he said.

○ ○ ○ ○

Agriculture secretary Henry Wallace approved Meier's transfer to the Bureau of Plant Industry, with the new title of principal scientist. The department agreed not only to pay his new salary as an aerobiologist but to hire him a team of researchers and to buy the equipment they needed for their labs and planes. Meier persuaded the Carnegie Corporation, a philanthropic fund set up by Andrew Carnegie in 1911, to pitch in thirty-five hundred dollars for the first year's work. If the results lived up to Meier's promises, the charity would consider donating as much as one hundred twenty thousand dollars for an even bigger effort that would last for years.

In June 1938, Meier laid out an ambitious plan for the first year of aerobiology. Meier would fly over farms to collect a menagerie of floating crop diseases, from tomato leaf spot to tobacco blue mold. He would soar across melon fields again and again to discover how the life in the air was influenced by clear skies versus clouds, humid weather versus dry, day versus night. He would inject the spores he caught on his flights into greenhouse tomatoes and observe any strange new diseases they caused. And Meier would launch the year's work with a spectacular voyage across the Pacific. In 1936, Pan Am had opened a route across the ocean, sending flying boats from San Francisco to Manila. Now the airline agreed to have Meier bring his sky hook aboard the *Hawaii Clipper* on a July 1938 flight. He would be accompanied by McKinley, who would disembark in Manila to work on an experimental test for leprosy.

"The Pacific Ocean offers opportunity to check distances to which spores of fungi and pollen grains may be carried by winds," Meier wrote. His first flight on the *Hawaii Clipper* would lead to a vast, ongoing survey of the Pacific. He would teach the pilots and flight hostesses to use his sky hooks to catch spores like Lindbergh and Earhart had. Before long, they were sending weekly shipments of fresh slides to Washington for Meier's team to study.

Closer to home, Meier had even more ambitious plans in store beyond the first year. Aerobiologists would "investigate the feasibility of systematically charting the air content of micro-organisms over significant areas of the United States and contiguous land and water areas," he wrote. Meteorological weather flights at twenty points across the country would use sky hooks to sample the air, ground weather stations would make surface collections, four commercial airlines would volunteer their services, a navy weather ship in the mid-Atlantic would help as well, and special flights would be arranged by the army, navy, and coast guard. Ultimately, Meier hoped the efforts would lead to the creation of a permanent government aerobiology lab.

Meier's ambitions reached out as far as the atmosphere itself. Aerobiology would make sense of life not just in the upper air, not just across vast expanses of wheat fields, but in the intimate spaces where humans lived. The concepts aerobiologists were going to develop would apply to all things airborne, including floating germs that made people sick.

The concept of floating germs was not popular among doctors in 1938, of course: they had been trained by experts like Charles Chapin to believe that airborne infections were largely a matter of miasmic superstition. But Meier suspected that he could unify aerobiology, especially with the help of the Doctors Wells. He was so intrigued by what he had heard at the November 1937 meeting that he followed up with a letter to Long. He asked for a picture of William Wells at work in his lab, in order to impress potential donors with their promising work on "diseases of Man."

There's no evidence that Meier met the Wellses before he left Washington for his trip on the *Hawaii Clipper.* But they already seemed like natural partners.

five

A PERFECT CYCLE

William Firth Wells showed few signs as a young man that he would go on to become one of the earliest aerobiologists. The air meant little to him at first. When he went to MIT in 1906, he joined a crusade to give the world clean water to drink.

The crusade's leader was William Sedgwick, a microbiologist famous for stopping a typhoid fever outbreak in the Massachusetts city of Lowell in 1891. When he traveled to Lowell, Sedgwick found that many locals blamed the disease on bad weather. But he traced the outbreak to *Salmonella*-laced sewage that was contaminating the city's drinking water. When Lowell cleaned up the city water supply, the typhoid went away.

At MIT, Sedgwick set about creating a new kind of sanitarian: engineers who would apply the germ theory of disease to public health. One of his students later recalled how he would hold up a glass of water during his lectures and "scare us to death by saying that it contained enough germs of typhoid fever to give the disease to a thousand people, and then go on to show how sanitary engineers could make the water safe to drink."

Wells felt an awe around Sedgwick he would still remember late in life. "He strove to inspire us for the coming battles which lay ahead," Wells said.

As part of his students' training, Sedgwick would send them on a train down to Providence, to pay a visit to Charles Chapin, the city's superintendent of health and, in Sedgwick's opinion, "the most thoughtful and scholarly health officer in the United States." The silver-mustached teetotaler would lecture Sedgwick's students about protecting people from germs. "I consider it a very great thing for young fellows meaning to go into Public Health work to be able to sit at your feet for a couple of hours and imbibe some of your ideas and inspiration," Sedgwick told Chapin.

Chapin and Sedgwick both instilled in their students the conviction that germs were largely spread through contaminated water and milk, direct contact, and short-range spray from coughs and sneezes. They shared an adamant loathing of what they called filth theory—the idea that waste left on the ground could corrupt the air and spread diseases over long distances. "Any thoughtful student knows that the filth theory as ordinarily understood is as dead as a door nail," Sedgwick told Chapin. "Your repeated blows at the old filth disease theory are sound and timely."

Sedgwick's skepticism about the air came in part from his own research. In 1887, he and his student Greenleaf Tucker designed an aeroscope to sample the air around Boston. To operate it, Tucker pulled a lid off a vacuum-sealed glass cylinder and let air surge in. Any germs sucked in could get stuck in a layer of sugar at the bottom. Later, Tucker would pour warm gelatin over the sugar and then turn the cylinder to coat it with the sticky film. Over the next few days, he checked the film for colonies of bacteria.

On one outing, Tucker took the aeroscope to Boston City Hospital. He collected surprisingly few bacteria inside the building—about as many as he got when he stepped out of the north side of the hospital to capture fresh air. He credited the meager catch to the cleanliness of the hospital staff. "This is as it should be," Tucker declared. "Bacteria, in a way, represent so much dirt."

○ ○ ○ ○

At MIT, Wells studied how to filter bacteria out of drinking water. He invented a portable kit to test rivers and lakes for contamination, by detecting the harmless gut microbe *Escherichia coli*. After graduation,

Wells shuttled between low-level jobs in North Dakota and Illinois before settling in Washington, DC, in 1913, and joining the US Public Health Service as a sanitary bacteriologist. His first job was to study outbreaks of typhoid fever caused by sewage spills in Maryland and Virginia. Wells walked the banks of the Potomac River, measuring the contamination.

People could get typhoid not just by drinking tainted Potomac water. Oysters could also turn into lethal meals. The estuaries and coasts of the mid-Atlantic were lined with dense beds of oysters that filtered dozens of gallons of water a day in order to trap particles of food. They sometimes ended up with *Salmonella typhi* in their tissues. "Grave danger exists in the possibility of disease germs lurking in the oysters taken from the Potomac River," the *Washington Times* warned.

Wells was dispatched to study how oysters became carriers of typhoid. He traveled to a desolate island in the Chesapeake Bay where the Public Health Service had a laboratory. Over the course of a year, he became intimately familiar with the shellfish. He worked out details of their life cycle and figured out how to purify oysters with chlorine and the tidal currents of the Chesapeake Bay. A contaminated bivalve could flush itself clean in a matter of hours.

"He did excellent work of considerable originality," the public health expert Wade Frost later recalled. "He is, I should say, a rather brilliant man whose thinking is perhaps more that of an engineer than a biologist, with a mind which is active and original but somewhat erratic."

Yet his original work did not lead to an offer of a stable job. Wells grew frustrated and began to distrust his fellow scientists. Earle Phelps, the director of the Hygienic Laboratory at the Public Health Service, called Wells "paranoiac." Phelps was speaking as a friend. He watched in dismay as Wells undermined his own achievements by leaping impatiently from experimental evidence to sweeping conclusions.

Phelps was not just a mentor to Wells but also a matchmaker. In 1915, Phelps hired a young doctor named Mildred Washington Weeks to work in his lab. Within two years, Mildred and William married in a small April ceremony in Virginia. Four months later, Mildred was pregnant. By the time she gave birth to William Wells Jr. in April 1918, William Sr. was gone. He had been drafted into the army, where he would deploy the germ theory of disease in the Great War.

○ ○ ○ ○

As the US Army prepared to enter the war, it hastily threw up camps to train raw recruits. Many of them turned to seas of mud. Some barracks leaked so badly that conscripts had to wear raincoats to bed. Scarlet fever, diphtheria, mumps, and measles swept the camps, and the sick soldiers got poor treatment in unheated, understaffed military hospitals.

After the families of dead soldiers complained to Congress, the War Department assembled three thousand officers with what it described as "special skills in sanitation, sanitary engineering, in bacteriology, or other sciences related to sanitation and preventive medicine." The new unit was dubbed the Sanitary Corps, and it soon welcomed Lieutenant William Firth Wells.

Wells traveled to the newly built Camp Meade in Maryland to provide clean water to the tens of thousands of soldiers stationed there. The only nearby source was the Little Patuxent River, a meager, muddy, sewage-laced stream just three feet deep. Wells converted a truck into a mobile purification plant. His team—including a chemist, a bacteriologist, and a pump-and-engine man—parked the truck by the river and pumped water through a sand-filled tank to filter out bacteria. They then doused it with chlorine to kill the rest. Every two hours the team checked the quality of the water to make sure a surge of sewage could not sicken the soldiers. When a reporter visited, Captain Wells—newly promoted—demonstrated his confidence in his system by drinking a treated glass from the foul Little Patuxent.

Under Wells's watch, waterborne diseases caused no notable harm at Camp Meade. But in the fall of 1918, the camp was hit by a devastating outbreak. It was caused by a germ that did not travel through water to reach its victims. Camp Meade soldiers began to suffer from scorching fevers. They grew delirious. Influenza had arrived.

○ ○ ○ ○

Before World War I, the flu did not rank high on the army's list of worries. It was just one of many infectious respiratory diseases. A five-hundred-page Sanitary Corps handbook published in 1917 included just a brief mention of influenza, wedged between pneumonia and tonsillitis.

When the flu arrived in camps the following year, its savagery took the army by surprise. At Camp Meade thousands of soldiers fell ill in one week. The division surgeon issued an order prohibiting "massed singing as it was observed that men singing in large groups frequently held their heads close together." The outbreak kept growing, forcing him to shut down buildings where soldiers gathered socially. But the cases kept climbing anyway. On September 24 the entire camp went into quarantine as masked nurses and doctors worked fourteen hours a day or more in packed wards.

It wasn't until early October that the caseload started to ebb. But then a new catastrophe struck: the lungs of a quarter of the flu patients filled with pus and blood. "The heart-breaking feature of this complication was the ghastly constancy of the incidence of pneumonia," the division surgeon reported. So many soldiers died that four new morgues had to be built in a day.

In the midst of this stateside carnage, Wells was shipped to France. There he was assigned to the 301st Water Tank Train. Wells and nineteen other officers commanded four hundred twenty men transporting water in a convoy of trucks from pumping stations to the front. He continued using the science he had learned from William Sedgwick to protect the army from waterborne diseases. But for all the good he could do for the army, Wells could do nothing about the influenza that American soldiers carried from stateside bases to Europe. While 227,000 American soldiers were wounded in battle during World War I, 340,000 were hospitalized with influenza.

○ ○ ○ ○

The Great Influenza started among soldiers, but it quickly spread to civilians. By the fall of 1918, it had even reached remote villages in Alaska. Most victims eventually recovered, but the rate of infection was so high that a staggering number of people died. It wasn't the old or the very young who were the flu's most likely victims, as in earlier pandemics.

Now young adults became its prime target. While Wells was in France, his twenty-six-year-old sister, Martha, died of pneumonia in Wilmington, Delaware: one life among the millions lost in 1918.

Public health authorities were badly prepared for the fight. Robert Koch had been able to battle cholera effectively because he knew exactly what his enemy was. But even in the early 1900s, influenza researchers still chased mirages. Like other diseases, influenza had once been thought to be caused by toxic air. The British physician Thomas Bevill Peacock wrote in 1848 that it was caused by "some generally-diffused poison." In 1892 a disciple of Robert Koch named Richard Pfeiffer seemed to have discovered its true nature, announcing that he found rod-shaped bacteria in the noses of influenza patients.

Following Koch's postulates, Pfeiffer used the bacteria to inoculate mice, rats, rabbits, guinea pigs, pigeons, and even apes. Only the apes and the rabbits got sick, and their lung infections did not resemble influenza in humans. Nevertheless, Pfeiffer declared victory. "I consider myself justified in pronouncing the bacilli just described to be the exciting causes of influenza," he claimed in the *British Medical Journal*.

When the pandemic surged across the United States in late 1918, US surgeon general Rupert Blue confidently declared that it was caused by "the bacillus influenza of Pfeiffer." Microbiologists brewed fresh stocks of the bacteria, killed them, and used their dead husks to make vaccines. But the vaccines proved useless, and it slowly became clear that the bacteria Pfeiffer had isolated did not cause influenza. They infected people only after their lungs had been weakened by the flu.

Mired in ignorance, public health officials argued with one another about how the germ spread. For his part, Charles Chapin rejected the idea influenza traveled long distances through the air. "The air has no part in its extension from place to place," he wrote in 1914. When influenza raced into Rhode Island in 1918, the *Providence Journal* offered Chapin's advice on its front page. "Influenza is chiefly sprayed by droplets from the mouth and nose in talking, coughing, and sneezing," he said. "Keep at arm's length from everybody and the chance of thus getting it is small. Don't go where you have to crowd close to others. Don't let people talk in your face."

Still, Rhode Island's death toll grew, and Governor Robert Livingston

Beeckman came under pressure to do more. Chapin urged the governor to resist calls to shut down public meeting places. He assured Beeckman that this huge sacrifice would not stop the spread. But Chapin lost the fight, and Beeckman went ahead with closing dance halls, schools, theaters, and churches—except for Sunday services. "One may fool the people," Chapin grumbled, "but one cannot fool disease."

○ ○ ○ ○

Inside Providence's hospitals, doctors and nurses took another measure against the pandemic: they put on masks. Elsewhere in the United States, entire cities ordered that people wear masks when going out in public. San Francisco imposed a jail sentence of ten days for a bare face. Masked police directed traffic. Baseball players hid their faces as they came to bat.

Two decades had passed since Carl Flügge had first persuaded surgeons in Germany to wear masks during operations. But using masks to stop an outbreak was a much younger practice. It started at the end of 1910, when an inhaled form of plague emerged in China. People in the Manchurian city of Harbin suddenly started turning purple, coughed bloodstained sputum, and died in as little as a day. Before long, corpses were laid out for miles on the frozen ground.

Wu Lien-teh, a doctor who had trained in microbiology labs in Europe, was put in charge of stopping the outbreak. Inspired by the beaked masks of plague doctors, Wu designed two-layer cotton masks and made them mandatory around Harbin and the vicinity. "Otherwise, the germs present in large numbers with every spit, would be carried into the throat, and then in the lungs of a healthy individual nearby," Wu wrote. Reports spread across the world of Harbin being filled with masked faces. When the plague retreated in March 1911, having killed more than sixty thousand people, Wu was hailed for preventing an even bigger catastrophe.

But Wu had not based his mask mandate on any experiment he had run beforehand. And when Wu wrote about the plague in later years, he never explained how the plague had killed almost three hundred members of the anti-plague team in Harbin, despite the fact that they were all supposed to wear masks.

Only after the epidemic did a scientist test Wu's hypothesis. Oscar Teague, who worked at the Bureau of Science in Manila, had his assistants put on masks and then sprayed harmless *Serratia marcescens* into the air next to them. Later, they took off the masks and spit into Petri dishes. The bacteria's red colonies emerged, revealing that the bacteria had gotten into their mouths. They had either slipped around the edges of the masks or penetrated straight through. "Their use during the recent epidemic of pneumonic plague lent a *false* sense of security which may have led to the taking of unnecessary risks," Teague and his colleagues warned.

Despite those warnings, masks became a more familiar sight in the United States. In 1916, George Weaver, the director of the Durand Hospital in Chicago, ordered his staff to wear masks for their own protection. Weaver claimed that colds and other respiratory infections almost disappeared from the staff, and he gave the masks credit. "The mask not only protects the healthy person from infection and from becoming a carrier, but also prevents a carrier from spreading infection to others," he wrote.

Weaver never ran a trial comparing masked workers to unmasked ones, so he could not say how effective his masks had actually been. And when the influenza pandemic emerged two years later, cities put new mask rules in place without any hard data to justify the decision. A debate broke out among public health experts as to whether masks did any good, and the American Public Health Association formed a committee to come up with guidelines. They ended up recommending only that people wear masks in hospitals and barbershops. "The evidence before the committee as to beneficial results consequent upon the enforced wearing of masks by the entire population at all times was contradictory," they wrote.

Wilfred Kellogg, the executive of the California State Board of Health, went even further: he declared that universal mask mandates provided no benefit at all. Too often, people wore crude handmade masks, which were too thin to stop droplets. Even well-made masks could fail if people didn't tie them tightly enough to keep droplets from sneaking in around the edges.

Worst of all, mask ordinances were often toothless. "During the compulsory universal wearing of the mask in San Francisco it was generally observed that the masks were worn carefully under circumstances of least necessity, as upon the public streets and in the open air," Kellogg said. "They were just as conscientiously laid aside in private offices, and among gatherings of friends, the very places where the chances of contact with an early case of influenza and where the conditions for the transfer of droplet infection were the most favorable."

○ ○ ○ ○

In November 1918, the influenza pandemic was at its peak when Germany and the Allies signed an armistice. Mildred Wells spent the winter in Austin with her mother, caring for William Jr. Meanwhile, William Sr. stayed on in Europe, working with other American sanitary engineers to rebuild France's ravaged water supplies and sewage systems.

By the time William returned to the United States in July 1919, the pandemic was largely over. It had killed somewhere between 50 and 100 million people—an estimated 3 to 5 percent of humanity. The world recovered from the carnage in later years, but it did not shake off influenza. The disease settled into a seasonal cycle, killing hundreds of thousands of people each year. No one could say if a new pandemic would strike. If it did, no one knew if the world would be any better prepared.

The pandemic had raged through the army during Wells's service, but he escaped unscathed, at least physically. Yet he would be haunted by "the failure of public health to deliver mankind from the 1918 pandemic of influenza," as he later wrote. Sedgwick, Chapin, and his other mentors had trained him to use the germ theory to protect people from death. They told him to clean water and dismiss air. And he had been helpless to halt the dying that surged around him.

○ ○ ○ ○

With the war now over, William needed a civilian job to support his new family. In 1919, he found work with the New York State Conservation Commission. He returned to saving oysters from the rising tide of twentieth-century pollution.

New York's oysters had a long, spectacular history. For thousands of years, the Lenape people had feasted on the massive beds that carpeted New York Harbor. Early European settlers wrote of oysters that grew over a foot long. By the nineteenth century, residents of New York were eating a million oysters a day. They ate them standing at street carts or packed into oyster houses. Their ravenous appetite wiped out the foot-long giants, and the Industrial Revolution drove down the remaining population. "Oyster culture, the most valuable fishery in the State of New York, is rapidly declining and threatens to become extinct," Wells warned.

Wells moved his family to Long Island, home to some of the state's densest surviving oyster beds. North Atlantic Oyster Farms gave him space at their West Sayville headquarters to open a laboratory. He set out to restore New York's dying beds by breeding millions of larvae. He filled jugs of salt water with fertilized eggs and let them grow into a thick soup of larvae. The larvae needed their water changed to keep growing, but when Wells tried to dump out the old water, he dumped the larvae as well.

"I was completely stumped," Wells later recalled. "Then one afternoon I saw a milk separator—one of those centrifugal machines which separate the milk from the cream. And I began to wonder, 'Why won't that do the stunt?'"

Wells poured his oyster-laden water into a milk separator and switched on the power. As it spun, the centrifugal force hurled the larvae to the sides of the machine, where they could be pumped out. Noticing that some of the fragile oysters died during the merry-go-round, Wells tinkered with the separator to make it a gentler ride. "It doesn't bruise the little fellows quite so much," he said.

With his new oyster centrifuge, Wells could now separate larvae safely from old water and rear them to maturity. By the fall of 1923, Wells had raised millions of oysters, which he released into Long Island Sound. "I have had to be both father and mother to these youngsters and have sat up with them and watched over them both night and day," he told a reporter.

Wells not only started rebuilding the oyster population but figured out how to disinfect their stocks on an industrial scale. Newspapers hailed his work, and he used that attention to protect the interests of oyster companies. When a typhoid outbreak killed one hundred fifty diners in New York and Chicago, newspapers received photographs of Wells, the trusted authority, "eating with obvious gusto a plate of New York oysters on the half-shell." Wells put on a press-friendly performance at the Hotel Pennsylvania in New York in October 1925 for the annual meeting of the National Association of Fisheries Commissioners. He served the commissioners the first banquet of artificially propagated oysters. Each of them was handed a menu that included a birth certificate for the oysters he was served. *Date of birth: July, 1923. Father's age: 4 years. Occupation: giving pleasure.*

To save oysters was miracle enough. "Artificial oysters! Who would ever have supposed it possible?" the *Pittsburgh Gazette Times* gushed. But Wells had gone even further: he had tamed the wild bivalve and could now pick the most desirable oysters from which he would harvest egg and sperm and breed a new race.

"The super oyster, sporting a pedigree if not a pearl, has arrived," the Associated Press announced in 1927. "The aristocratic bivalve has not yet left its laboratory nursery, but probably will be seen at the dinner table soon. The blue-blooded mollusk is a direct descendant of the first families of artificially propagated oysters. Its stepmother was a cream separator. Its godfather is William Firth Wells, biologist of the New York State conservation department."

The Roaring Twenties were a good time to be an oyster godfather. Wells left his job with the conservation commission to work as a biological engineer for the North Atlantic Oyster Farms Company. They paid him a comfortable salary of thirty-six hundred dollars a year. Wells improved his oyster-rearing techniques, experimented with chlorinating seawater, and even filled an auditorium at the American Museum of Natural History for his movie about developing oyster embryos.

And then the oyster days came to an end. The stock market crash of 1929 led North Atlantic Oyster Farms to be sold to General Foods, which no longer wanted to pay Wells for his research. Wells found a little work here and there, such as consulting for the city of New York and writing a report for a paper company on how its pollution might poison oysters. But he could not find any permanent oyster work and abandoned his career. In later years, shellfish biologists would credit Wells as a pioneer in conservation, but his efforts in the 1920s did not halt the decline of New York's oysters. New York Harbor's beds became so badly polluted that the city closed them for good. Long Island's beds stayed open, but they declined as its towns—and their sewage—expanded eastward.

It would not be the last time that Wells failed to deliver a miracle.

◦ ◦ ◦ ◦

Mildred Weeks Wells left behind few clues about the first four decades of her life. Archives hold none of her early letters. The surviving correspondence from later years includes no mention of her upbringing in Texas. Only one picture from her entire life appears to have ever been published. A twenty-four-year-old woman wearing a black gown and mortar board smiles in profile. The longer you look at her smile, the sadder it becomes.

That portrait of Mildred Washington Weeks, MD, appears in the 1915 *Cactus*, the yearbook of the University of Texas at Austin. It preserves a few clues about a strong-willed woman who was already set on a future in medicine. Mildred's medical school class was composed of three women and thirty-one men. "Mildred has a widespread reputation," the yearbook declared, "already having practiced from Washington D.C. to Austin, Texas, and to Galveston. My! How we envy her—and we know her already established success will continue."

Mildred belonged to a Texan family famous for wealth, violence, and scandal. Her great-grandfather John Bunyan Denton was an itinerant Methodist preacher who got rich as a land speculator and a lawyer. He then joined a volunteer militia that attacked the K'itaish people, and he was killed in the battle. Both the city and county of Denton were named in his honor.

Denton's son Ashley—Mildred's grandfather—joined the Confederate Army as a surgeon. After the Civil War he was appointed as the director of the State Lunatic Asylum, a vast complex that housed more than five hundred inmates. In 1885, Ashley beat a newspaper editor with a cane for accusing him of mismanagement. He was accompanied by his bookkeeper, William Weeks, who pulled out a pistol during the fight. The editor's wife grabbed it before Weeks could shoot.

Even after Denton and Weeks resigned from the asylum, rumors flew that they had embezzled funds. The scandal subsided long enough for Weeks to marry Denton's daughter Mary Alice in January 1888. But the following year, a grand jury indicted him for embezzling fourteen hundred dollars from the asylum. Weeks was arrested and returned to Austin to stand trial. When the *Austin American-Statesman* asked Denton about his former bookkeeper and new son-in-law, Denton replied "that he believed the case against him was a very bad one."

The case was dismissed, but Denton and Weeks did not patch things up. After William and Mary had their first child, Marion, they moved to Oklahoma Indian Territory, where Mildred was born in 1891. Weeks seems to have vanished at that point. Mary returned to Texas, left her daughters in the care of her parents in Austin, and moved away to San Antonio. In 1901, shortly after Ashley Denton died, Mary Weeks died as well. Her obituary made no mention of Mildred's father.

With one parent dead and the other vanished, Mildred spent the rest of her childhood in Austin with her grandmother and sister. She appeared in Austin society pages when she attended a Halloween party or played a game of bridge. Mildred enrolled at the University of Texas, where she studied medicine as her grandfather Ashley had six decades earlier.

She then moved to Washington to work for Earle Phelps in his bacteriology lab at the US Public Health Service. Within three years, she was married to William and had become a new mother. Mildred does not appear to have ever worked as a doctor. In 1925, a New York census worker recorded her job as "housework." As William was celebrated in the national press as the oyster godfather, Mildred appeared from time to time in the *Suffolk County News*, listed among the ladies who attended meetings of the Sayville unit of the South Shore College Women's Club. "After the business meeting, bridge was enjoyed for two hours, followed by delicious refreshments," read a typical report.

What the newspapers didn't report was that Mildred likely spent much of her time caring for William Jr., whom she and William had nicknamed Bud. A colleague later recalled that he was "confined at home as he had never been able to get along in schools." Bud would live with Mildred and William until their deaths.

In 1929, with William out of a job, the Wellses left Long Island and moved into an apartment in Harlem on 123rd Street. While William searched for a job, Mildred found work. She became a polio detective.

° ° ° °

In 1928, the International Committee for the Study of Infantile Paralysis organized a team to review all the published research on polio. They invited Mildred to join the effort. Her task would be especially challenging: she would have to read every published epidemiological study on polio and try to make sense of its spread.

Ever since polio had emerged in the United States in the 1890s, it had terrified parents by striking children seemingly at random, leaving them lame, paralyzed, or dead. "The medical profession is entirely in the dark regarding the epidemic," the *New York Times* reported in 1899. One glimmer of light came in 1908, when the German physicians Karl Landsteiner and Erwin Popper managed to transmit polio to new hosts. They extracted spinal fluid from a boy who had died of the disease and injected it into two monkeys, which both became paralyzed and died.

Landsteiner and Popper autopsied the monkeys but found no sign of bacteria in their nervous system. They correctly guessed that a virus, far smaller than bacteria, was the cause of polio, but decades would pass before scientists isolated it. In the 1920s polio still moved like a ghost from host to host.

Mildred pored over reports of polio outbreaks and wrote a 172-page review packed with tables in which she broke down cases by state, season, age, and weight. She judged the evidence that the virus might be carried by flies, by dust, by milk, by feces. She concluded that polio defied conventional ideas for how a germ spread. "The lack of obvious connection between cases of poliomyelitis is one of the striking and constant features of the epidemiology of the disease," Mildred observed. Polio, she wrote, "does not behave epidemiologically in accordance with the concepts that have become crystalized as to how a contagious disease should behave."

Mildred presented the leading theories of the disease, but she paid particular attention to the idea that polio could spread through the air. Entertaining that possibility put her at odds with Charles Chapin. "Absence of infection in hospitals indicates that the air is not an important vehicle of infection," Chapin had said of polio.

But Mildred suggested that the air might be an important vehicle after all. She pointed to studies showing that mucus from the nose could transmit polio. She suggested that the disease should be classified among "the various air-borne or droplet infections." In other words, people might spread polio by expelling heavy droplets in their coughs, or by releasing smaller ones that could become airborne. Most people infected with polio never developed symptoms. But they might surreptitiously spread the disease onward from their nose or mouth, in some cases with devastating results.

"It is conveyed from the sick to the well by these secretions, primarily through infective droplets sprayed about in talking, breathing, coughing, etc.," Mildred wrote.

In 1932, her survey appeared in *Poliomyelitis*, a cinder block of a book. At forty, she was at last a published author. But Mildred got little credit for her labors. The *New England Journal of Medicine* praised the book's wide scope, but warned doctors they wouldn't get any clear guidance for treating patients. The journal also gave Mildred and her colleagues a backhanded compliment for their thoroughness: "It is interesting to note that this book is entirely the product of women in medicine and is the first book, so far as the reviewer knows, by a number of authors, all of whom are of the female sex." The journal added that "no one is better fitted than a woman to collect data such as this book contains."

Overall, Mildred's reviewers seem to have missed her provocative idea about how polio spread. One writer brushed aside all her work: "No new conclusions are drawn."

But at least one person recognized a new conclusion: William Firth Wells. In later years, he would look back at her insights about polio as a turning point in his own career.

○ ○ ○ ○

While Mildred investigated polio, William found what seemed like stable work. In 1930, he accepted a job at Harvard as an instructor in sanitary science. But despite the university's prestige, he faced a dire pay cut. William had made thirty-six hundred dollars a year breeding oysters, and now his instructor salary would come to only twenty-five hundred dollars—barely more than what he had made right out of college. A colleague at Harvard wondered how the Wellses managed to survive on so little. In the midst of the Depression, William—now forty-three—was lucky to find a job at all.

When Wells arrived on campus, he fit in comfortably, at least in appearance. "This slender, middle-aged man with black hair, dark eyes and a black mustache had a quizzical expression and a diffident manner," his assistant Edward Riley later recalled. "Puffing on the ever present briar pipe and gazing off into the distance, he was the picture of an absent-minded professor."

At first Wells seemed like an acceptable new hire. "He has given the course this year in an admirable way," said Wilson Smillie, a Harvard professor who oversaw his teaching. But Smillie could see that Wells had some shortcomings. He only had a bachelor's degree, and his career had been limited to water pollution and oysters. Smillie sent Wells on a summer journey across the southern United States to fill in the gaps in his understanding of public health. On his trip Wells learned about building outhouses, digging wells, and killing mosquitoes.

In the fall, Wells returned to Harvard, where his new colleagues at Harvard got to know him better. Many of them ended up choosing the same word to describe him: peculiar. Although he had been hired to teach, Wells was indifferent about his class. What he cared about, intensely and obsessively, was making a name for himself as a scientist. Wells felt as if once again he was walking in the path of his old professor William Sedgwick, who had helped found Harvard's School of Public Health.

But Wells had trouble making a place for himself in his new home. For one thing, he didn't get along with his supervisor, the sanitary engineer Gordon Fair. Fair was only thirty-five when he met Wells, and he had already been appointed as Harvard's dean of engineering. The successful young scientist might have had no idea what to do with Wells, a difficult man eight years his senior who lacked a PhD and had gained fleeting fame for breeding oysters with a milk separator.

Wells got a friendlier reception from the scientists who worked at the Harvard School of Public Health, although they mostly studied air rather than water. The age of miasmas had ended some three decades earlier, but it was now clear that the atmosphere contained other dangers, many created by modern life. Wells's new colleagues studied carbon monoxide poisoning and metal fume fever. They evaluated masks designed to protect factory workers from inhaling deadly chemicals. They studied the humidity in hospitals and demonstrated that premature babies got sick if they breathed air made too dry by heaters.

For his own work, Wells was assigned a small, dingy office in Pierce Hall. On the wall there was an old pen-and-ink picture of what looked like a champagne flute topped with cotton. Wells realized it was the aeroscope that William Sedgwick and Greenleaf Tucker had invented in 1887 to test the air at Boston City Hospital. Tucker's surveys with the aeroscope had strengthened Sedgwick's doubts about the danger of airborne germs. And yet, sitting in his office looking at the picture, Wells began to wonder if Sedgwick had been wrong.

After all, the aeroscope collected only what little air rushed into the cylinder after its vacuum seal was broken. People breathed in thousands of gallons of air every day, and it might take just one gulp to inhale a disease-causing germ. Mildred likely fueled her husband's doubts even more by refusing to go along with Chapin's presumptuous claim that polio could not spread through the air. Wells now decided to study the air for himself.

He began by mulling how he could sample more air. It dawned on him that he could use the milk separator he had converted for harvesting oysters. Now he turned it into an air centrifuge.

From the outside, his new invention looked like a wooden box trailing an electric plug. A door opened on the side to reveal a metal shaft. When Wells wanted to catch germs, he slipped a hollow glass cylinder inside the shaft and turned on the power to make it spin. In addition to turning the cylinder, the motor also powered a fan that drew air up through the box. Traveling through the cylinder, it began spinning as well. The centrifugal force flung dust motes, pollen grains, droplets, and other airborne particles against the cylinder's inner wall. Wells could set the fan to spin at different speeds, letting him control exactly how many gallons of air he pulled through the device.

When he was done sampling the air, Wells would shut off the box and take the glass cylinder out. The inner surface of the cylinder was sticky with a nutrient-rich goo, and organisms that could eat it grew into visible colonies. Wells perfected recipes to feed different kinds of bacteria. For example, microbes that lived harmlessly in the mouth, called alpha streptococci, could multiply in the cylinders if Wells provided them with a few drops of human blood.

Wells's colleagues at Harvard looked on skeptically as he built his air centrifuge. One of them even bet him that it would fail. But then one day, Wells turned on the air centrifuge, took in some air, and got bacteria growing on his tubes.

"He won," said the Harvard epidemiologist Edwin Wilson, "not by accident, but because he knew it would work."

o o o o

Shortly after Wells invented the air centrifuge, Gordon Fair informed him he had a new assistant. Edward Riley, a twenty-four-year-old engineer, had become a fellow victim of the Great Depression. After losing a job turning steel into fans and forges, he came to Harvard to get a master's degree in sanitary engineering. When Riley's money ran out, Fair sent him to work for Wells.

Wells dispatched Riley to test his new air centrifuge at sewage treatment plants around Boston. He wondered if it could capture germs kicked up by the wind. Riley found nothing noteworthy. After Riley finished his master's degree in 1934, Wells hired him back to help run more experiments. They poured a broth of *E. coli* into an air conditioner in the basement of the School of Public Health. The air conditioner sprayed out droplets laden with the bacteria. Wells and Riley climbed to the top of the three-story building. When they ran the air centrifuge, they captured a surge of *E. coli* at the end of every hallway.

The Commonwealth of Massachusetts then asked Wells to bring his air centrifuge to textile mills around the state, to measure bacteria that might harm the workers. When they pulled apart cotton or wool in carding rooms, they released dry bits of fibers on which bacteria might stick. Meanwhile, workers in the weaving rooms sprayed down the floors with water in order to make the air humid, improve the consistency of the fibers, and lower the risk that static electricity might spark a fire. But public health officials worried that the water might be contaminated with sewage and that its bacteria might theoretically end up in the air.

Wells and Riley visited fourteen textile mills around Massachusetts to sample their bacteria. They laid down Petri dishes in dusty carding rooms and humid weaving rooms to catch germs that fell to the floor. They also switched on the air centrifuge for fifteen minutes in every room to catch floating microbes. Back at Harvard, Wells and Riley waited for the dishes and glass cylinders to grow colonies.

A clear difference emerged between the factory rooms. In the dusty carding rooms, Wells and Riley found a comparable number of bacterial colonies in both the dishes and the centrifuge cylinders. But in the humid weaving rooms, the cylinders harbored far more. Wells suspected that the bacteria they caught in the carding rooms were being carried aloft by dust. The dust then settled quickly back onto the floor. But in the humid weaving rooms, droplets carried bacteria into the air and then floated for long stretches of time.

The trips to the mills seem to have ignited Wells's imagination. He wondered if people's lungs could also fill rooms with airborne droplets laden with germs. Perhaps those germs could float across those rooms and infect people inhaling them.

To find out, Wells hired Edward Riley's younger brother, Richard, as a second assistant. Richard had studied music at Harvard but decided to give up a career as a pianist to enter medical school. Both brothers were dazzled by their peculiar mentor. Wells was "a wizard whom I found irresistible," Richard later said. "The many hours I spent with Wells did not improve my scholastic standing, but did give me the privilege of keeping pace for a while with a man whose mind took giant strides from peak to peak and quickly encompassed a new and fertile territory."

Wells took his disciples along on his exploration of airborne droplets.

In the basement of the public health building, he discovered a giant metal chamber the size of a dentist's operating room that had been used to test poison gas during World War I. Wells figured out how to spray a mist of water droplets inside the tank.

In ordinary light, the mist was invisible. But when Wells turned out the overhead lamps and pointed a bright beam of light into the chamber, he could see a haze. To Wells's surprise, it lingered for hours. Even four days later, some remnants survived.

In later years, Wells would credit Richard Riley for helping him figure out how the mist defied gravity for so long. A big droplet of water is so heavy that it falls quickly, like a tossed marble. But as the droplet falls, it also evaporates. As it shrinks, it becomes lighter. If the droplet is small enough, even a faint current of air is sufficient to push it back up and keep it floating.

Wells then made some rough calculations about the fate of droplets. In humid air, a big droplet will evaporate too slowly to resist gravity. In drier air, its chances of shrinking into an airborne core are better. In dry air, he concluded, droplets bigger than a hundred microns are doomed to splat. Below that diameter, a droplet can potentially have enough time to shrink down to a floating core—what Wells called a droplet nucleus.

Mildred appears to have helped William think through what all that meant for medicine. As people coughed or talked, they might exhale a mix of droplets of different sizes. The ones bigger than a hundred microns would land on the floor after a brief flight. Any germs they carried could threaten people only at close range. A heavy droplet might hit people in the face or their clothes. If they wiped it into their mouth or nose, a germ could start to replicate. But droplet nuclei—released by coughing or perhaps just by talking—could drift over longer distances and then be ingested by people breathing in the air. "The contrast between droplet and ingested infections was first impressed upon us by an epidemiologic study of poliomyelitis made by Dr. Mildred Weeks Wells for the International Committee for the Study of Infantile Paralysis," William later recalled.

William gave his first public talk about his new work at a Boston conference in December 1933. He then started work on a paper to lay out his concepts in more detail. "It appears," he wrote, "that transmission of infection through air may take one of two forms depending on the size of the infected droplet." The first was caused by heavy, fast-falling drops, which he called droplet infection. "The second form may be called air-borne infection," he wrote. He used that name to convey how droplet nuclei could drift over long distances, for long stretches of time, filling a room like smoke.

○ ○ ○ ○

On a chilly day in late 1934, Wells used the students in his hygiene class to demonstrate his new theory. The students, sitting scattered across the lecture hall, were likely experiencing a mix of boredom and frustration brought on by Wells's incompetence as a teacher. "This is the type of work Wells does extremely badly," the dean of the School of Public Health said.

Next to Wells sat one of his air centrifuges. Twenty minutes before the end of class, he stopped his lecture to insert a glass cylinder inside the wooden box and switch on the power. As the box whirred, Wells took out a jar of sneezing powder. He dropped a pinch into the palm of his hand, which he held near the top of the box. The breeze from the fan wafted the powder toward one side of the room.

On that cold day, the windows were shut, letting little fresh air into the unventilated hall. Gentle currents spread the powder around the room as Wells went on talking. Soon, the students began to sneeze. After a couple dozen sneezes, Wells shut off the air centrifuge and replaced the glass cylinder with a fresh one. Once it started spinning again, he lectured for another ten minutes until the end of class. His students pushed open the doors and filed out, allowing fresh air to rush in. Wells stayed behind and swapped the second cylinder for a third one. He let it whir for ten minutes more, then packed up and left.

"A practical joke was turned into a conclusive experiment," William and Mildred later wrote. In the days that followed, Richard Riley incubated the glass tubes at body temperature to see if any alpha streptococci grew on their walls. The first cylinder, which Wells had spun as his students sneezed, developed none. But the second, which spun after the sneezing stopped, grew twenty-two hundred colonies, and the third, which Wells spun after class, produced nine hundred twenty.

Here, the Wellses argued, was evidence of long-distance airborne infection. When the students sneezed, they unleashed droplets laden with bacteria from their mouths. While some of the droplets fell to the floor within a few feet, others had turned into floating droplet nuclei and wandered all the way to the front of the room, where Wells scooped them up in his air centrifuge. Even after the end of class, enough droplet nuclei still drifted through the hall that Wells could capture some more.

"In this experiment, contamination of the air by the machine was evidenced by the sneezing, and contamination of the room by the sneezers was evidenced by bacterial colonies in the tubes," William and Mildred wrote. "A perfect cycle was thus completed."

It's unlikely that many students in the lecture hall on that day in 1934 recognized what a radical case William was making. Few of his colleagues recognized it either. "'Air-borne infection' might revive the ancient and exploded theory of miasms," William and Mildred acknowledged. The distinction between airborne infections and miasmas was easy to miss, because droplet nuclei spread out invisibly in the atmosphere. "They may be considered to float or drift with the slightest air currents," William and Mildred wrote, "and therefore to be in effect a part of the atmosphere itself."

But the Wellses insisted that their theory was firmly grounded in the germ theory of disease. The air did not become corrupted by filth. It was simply an effective route for germs to reach new hosts. If the Wellses were right, airborne infections might be a major factor in human health. As Max von Petterkofer had observed, we live in an ocean of air, inhaling thousands of gallons each day. To William and Mildred, we are aerial oysters.

○ ○ ○ ○

With her review of polio behind her, Mildred joined William to work on his theory of airborne infection. While he ran experiments and engineered new instruments, she read through the medical literature, analyzed disease records, argued with William about his experiments, and wrote up his results. In the opinion of some colleagues, Mildred had the scientific rigor William needed to keep from lurching too far ahead of his data. As a couple, it was said, the Wellses made a much greater contribution than William would have alone.

Mildred gained a reputation at Harvard as the more aggressive of the two. The professors who made that judgment were, of course, overwhelmingly men. They were also probably ill-equipped to deal with a strong-willed Texan woman who had endured a traumatic childhood and earned a medical degree in a class full of men. Nevertheless, Mildred earned their grudging respect. In 1935, the School of Public Health appointed her as a research associate. Her appointment was recognition of her importance to William's work, albeit one without a salary.

William tinkered with the chemical weapons tank, connecting the air centrifuge to it through a tube. That change allowed him to see how long bacteria could stay alive floating in the air. He injected germ-laden mists into the tank, and then periodically switched on the air centrifuge to capture some of the droplet nuclei.

The bacteria that live in the intestines, such as *Salmonella*, could grow into colonies if he caught them in the first few hours of their life in the air. After that, they died off. On the other hand, alpha streptococci, which infected the airway, could stay viable even after floating for two days. It looked as if germs that relied on being inhaled were adapted to survival in the air.

Leaving the basement with the Riley brothers in tow, William went on bacteria-catching expeditions in Boston's parks, theaters, hospitals, and schools. When they ran the air centrifuge outdoors, ten cubic feet yielded not a single alpha streptococcus. The droplet nuclei that people expelled as they walked down the streets of Boston were diluted into the vast soup that is the outdoor atmosphere. Only indoors, Wells found, did people concentrate their breath.

° ° ° °

As Wells began publishing the results of his surveys, other scientists decided to try out his invention for themselves. In New York City, a team of Columbia University researchers took the Wells centrifuge into subway cars, classrooms, and movie theaters. All told, they captured thirty-six thousand cubic feet of air. More than two-thirds of the air samples they collected in subways contained bacteria; only a quarter of the ones from Central Park did. It was a striking confirmation that people filled indoor spaces with floating germs, even as the outdoor air diluted them.

But trapping bacteria did not prove that airborne infections were a serious threat to public health, and Wells's first batch of papers left most experts unswayed. Milton Rosenau, who directed Harvard's epidemiology program, was one such skeptic. In a 1935 medical textbook, he wrote that respiratory diseases were "the most prevalent and damaging of the infections to which flesh is heir." And yet, in those same pages, Rosenau still dismissed airborne infections as miasmic myth. "The hazard is much less than formerly thought," he wrote. "The radius of danger is limited."

William and Mildred believed this was a catastrophic error. They even raised the possibility that public health experts had made the same error during the influenza pandemic. Chapin and others had claimed that the flu spread mainly through close contact, but the Wellses doubted that it could have overwhelmed the world so quickly by this route alone. "If, on the contrary," William and Mildred later wrote, "the sneeze-infected air of a room were breathed by many persons over a considerable interval of time, simultaneous infection could be realized, and the velocity of infection would be ample to explain even the explosive pandemic of influenza in 1918–1919."

During the pandemic, doctors did not know what caused influenza. But now that mystery had been solved. In 1933, British researchers collected throat gargles from people sick with the flu. They filtered bacteria from the fluid and then put drops into the noses of ferrets. The ferrets got sick and could pass on the disease to other ferrets. Here was proof that a virus caused influenza. A Harvard researcher named Harold Brown taught himself how to isolate the flu virus, and soon he was collaborating with Wells to see if the virus could survive in droplet nuclei.

They began by creating a broth of flu viruses, which they sprayed into Wells's tank. Wells and Brown then used an air centrifuge to capture some of the floating droplet nuclei. When they transferred the harvested water into the noses of ferrets, the ferrets got sick. Wells and Brown found that influenza viruses could float in the air for up to half an hour and remain viable.

If diseases like influenza could spread like smoke, they could make life difficult for public health authorities. Chapin had merely offered precautions that people could take to avoid getting hit with big droplets, along with courtesies to show others. But that sort of personal advice would not be enough to stop airborne pathogens. Governments would have to intervene to ensure people could enjoy safe air, just as they did to provide safe water and food.

"The complacency of those who would rather blame the victim than vested authority for contagious epidemics was rudely jolted," Wells later wrote.

The news of the influenza experiments returned Wells to the kind of fame he had known seven years before as the oyster godfather. But he was reincarnated as an even more intriguing hero: an explorer of the air. Waldemar Kaempffert, the science editor of the *New York Times*, recounted how physicians had been taught for decades that germ-carrying droplets plummeted to the floor. Wells and his students "shatter this comforting doctrine with discoveries which will make it necessary to re-examine the possibility of transmitting diseases by means of the air," Kaempffert wrote.

Two days later, Wells appeared again in the *Times*, this time in an editorial. The writer linked him to the other prominent aerobiologist, Fred Meier, in a piece entitled "Germs and Winds."

"What happens to them in the open air where they are caught by every wind?" the writer asked. Some hints of an answer were coming from Meier, who had recently announced the results of the Lindbergh expedition. "A truer picture of the distances to which minute forms of life can be wafted is thus obtained," the writer suggested. Wells might toil in his basement while Meier flew through the clouds, but they were both seeking to reveal the same living atmosphere.

○ ○ ○ ○

Wells came to believe he could do more than just paint a picture of the air. He could control it. As a young sanitarian, he had been a soldier in Sedgwick's campaign to purify water and food. Now Wells wanted to do the same for the air. "The indoor air we breathe is perhaps the last great frontier in the environmental control of infectious disease," he later said.

Wells tried to adapt the methods that Sedgwick and others had used to conquer waterborne diseases. He knew that chlorine killed off bacteria in swimming pools and water treatment plants. But he couldn't use it to disinfect the air: chlorine gas was so deadly that both sides in World War I had used it as a chemical weapon. Wells instead tried out ozone, which disinfected water fairly well and didn't become a lethal gas. But when Wells injected ozone into his tank, it managed to kill off only some bacteria floating in droplet nuclei. He needed a weapon that was both potent and safe.

It occurred to Wells that Gordon Fair had tested ultraviolet radiation years before. He had trained mercury lamps on tainted water and killed the bacteria that swam in it. But Fair hadn't thought that UV radiation would work on a large scale, so he had abandoned the work. The lamps still sat in his lab. Wells asked to borrow them, and Fair let him. Fair seems to have had no interest in Wells's idea. Perhaps he thought it didn't have a prayer of working.

Wells rigged a lamp to the tank so that it bathed droplet nuclei in ultraviolet light as they floated into the air centrifuge. When he sprayed mists of bacteria into the tank, the radiation killed the germs. No colonies grew on his glass cylinders. Next, Wells tried the lamps out on influenza. He sprayed flu viruses into the tank and hit them with ultraviolet rays. He then collected the droplet nuclei and put them up the nostrils of ferrets. None developed the flu.

When Wells announced the results of his experiments, doctors flooded him with requests to help put ultraviolet lamps in their hospitals. In Boston, Charles McKhann set up lights at Infants' Hospital and found infection rates of 12.5 percent in a ward without the lights and just 2.7 percent in one with them. In Philadelphia, Joseph Stokes, the physician-in-chief at the Children's Hospital, installed curtains of ultraviolet lights in doorways to stop viruses wafting into rooms. Doctors weren't the only ones to ask for help. The Pullman Company asked Wells to help them install lights in their trains on the Bangor and Aroostook line in Maine.

In the 1920s, Wells had dazzled the public with a technological solution to the country's pollution crisis. Now he tantalized them with a far more profound fix. "Ultra Violet Air Dooms Germs," one newspaper announced; "Scientists Fight Flu Germs with Violet Ray," declared another. *Newsweek* predicted that ultraviolet lights would become a familiar sight in the modern world. "In time the lamps might serve to give hospitals, clinics, children's homes, theatres, and offices daily microbe housecleanings." Wells posed heroically for photographs next to his hulking test chamber, in which he had uncloaked the invisible threats of the air. "Chief among his aids, Wells said, was his wife, Dr. Mildred Wells," the Associated Press noted. She never appeared in the pictures.

Time published a profile of Wells that cast him as both wizard and savior. "In his fast scientific stride Biologist William Firth Wells, industrious instructor of sanitary science at the Harvard School of Public Health, has made oyster eggs germinate artificially and by means of artificial sunlight made germs vanish from thin air," the magazine declared. "Last week after working persistently against smaller & smaller forms of life, Biologist Wells was able to announce that by means of ultraviolet light he destroys the minuscule cause of influenza as it floats in air."

A few weeks after the profile appeared in *Time*, Biologist Wells presented his theory to an enormous gathering. Harvard celebrated its three hundredth anniversary by assembling some of the world's leading scholars, along with President Roosevelt, to deliver speeches to vast audiences. People across the country listened in by radio. For three weeks, a reporter wrote, the Harvard Tercentenary Conference of Arts and Science became "the intellectual center of the world."

William was invited to present a talk about airborne infection. Mildred helped him prepare his speech, adding epidemiology to his experiments and bringing a sweep and clarity to the text. The *Journal of the American Medical Association* would later publish the speech as a two-part essay.

Delivering the lecture in August 1936, William started by evoking the floating life that Fred Meier had recently revealed overhead—seeds, pollen, fungal spores, bacteria, and viruses. "The theater of operation of these phenomena is vast," Wells said. "Microorganisms have been found in polar flights and in the highest penetrations of the stratosphere."

After his nod to Meier's discoveries aboard *Explorer II*, Wells brought his audience back down to Earth. "We are not concerned with this outer atmosphere except as a source of infinite dilution," he said. Instead, he turned his attention to "those semienclosed atmospheres wherein we live."

William placed the research he and Mildred had been carrying out in the history of biology. That history started with Pasteur capturing germs from the air, and moved forward to Flügge's experiments with dishes on a laboratory floor. He condemned Chapin and others for rejecting those findings and claiming there was practically nothing beyond the short-range spray of heavy drops. "The theory of air-borne transmission was well-nigh abandoned," William told his audience.

William argued that his experiments proved that doing so had been a mistake. He ended the lecture by urging his audience to consider clean air as vital to their well-being. "The great reduction of intestinal disease through water purification since the turn of the century might prompt us to hope that some of the diseases transmitted through discharges from the respiratory tract may be checked by methods of controlling air supplies," William said.

For weeks, William basked in the afterglow of the lecture. He marveled at how far he and Mildred had come in six years. "The growth of these ideas has been so gradual, and the steps so simple, and short, that I could not grasp the distance we had wandered from established lines," William told the Yale microbiologist Charles-Edward Amory Winslow. "And now I must carefully determine how far I can venture into this new territory, which seems so fertile and inviting, without risk of my scalp."

○ ○ ○ ○

William was right to worry. Despite their accolades, he and Mildred were in dire professional trouble. Few people involved in the trouble wanted to talk about it much in later years. But it seems to have started with Gordon Fair.

Fair had been indifferent about Wells borrowing his lamps, but once they proved able to kill the flu, he had a change of heart. He demanded to be included as an author on the paper Wells was writing about the UV experiment. He also insisted that he share a patent for using ultraviolet lamps to purify indoor air. Fair then reportedly ordered that all further research take place in his own lab, under his control.

The Wellses pushed back—Mildred pushing harder than William, it was said. They demanded complete scientific independence from Fair. But William and Mildred were in a weak negotiating position, thanks to their difficult personalities, their huge claims about the significance of their work, their lack of PhDs, and William's wretched teaching.

If the Wellses no longer answered to Fair, they would have to find a new home at Harvard. But none of their colleagues offered them one. Harvard officially terminated William and Mildred in September 1937, just over a year after it had celebrated their accomplishments at its tercentenary.

As the Wellses scrambled to find new jobs, their reputation preceded them: impressive research, difficult people. Wade Frost, who supervised William at the US Public Health Service before moving to Johns Hopkins University, said he wasn't sure how solidly William "has his feet on the ground." Frost also wondered what had happened at Harvard that led to his firing. "I am a little puzzled by the fact—if it is a fact—that his work seems to have found so little favor at Harvard, where I understand that the special equipment needed for his experimental work is already available," he said.

Glimmers of hope appeared, only to wink out. Earle Phelps, who had given Mildred her first job as a bacteriologist twenty-two years earlier, tried to hire her and William at Columbia University, where he was now running a lab. But the grant he was counting on to pay them fell through. Royd Sayers, the director of the Division of Industrial Hygiene at the US Public Health Service, contemplated bringing William back to Washington. But when Sayers made a few inquiries, he learned that Mildred carried out much of the research. Sayers believed that in government service, it was not possible for a man's wife to be responsible for much of his job. It apparently did not occur to Sayers to give Mildred a job as well.

Their luck finally turned when Charles McKhann contacted Joseph Stokes. McKhann suggested that the University of Pennsylvania hire the Wellses, and Stokes made the idea his mission. "I had an extraordinary amount of respect for his scientific insight and his tenacity of purpose," Stokes later recalled of William.

To bring the Wellses to Penn, Stokes enlisted two influential professors to pull strings. One was Stuart Mudd, the founder of Penn's microbiology department. Another was Esmond Long, who ran the Henry Phipps Institute, a hospital for tuberculosis. Long suspected that tuberculosis was airborne and hoped the Wellses might confirm his hunch.

Stokes, Mudd, and Long had all heard about Harvard's difficulty with the Wellses. But the three men agreed their research had such far-reaching importance that it would be worth the trouble. Mudd said that he and Stokes and Long "have all had a good deal of experience in dealing with queer people, and none of us feels any particular apprehension on this score."

The university agreed to create the Laboratories for the Study of Air-borne Infection, where William and Mildred would conduct research without any requirement to teach. The money for their work came from the Commonwealth Fund, a private philanthropy that had grown into a major supporter of American medical research.

Before handing over the money, the officers at the Commonwealth Fund investigated the Wellses. They agreed with Mudd that the importance of their work overrode their peculiar personalities. Roderick Heffron, who would go on to oversee the Wellses for years, once said that "there is a brilliant and genius-like quality to Mr. Wells that forces respect."

In their new jobs, William would earn forty-five hundred dollars a year, while Mildred would receive fifteen hundred dollars. Their combined salary more than doubled their income at Harvard. While William earned far more than Mildred, she was an equal partner. Even before the labs opened, Mildred sent Mudd a signal to that effect. She wrote him a letter in which she described the technician who had worked for them at Harvard and who had agreed to come with them to their new labs.

> So far as I can see the only real drawback to her is the question of her race. Her name, Ruth Blumfeld, is certainly Jewish. . . . Will and I have talked it over and think that if it is agreeable to you, we would pay her personally for the next year. . . . I hope you will be absolutely

frank about this. But I also hope you will see your way clear to consent.

Mildred and William reported for duty at their new lab on September 22, 1937, in Philadelphia. So did Blumfeld.

° ° ° °

The Wellses had been at Penn for just two months when Fred Meier held his first aerobiology meeting in Washington. They were too busy launching the next chapter of their work to attend.

They arrived in Philadelphia entirely convinced of their theory of airborne infection. "I am afraid we are so completely convinced of the superiority of the bacteriological methods we are using that we do not sufficiently hide it," Mildred confessed to Winslow. Now she and William began working on experiments that they hoped would persuade others of their theory too.

Leaving behind his chemical warfare tank at Harvard, William invented a new kind of chamber for his research. This one would let him directly expose animals to droplet nuclei. It came to be known as the Infection Machine.

At its heart was a bell jar big enough to house a cage of mice. It was connected to a narrow glass tube that ran fifty feet around the lab. At the far end of the tube, William sprayed a mist laced with bacteria or viruses, and the droplets drifted along its length. Along the way, the heavier blobs of water crashed into the walls, while the lightweight droplet nuclei continued to float, finally reaching the bell jar, where the animals inhaled them. The air then exited the bell jar, passing through a flame that incinerated the germs. By adjusting the density of the mist and the height of the flame, William could precisely control the number of droplet nuclei that the animals breathed. After the animals' exposure, William and his colleagues could then observe whether they became sick from what they had inhaled.

The Wellses also wanted to demonstrate ultraviolet light's germ-killing power outside a hospital. If they could set up lamps in a school, they could shield children from airborne infections. Joseph Stokes made that experiment happen by introducing the Wellses to the Germantown Friends School, where he served on a committee. It was a savvy choice. The venerable Philadelphia establishment, founded nearly a century before, attracted some of the wealthiest families in the city. It also had a fiercely progressive culture. Instead of listening to old-fashioned lectures, the students learned how to type and put on a pageant in celebration of Virgil's two thousandth birthday.

Germantown Friends School also followed the latest developments in medicine. Theodore Wilder, the school doctor, consulted epidemiologists in order to keep diseases from spreading through the school. "Pupils come to this school from the better type of family, judged by material and intellectual standards," Wilder boasted. "Most parents cooperate in keeping at home children with obvious colds or symptoms."

The school welcomed the Wellses and their experiment, and they got to work in late 1937. They measured the circulation of the air in the classrooms and picked out different ultraviolet lights to install in each space. Some rooms got bowl-shaped lamps that hung from the center of the ceiling. Others got long, thin lights bolted high on the walls. All the lights were shielded on their undersides so that their rays shot upward rather than down at the children. The Wellses expected that the ultraviolet radiation would keep the rooms clean by destroying germs in the droplet nuclei that escaped the mouths of the children and rose toward the ceiling.

Philadelphia's newspapers gloried in the scientific advances now unfolding in their city. The *Philadelphia Bulletin* published a story entitled "Black Light Kills 'Flying' Germs," which cast the Wellses in a war against diseases. "Directing this air-fight are two pioneers in the field, William F. Wells, biologist for many years connected with the Harvard University School of Public Health, and his physician-wife, Dr. Mildred W. Wells," the journalist Steven Spencer wrote.

The Wellses told Spencer that they dreamed of safeguarding every indoor space where people congregated. They would snuff out clusters of infections before they could unleash epidemics. "The hope of all this work is to draw up blueprints for a practical campaign which may some day mop up the air of all schools, trains and other such germ-swapping places as trolleys, movies and stores," Spencer wrote.

By 1938, the path that the Wellses had taken to scientific prominence bore a striking resemblance to the one taken by Fred Meier. They had all started off two decades earlier as outsiders—Meier working on rotting watermelons, William Firth Wells on sewage-tainted oysters, Mildred Weeks Wells raising Bud. They lacked PhDs to establish their authority. None enjoyed the comforts of tenure.

Nevertheless, Meier and the Wellses were now famous for inventing simple but powerful new instruments—sky hooks and air centrifuges—that let them plumb the living air. Meier was celebrated for discovering life as high as the stratosphere, while William and Mildred Wells were revealing the microbial soup in which humans lived, inhaling germs as if they were aerial oysters. Together, they seemed poised to realize Meier's dream of aerobiology—to establish a science that would make sense of airborne life, whether it dwelled in the close quarters of home or at the border of outer space.

six

THE SCATTERED WORKERS

On July 23, 1938, Fred Meier boarded the *Hawaii Clipper* as it floated in San Francisco Bay. The propellors on the elephantine flying boat began to turn, and the plane gradually lifted itself off the water. The *Hawaii Clipper* then did something new and strange. It turned west and flew out over the Pacific.

The idea of crossing the Pacific by plane was still so exotic in 1938 that Meier's flight was the stuff of newspaper reports. "The *Hawaii Clipper* was to hop for Honolulu today on a 'mercy mission,'" the Associated Press announced. One of the passengers, a wealthy manufacturer from St. Louis named Henry C. Hutchinson, was escorting a neurologist to Honolulu to operate on his brother. Among the other passengers that day, the AP reported, was Wah Sun Choy, a Jersey City restaurateur headed for Hong Kong. Earl McKinley, the dean of the George Washington University School of Medicine, was bringing an experimental leprosy test to the Philippines. The article also briefly mentioned that a plant pathologist named Fred Meier was on board, but newspaper readers were left to wonder how a plant pathologist had ended up in such glamorous company.

For Meier, the flight was a new chapter of his life. "Fred had left Washington in high spirits," two of his colleagues later recalled. "His dreams of the development of a far-reaching field of science, which he designated 'aerobiology,' were beginning to be realized."

Meier kept his colleagues up-to-date on his progress. When the

Hawaii Clipper landed at Pearl Harbor on July 24, he sent a telegram to the Bureau of Plant Industry in Washington: ARRIVED HONOLULU 7MORNING LEAVING FOR MIDWAY TOMORROW.

The next day, the *Hawaii Clipper* took off again. It flew thirteen hundred miles west to Midway, where Meier telegraphed again: ARRIVED THREE.

The plane took a third leap, from Midway to Wake Island, and then from Wake Island to Guam. Meier sent an update: ON SCHEDULE.

Six days after Meier had departed from California, the *Hawaii Clipper* took off on July 29 from Apra Harbor, headed for Manila. It cruised that morning at an altitude of ten thousand feet. Rain began to fall, and the plane encountered headwinds. As the raindrops pelted the plane's antenna, they created static on the radio. William McGarty, the radio officer on board the *Hawaii Clipper*, struggled to hear the messages from operators on the ground.

At noon, with six hundred seventy miles to go to Manila, McGarty signaled to Eduardo Fernandez, the radio operator on the Philippine island of Panay. McGarty told Fernandez that the *Hawaii Clipper* was flying through the tops of cumulus clouds, bumping through rough air. Fernandez asked if McGarty was ready to receive the latest weather report. "Stand by for one minute before sending," McGarty replied. "I am having trouble with rain static."

Fernandez waited for McGarty to get back on the radio. A minute passed in silence. Fernandez called to the flying boat again, but got no response. Hours passed, and at five p.m.—the *Hawaii Clipper*'s scheduled arrival time—it failed to appear in Manila. When night fell, it became clear that one of the world's most famous airplanes had vanished.

At the time, the Philippines was a colony of the United States, which operated a constellation of naval bases on the islands. At midnight, the naval commandant dispatched a fleet in search of the *Hawaii Clipper*, accompanied by six submarines cruising underwater and six long-range bombers flying overhead. It was possible that they'd find the plane bobbing peacefully on the ocean. The crew and passengers had a two-week supply of food they could eat as they waited for a rescue. But a day of searching did not lead the search party to the flying boat. The only clue was found by the US Army transport ship *Meigs* close to the *Hawaii Clipper*'s last reported position. It encountered a floating patch of oil.

By then, the disappearance had come to the attention of reporters. Newspapers around the world put the missing plane on their front pages, and they followed the search day after day. Some papers even ran a photograph of Meier's elderly parents gazing at a radio, hoping for an update. Meier's wife was reported to be waiting by her telephone.

The crisis suddenly pulled back the veil on Meier's research. "The present problem, undertaken through the sponsorship of the committee on aero-biology of the National Research Council, was concerned with the possibility of the transmission of disease germs and other organisms by trade winds blowing over the ocean," the *New York Times* reported on July 30. Aerobiology made its debut as a nightmare rather than a celebration. The *Washington Evening Star* noticed an odd, dark coincidence in the science's short history: "Strangely enough, Dr. Meier once received assistance in his investigations from another air voyager who disappeared on the Pacific—Amelia Earhart."

For a week the search party scoured 160,000 square miles of ocean between Manila and Guam. They found nothing beyond the oil slick. As the days passed, Meier's family and friends began to surrender hope that he was alive. "The loss to our Committee, to aerobiology and to Science is immeasurable," Robert Coker wrote to his fellow committee members. "Most of us too have lost friends singularly high in our personal esteem. I have known no one who has impressed me more strongly than Fred Meier for his greatness and fineness of spirit, his strength and lovableness of character, his keenness of mind, his courage and his effective devotion both to regular duty and to the study of aerobiology, to which he had committed himself at great personal sacrifice for years, and ultimately the greatest sacrifice."

As an air safety board began a formal investigation, people were left to guess at how a huge flying boat had gone missing without a trace. It left no debris, no radio signal, no flares, no search-marker balloons. Stories about the disappearance of the *Hawaii Clipper* turned into fever dreams. Wah Sun Choy was actually the chairman of the Chinese War Relief Fund according to one rumor, and he had been bringing gold on the *Hawaii Clipper* for Chiang Kai-shek. Before the flight, he had reportedly said that he was being followed. Perhaps the Japanese had put a bomb on the plane.

On September 7, a *Philadelphia Inquirer* columnist named Paul Mallon even floated the idea that the aerobiologists were to blame. "Fantastic tales are being passed around suggesting the government hushed investigation of the Hawaii Clipper disaster," Mallon wrote. "Two scientists were aboard, studying transportation of microbes in upper air currents. They had opened the hatch of the ship and had placed their instruments outside. The experts agree this unusual circumstance, which created a strong draft through the ship, probably initiated the cause of the disaster, but they have not been able to prove just how."

The insinuations came as a shock to the aerobiology committee. They conferred with an attorney. "He has of course advised keeping the matter in as close a circle as possible," Albert Barrows, the council's executive secretary, told a committee member. Barrows quietly tried to figure out what had happened on the *Hawaii Clipper*. He got Meier's drawings for the sky hook and asked Karl Compton, the president of MIT, to help him figure out whether it might have led to a disaster. Compton speculated that Meier might have lost control of his sky hook, which then hit a rudder control.

Two months after Meier's disappearance, the aerobiology committee gathered in the Washington boardroom of the National Research Council to discuss Meier's disappearance. Elvin Stakman—the man who had encouraged Meier to fly in the first place—was named the new temporary chairman. "I think probably this is one of the hardest jobs I have ever had to do," he said.

The conversation quickly turned to the crash of the *Hawaii Clipper*, and from there to Mallon's insinuation that Meier had caused it. "It seems to me," said Samuel Prescott, a professor at MIT, "that the story shows that a certain newspaper reporter has a very keen imagination, but I do not see how he has any proof whatever that it did so happen."

Yet the committee members could not say what exactly Meier had been up to. Barrows laid out the blueprints of the sky hook on the conference room table.

"Did they have to open the hatch to work the instrument?" asked Warren Vaughan, an expert on influenza.

No one had an answer.

The air safety board went on with its investigation. When they presented their report in November, they offered no new evidence. In fact, they took away what little evidence there had been. When the crew of the *Meigs* had encountered the oil, they scooped up a sample, which was analyzed by chemists. They concluded that it had not come from a plane. "Pending the discovery of some concrete evidence as to the fate of the Hawaii Clipper, the investigation remains in an open status," the board wrote. It cast no suspicion on Meier in its report. It didn't even mention his sky hook.

After the scandalous rumors blew away, Meier received a few posthumous honors. Nathan Smith, a member of the aerobiology committee, wrote an obituary for the journal *Science*. "If it had to be," Smith wrote, "this was a most fitting manner for him to die, engaged in the work he liked best, pioneering a field that has hardly been touched." Years later, the British scientist Philip Gregory called Meier the martyr of aerobiology.

But Meier's scientific legacy dispersed like a cloud. He had not managed to publish a body of work that could live on after his death at the age of forty-two. He had spent his eight years as a budding aerobiologist holding down a day job, and he had filled his free time leaning out of cockpits in storms, networking with scientists and aviators, and toiling late at night in his home lab. From time to time, he managed to write an abstract about a few of his trips or a short account for an obscure magazine. The paper he published on Lindbergh's flights gained the attention of newspapers, but it was a thin piece of scientific scholarship. It consisted of little more than a brief catalog of spores. Meier never published large-scale thoughts about what determined how high life could fly or the forces that steered its spread across the planet.

In the weeks before his Pacific flight, Meier had been working on a new article for the *Botanical Review*. Perhaps he was writing a manifesto for aerobiology that would attract scientists to the new field. We'll never know. Agnes Meier was aware of the article, but when she searched for the manuscript, she found only a bibliography Meier had put together. His colleagues at the Department of Agriculture worked their way through the slides and samples he had collected over the years to see, as one of them said, "if anything of value can be salvaged." What little they found was filed away in government archives. There's no evidence that anyone studied them again.

Before he died, Meier managed to give aerobiology its name, but his dreams for the science did not survive without him. The Carnegie Corporation refused to support the project with Meier gone, and his committee lost momentum. They never settled on sky hooks or air centrifuges or another device as the standard tool that all aerobiologists should use. They never built Meier's central laboratory. They never established national surveys of the sky. Instead, the world's few aerobiologists went on working as they had before. Some investigated how living things could soar high into the atmosphere and travel thousands of miles. Others, like Wells, stayed inside schools and hospitals, where they followed germs circulating in closed spaces.

Warren Vaughan recognized this schism early on. At the October 1938 meeting, during which the committee mourned Meier, Vaughan warned that the two branches of aerobiology had to be unified if the science was to survive. "We are existing as a committee on the assumption that there is a common basis," Vaughan said. "And it should be one of our functions to show how those apparently unrelated phases of investigation can be brought together."

The committee members tried over the next few years to bring those phases together, but without Meier's charm and relentless focus, their efforts fell short. Their one important achievement was organizing a symposium in October 1941. William Firth Wells, Elvin Stakman, and dozens of other scientists gathered in Chicago to present their research. A book based on the meeting, *Aerobiology*, came out the following year. It was the first effort to tackle the biology of the air in a comprehensive way. But the British biologist R. C. McLean wrote in *Nature* that it contained some telling flaws.

Although he admired the book, McLean spied a fault line that cleanly split it in two. It was a collaboration between two groups of scientists who existed in separate worlds, speaking separate languages. In the first part of the book, Stakman and like-minded scientists took turns talking about the life of the great outdoors: about clouds of pollen, bacteria drifting inland from the oceans, spores snowing down on farms. Then came the medical researchers such as Wells who explored the cramped confines of modern life where they believed droplet nuclei were spreading disease. Neither group made any reference to the other; neither found any inspiration across the divide for their own ideas about aerobiology. It was, McLean wrote, "not quite the book for which the scattered workers in this field have hoped."

McLean was writing not only as a book reviewer, but as a prophet. Aerobiologists would remain divided for decades, and their disarray would lead to much of their most significant work being ignored. And what wasn't ignored was hijacked for sinister uses that almost none of Meier's mourners could have imagined.

PART 2

The Suicide of Bacteriology

seven

WAR AT HOME

Meier's sudden death in 1938 left William and Mildred Wells as the most prominent aerobiologists in the world. They were the only scientists running a lab dedicated entirely to life in the air. Having built the Infection Machine, William was now conducting carefully controlled experiments to prove whether droplet nuclei could indeed spread airborne infection. At Germantown Friends School, Mildred oversaw the glowing ultraviolet lamps.

But when William traveled to scientific conferences to speak about their work, he could tell that many in his audiences were unimpressed. They had accepted Charles Chapin's doctrines about the routes of infection, and they couldn't see a reason to give them up. Still paranoiac, William suspected that enemies were spreading false stories about his theory and how he planned to use it to save lives. "Misunderstanding, or even misrepresentation, are real dangers to any project whose initial stages of development upset prevailing dogma," he said.

William and Mildred knew that one of their biggest challenges was that droplet nuclei were invisible. If people could only see the microscopic beads of water that they released with every breath, they might take the possibility of airborne infection seriously.

And then, one day in 1938, droplet nuclei became visible. It was not the Wellses who unmasked them. They were revealed in photographs that emerged from Nazi Germany.

○ ○ ○ ○

Before Adolf Hitler's rise to power, Friedrich Weyrauch was a little-known hygiene scientist at the University of Jena. He studied lead poisoning. He investigated contaminated milk. When the Nazi Party seized control of Germany and perverted the country's public health establishment, Weyrauch fell in line.

The Nazis declared that the chief aim of public health should be the protection of the Aryan race. They were especially concerned about Jews and Slavs spreading diseases throughout the Reich. Nazis claimed that inferior races were rife with diseases like typhus, to which they were mysteriously immune. As a result, they could secretly spread infection to good Germans. This racist myth provided some of the fuel for the genocide of the Jews, as well as for the invasion of the Soviet Union. In both cases, Nazis believed that they were shielding the Aryan race from epidemics spreading from the degenerate East.

Weyrauch didn't merely join the Nazi Party. He went so far as to enlist in the SS, Hitler's paramilitary organization. Although it was best known for terrorizing political opponents and building concentration camps, the SS also created a medical corps, a sprawling network of laboratories, and its own Hygiene Institute, which was led by Weyrauch's close friend Joachim Mrugowsky.

Mrugowsky was responsible for the health of Nazi troops fighting the Soviet Union, as well as the SS guards overseeing Hitler's concentration camps. Outbreaks of diseases such as typhus were common in the camps, and Mrugowsky took steps to keep the germs from spreading to the Nazi staff. Trained in the tradition of Robert Koch, Mrugowsky knew very well that typhus was not jail fever caused by bad ventilation. It was caused by bacteria spread by lice. Mrugowsky fought typhus by fumigating clothing, using a potent gas called Zyklon.

In 1938, Weyrauch was summoned (possibly by Mrugowsky) to Buchenwald. The camp had opened the previous year, receiving more than twenty-five hundred political prisoners, homosexuals, and Jehovah's Witnesses. Now it swelled with thousands of newly arrived Jews who had been arrested after Kristallnacht, a night of savage destruction of synagogues and Jewish businesses.

Not long after the Jews arrived at Buchenwald, an outbreak of an intestinal disease called paratyphoid began wreaking havoc. Transmitted in contaminated food and water, it spread not only among the prisoners but among Buchenwald's SS officers. Weyrauch traveled to the camp to examine the sick and to test the sewage for bacteria.

As dangerous as contaminated food and water could be, Weyrauch worried more about the diseases of the air. A follower of Carl Flügge, he believed that people could release germ-laden droplets from their mouths and noses. Airborne infections were not just a threat to concentration camp prisoners, but could spread through German offices, factories, schools, and streetcars.

"It is often forgotten that not only coughing and sneezing, but also loud and even normal speaking disseminates droplets," Weyrauch wrote. "In general, humans keep too little distance from one another."

To encourage his fellow Germans to keep their distance, Weyrauch set about showing them the droplets they were spraying. He happened to live in the perfect city for such an undertaking. Jena was the home of Carl Zeiss, a lens maker who had founded a company that supplied microscopes to early microbiologists. "A large part of my success I owe to your excellent microscopes," Robert Koch once told Zeiss. Zeiss went on to manufacture some of the world's finest photographic cameras. Weyrauch believed they would be powerful enough to capture fast-moving droplets on film.

Weyrauch enlisted a Jena photographer, Johannes Rzymkowski, to help set up a darkened studio illuminated with a single beam of light. He brought in volunteers, asked them to stand with their mouths grazing the edge of the beam, and gave them a pinch of snuff to make them sneeze. Rzymkowski snapped away.

When Rzymkowski and Weyrauch developed the photographs, they saw fiery streaks flying from scrunched faces. Rzymkowski took more photographs of people coughing and talking. Those pictures revealed swarms of droplets as well. "In general, the droplets sink fast to the ground so that they disappear from sight," Weyrauch wrote. "The main danger of a droplet infection will therefore in general have to be found in the direct vicinity of coughing, sneezing and speaking."

The photographs confirmed Flügge's experiments forty years earlier.

But one feature of the sneezes left Weyrauch mystified: he could see a glowing fog around the streaks. He wondered if water vapor in the surrounding air was condensing into tiny droplets. He tried heating up the studio to a sultry 86 degrees Fahrenheit to dissipate the fog. But when his volunteers sneezed into the heated air, the fog persisted.

In June 1938, Weyrauch and Rzymkowski published their photographs in *Zeitschrift für Hygiene und Infektionskrankheiten*, a journal founded by Koch and Flügge. News reports about atrocities in Germany had not yet stained the journal's scientific reputation. Mildred and William discovered Weyrauch and Rzymkowski's paper, and they marveled at the photographs. At last they could look at the things they had been studying for eight years. And while Weyrauch puzzled over the fog, the Wellses did not. It was the same haze that William had seen when he cast a beam of light into his basement tank at Harvard.

"We immediately recognized the fog in these pictures as photographic evidence of dispersion of myriads of minute droplets into the air," William later wrote. Weyrauch had unwittingly taken pictures not just of large droplets, but also of smaller droplet nuclei—the key to airborne infections.

Mildred promptly wrote to Weyrauch to ask for copies of the photographs. She received them in a letter from Switzerland, which she and William took as a sign that Weyrauch was trying to hide his correspondence from the Nazis. The Wellses apparently didn't know that he was a member of the SS who paid official visits to concentration camps. They cared only about Weyrauch's pictures.

To bring them to a broader audience, William and Mildred added four of the images to a paper they were preparing, called "Infection of Air." It appeared in August 1939 in the *American Journal of Public Health*. The four photographs—one of a sneeze, one of a volunteer pronouncing the letter *p*, along with two pictures of people saying *t*—did not make much of a splash. It didn't help that the faces in the pictures were blurred and turned partly away from the camera. William and Mildred also weakened their impact by tucking the photographs near the end of their paper with almost no explanation about why they mattered.

Fortunately, one reader knew exactly why. "With our disclosure this led to even more dramatic photographs with new scientific techniques," William later recalled. Those more dramatic photographs came from Marshall Jennison, a biologist at MIT.

Jennison liked to tinker with new technology to address old scientific questions. In the early 1930s, he had learned about a remarkable new camera invented by Harold Edgerton, his colleague at MIT. Edgerton kept the shutter of his camera open in a darkened room and then flashed a strobe light to take a picture. The flash, lasting a hundred-thousandth of a second, froze fast-moving objects, making them look as if they were trapped in clear ice. Edgerton stopped a hummingbird in downstroke and a bullet emerging from a gun barrel. He took pictures of splashing milk drops so strangely gorgeous that they ended up in the Museum of Modern Art.

Jennison borrowed Edgerton's stroboscope and connected it to a microscope. He wanted to observe tiny hairs called cilia that clams use to filter food. (Cilia also line our airways, where they flick back and forth to trap incoming bacteria and particles. They then create waves that push the particles back up out of our lungs to where they can be swallowed or coughed out.) Jennison trained Edgerton's camera on a piece of gill tissue from a steamer clam, and snapped two hundred pictures a second, creating a movie. The cilia, Jennison discovered, beat back and forth every half second. Their movements—flicking stiffly forward and then relaxing as they moved back—reminded him "of a fly rod during the backward movement of a cast," he wrote in a 1934 report.

Five years later, when the Wellses reprinted Weyrauch's photographs, Jennison saw another opportunity to use the stroboscope. He rounded up volunteers with hay fever and colds and had them sit down in front of Edgerton's camera. When they sneezed, he flashed the strobe.

The images Jennison captured were spectacularly superior to Weyrauch's. Respectable men in bow ties stood in crisp profile against a pitch-black background, expelling sharp constellations of saliva and mucus. "We have been able, by means of high-speed photography, to 'stop' the motion of droplets given off in coughing and sneezing," Jennison and Edgerton announced in 1940.

The pictures were so precise that Jennison could use them to draw some scientific conclusions. He estimated that a single cough released thousands of droplets, and tens of thousands came from a single sneeze. Jennison took sequences of photographs of individual sneezes and then inspected the images to measure the speed of the droplets. "The muzzle velocity of some droplets is as great as 150 feet per second," he concluded.

Jennison calculated that larger droplets fell to the floor within two to three feet, just as Charles Chapin would have expected. But he also saw the finer spray wandering off through the air. When Jennison took pictures of people pronouncing consonants, those images captured fewer falling drops, but still revealed a mist glowing in the darkness. He declared that his photographs were "proof of air-borne droplet nuclei."

Jennison's photographs were effective not only at proving droplet nuclei. They also worked as public health messages. Newspapers began reprinting them, with headlines like "This Is Nothing to Sneeze At." The iconic images would go on to appear in textbooks and on posters. When people prepared to sneeze or cough, they could now picture the germ-filled blast they delivered to the people around them, and would perhaps cover their mouths to stop the spread of diseases.

○ ○ ○ ○

In 1938, William also gathered his own evidence for airborne infection. He began putting animals in his Infection Machine and sickening them with a series of diseases.

For his first experiments, William chose a blood infection caused by a strain of *Streptococcus* bacteria. Stuart Mudd, one of his champions at Penn, had spent years infecting mice with *Streptococcus*. He simply injected the bacteria into the animals, ensuring that they suffered a lethal disease. Now William invited Mudd to help him try infecting the mice by air.

The two scientists put cages of mice into the bell jar and pumped a *Streptococcus* mist into the long glass tube snaking around the lab. Just as Wells had predicted, mice inhaling the bacteria got sick. But Wells could also control just how sick they got. If he adjusted the machine so that only a faint mist enveloped the mice, they got mildly ill. If he swathed them in a denser microbial fog, they died from an overwhelming infection.

In May 1939, Roderick Heffron from the Commonwealth Fund traveled to Philadelphia to see the Infection Machine at work. An expert on pneumonia in his own right, he was impressed. "Here for the first time in medical history is clear experimental proof of air-borne infection," Heffron reported back to New York.

For his next experiment, Wells turned to tuberculosis. Now he needed help from Max Lurie, the Phipps Institute's leading expert on TB in animals. Lurie was a shy, mouselike man, except when a conversation turned to tuberculosis, which made him suddenly turn garrulous. He had spent twenty years perfecting techniques for infecting rabbits with *Mycobacterium tuberculosis*, injecting the bacteria into a vein or putting drops into a rabbit's nose or mouth. But Lurie knew those methods weren't the natural way that the animals got sick. He suspected tuberculosis was airborne, but he had yet to prove it. Lurie accepted Wells's invitation to run experiments in the Infection Machine, hoping that they might lift his research into a new league.

Wells and Lurie put rabbits one at a time in the bell jar and exposed them to a mist of bacteria. A dose of a thousand microbes caused the animals to grow nodules known as tubercules in their lungs. They died soon afterward. When Wells and Lurie ratcheted down the dose, the rabbits survived for longer. Every time they reduced the number of microbes they delivered to the rabbits, the animals took longer to die.

"Now we have a natural way of producing disease, and a way we can measure precisely," Lurie told the *Philadelphia Record*. "That makes the ideal scientific arrangement."

For his third pathogen, Wells turned to influenza. At Harvard, he had harvested flu-laced droplet nuclei from the air inside his tank and put them directly into the noses of ferrets. Now, working with the Penn influenza expert Werner Henle, he created a more realistic experiment. Wells and Henle put cages of ten mice at a time in the Infection Machine and exposed them to droplet nuclei loaded with flu viruses. The results were much like the ones Wells saw with *Streptococcus* and TB. A mist dense with influenza viruses could kill an entire cage of mice in ten days. A diluted broth produced milder bouts of the flu and fewer deaths.

By 1941, Wells had demonstrated that three pathogens could cause airborne infections in animals. He then tinkered with the Infection Machine to stop the germs with ultraviolet light. He rigged a lamp to his long glass tube so that the droplet nuclei would be bathed in UV rays for a few seconds before reaching the bell jar. That exposure was enough to defang flu viruses and enfeeble *Mycobacterium tuberculosis*. Lethal doses now caused only mild illnesses. In some cases, the pathogens caused no disease at all.

Stokes, Mudd, and Long were delighted with the results. They were no doubt relieved as well, because the Wellses were becoming notorious around Penn. "The Wells themselves seem to me to be extraordinarily difficult and touchy people to work with," Heffron observed. William in particular maddened other scientists with his inability to focus on the essentials. "Among Mr. Wells' gifts, which I regard as unusual, the habit of breaking right out in plain English and saying what he means is not included," Mudd said. "Wells is a queer dick," said Alfred Newton Richards, a vice president at the university—and one of his most loyal defenders.

William and Mildred also got into fights on a regular basis with anyone they saw as a threat to their work, even their closest colleagues. When Mudd tried to rein in their spending, William haughtily responded that they had turned down lucrative offers from lighting companies in order to keep their research above reproach. "I cannot overlook this opportunity of calling to your attention again the meagerness of our allowance for running expenses when measured against the opportunities which are continually being offered to us to advance the subject of airborne infection," he complained.

Mudd shrugged off the attacks. "Creative workers in any field are apt to be temperamental; the Wells are peculiarly so," he said. The university mostly gave the Wellses what they wanted. William demanded a new title, which he got: director of the Laboratories for the Study of Airborne Infection. When the Wellses accused their patrons Long, Mudd, and Stokes of getting meddlesome, they were allowed to switch to the Department of Public Health and Preventive Medicine, where William was made an associate professor. People got out of their way and let them do their work, at least for the time being.

○ ○ ○ ○

By the time the Infection Machine started delivering impressive results, the experiment at Germantown Friends School was still getting off the ground. The trouble began with the lamps: each brand the Wellses tried turned out to have its own quirks, making it hard to create a uniform level of light in all the classrooms. The paint on the walls proved unexpectedly dull and soaked up UV rays instead of bouncing them back into the air. Even the social lives of the students could make a mess of the experiment. The Wellses thought they had two cleanly divided sets of students to study, when in fact the children sometimes mingled together in common rooms—sometimes with UV lights overhead, sometimes not. Even if the lamps could protect against airborne infections, the Wellses were not sure they'd be able to prove it.

Once the lamps were finally turned on, Mildred scanned the school's medical records for evidence that they were protecting the children. Two years passed, and the only infectious diseases of note at the school were colds. The children in the rooms protected by the ultraviolet light seemed about as likely to get a cold as those without protection. Mildred suspected that they were getting infected from their families at home. For preventing colds, at least, keeping the air at school clean didn't make a difference.

While the Wellses waited for other diseases to emerge at the Germantown Friends School, they looked for other places to put up their lights. They persuaded the nearby town of Swarthmore to install lamps in two of its public schools. And in the spring of 1940, their first big test arrived: Philadelphia was hit with a wave of chicken pox.

"Probably no contagious disease is more volatile than chicken pox," Mildred and William later wrote, "and even those investigators who are reluctant to accept a general hypothesis of air-borne contagion concede the probability that this disease is spread through the medium of air."

The Wellses hoped that the ultraviolet lights would fare better against chicken pox than colds. Chicken pox appeared to spread explosively between children, which turned schools into hot spots for new infections. The Wellses suspected that schools were also places where chicken pox might be stopped. "No more severe test could be offered, and we naturally are awaiting the coming days with great expectancy and anxiety," William said.

Tallying the records, Mildred saw hints that the lamps were working. Sick children rarely managed to pass chicken pox to other children in the UV-equipped classrooms. In rooms without the protection of the lamps, the virus proved far more contagious: when an infected child came to an auditorium for a play rehearsal, fifteen students and a teacher fell ill.

The results were heartening, but it turned out that the chicken pox outbreak was just a trial run before a far bigger test. In October 1940, Philadelphia was slammed by a massive wave of measles.

The victims of measles—mostly children—were hit with runny noses, coughs, and a speckling of white spots in their mouths. After a few days, a rash spread across their skin, accompanied by a high fever. Most people who got measles recuperated and gained an immunity so powerful that they never got measles again. But a small fraction suffered devastating outcomes: pneumonia for some, brain-damaging encephalitis for others, and death for an unfortunate few. Because measles was so contagious—among the most contagious diseases doctors had ever encountered—an outbreak could leave a wake of devastation. In the 1930s, about 5 million people got measles in the United States every year, and about ten thousand died.

Measles left doctors and parents alike feeling helpless. Even in 1940, microbiologists could not say if it was caused by bacteria or viruses. Researchers could not make a vaccine. There was no serum to treat the sick. "Many parents have gotten in the habit of looking upon it as inevitable in every childhood," a doctor wrote in the *Philadelphia Inquirer.* After measles arrived in October 1940, it surged through the city schools for months, and by May 1941, the city had registered twenty-five thousand cases, the biggest wave it had ever experienced.

When Mildred added up the cases in their schools, she saw a stunning contrast. In the Swarthmore schools, just over half of the unprotected students caught the measles. But among the children protected by ultraviolet light, only 15 percent got sick. In Germantown, the difference was even starker. Mildred estimated six-year-olds protected by UV lamps were ten times less likely to get measles than those without the protection.

Newspapers celebrated the results. They predicted that ultraviolet lamps would be installed in many buildings in the future, halting the spread of many epidemic diseases. Heffron declared the experiment "perfectly extraordinary." But for Mildred and William, the most important validation came from Boston. Edwin Wilson, the Harvard epidemiologist, praised their new work, saying that it "proved as conclusively as need be that air-sanitation by light works."

○ ○ ○ ○

The Wellses were gratified but also exhausted. "I have been under the severest pressure during the last few months because of the enormous interest which has been aroused by the rather dramatic experience with measles in the schools," William wrote in October 1941. He and Mildred pushed out a cluster of papers on the measles outbreak, hoping the evidence would be enough to throw the old dogmas into question.

They failed. "This was more than the scientific community was ready to accept," Richard Riley later said. Many doctors didn't think the Wellses could draw firm conclusions from the schools. And in Toronto, the Canadian microbiologist Ronald Hare ran an experiment that raised even graver doubts about airborne infections.

Hare rounded up volunteers with sore throats and had them sit at a table. Before them sat an odd piece of lab sculpture. They faced a large, curved piece of cardboard with five Petri dishes attached to it. Hare positioned the highest dish straight ahead of his volunteers at eye level, while the lowest one sat just above the table, close to the volunteers' chests. Hare asked the volunteers to cough, and then to talk as if they were teaching a class of twenty students. Some sat close to the table, while others sat a few feet away. After each trial, Hare pulled the dishes off the cardboard arc and incubated them to see if the volunteers had managed to spread their bacteria through the air.

The volunteers who sat close to the arc, Hare found, could seed colonies in the lowest dishes, but they left the highest ones sterile. Hare concluded that the heavy droplets his volunteers released carried some bacteria, while the tiny droplet nuclei floating higher carried none. A similar result emerged when Hare had volunteers sit farther back from the table, so that the heavy droplets fell to the floor before they could land in the dishes. In these trials, hardly any colonies formed at all.

These results left Hare convinced that there was little reason to fear germs floating through the air, because droplet nuclei were too small to carry them. In later years, Hare would become an outspoken critic of the Wellses, attacking the idea that droplet nuclei could spread infection like smoke. "There is reason to believe that this may be a rare event," he said.

After four impressive years of aerobiology research in Philadelphia, the Wellses were frustrated to meet such skepticism. It was hard to know what to do next. If a few hundred schoolchildren protected from measles was not enough to shake the dogma, the Wellses would need to run an even bigger experiment. They would have to find thousands of people to protect with their rays.

In peacetime, that was a dream. But another war in Europe was about to make it a necessity.

○ ○ ○ ○

The threat of war had been growing ever since the Wellses moved to Philadelphia. In October 1937 President Franklin Delano Roosevelt delivered an ominous speech in Chicago in which he warned that a "reign of terror and international lawlessness" was on the rise around the world. The United States could not isolate itself from the growing threat to civilization from countries like Germany and Japan.

Charles Lindbergh spoke out against war, becoming one of Roosevelt's biggest enemies. Since his move to Europe, he had grown enchanted with Nazi Germany. When Lindbergh visited Berlin in October 1938, Hermann Goering awarded him the Service Cross of the German Eagle, which he described in a letter to Goering as "an honor which I shall always prize highly." The following year, Lindbergh told a Nazi general that it was essential for Germany and the United States to arrive at a better relationship, "not only for us, but for our entire western European civilization as well." Lindbergh assured the general that "I feel that I understand many of the problems you are confronted with."

Lindbergh returned to the United States to urge Americans to form an alliance with Nazi Germany so that they could work together against their common enemy, the Soviet Union. "A general European war would, I believe, result in something akin to Communism running over Europe and, judging by Russia, anything seems preferable," he said. To Lindbergh, this alliance was essential for the survival of the white race. "It is time to turn from our quarrels and to build our White ramparts again," he said. And he laid some of the blame for the push toward war on American Jews. "Their greatest danger to this country lies in their large ownership and influence in our motion pictures, our press, our radio and our government," he said.

Roosevelt lashed out at Lindbergh as a traitor and pushed ahead with supplying military aircraft to France and Britain. When Hitler attacked France in May 1940, Roosevelt took another step toward war by expanding the US Army from 188,000 soldiers to 530,000. The prospect of hundreds of thousands of new recruits coming together in army camps frightened the Wellses. They thought back to World War I, when the Sanitary Corps had protected American soldiers from waterborne diseases but could do nothing to stop influenza.

"Casualties from respiratory disease during the last war exceeded the casualty lists in action," William and Mildred warned in August 1940.

On the verge of a second world war, with years of research behind them, the Wellses hoped to use aerobiology to stop airborne illnesses like the flu. "Is it too radical to believe that we can, with the methods already available, accomplish, in the coming concentration of troops and civilians, with respiratory infection what was done in the last war with intestinal disease?" they asked.

Stuart Mudd agreed. "This, of course, is one of the situations which the last three years' program on air-borne infection has been designed to show us how to meet," he said.

In 1940, Mudd and William Wells joined dozens of infectious disease experts as advisors to the War Department. The scientists reviewed what was known about the spread of diseases and pinpointed the most urgent questions the military should answer with new experiments to keep their troops safe.

At first Wells took the lead in the conversation. In October 1940, at a meeting convened by the National Academy of Sciences, he delivered a lecture for the army about the threat of airborne infection. Wells proposed using ultraviolet lamps to purify the air of military buildings in the coming winter. "Dr. Wells feels that many of the so-called infectious diseases are air-borne, and that their germs or viruses can be kept down by flooding spaces where they float with ultraviolet radiation of germicidal wavelengths," one reporter wrote.

Wells went into the war with some powerful allies. Arthur Hitchens—a lieutenant colonel in the US Army Medical Corps and a professor of public health at Penn—was impressed by William and Mildred's experiments with the Infection Machine and in Philadelphia's schools. Speaking at the annual meeting of the American Public Health Association, Hitchens predicted that similar studies would soon be carried out in army barracks. Hitchens believed that ultraviolet lamps, combined with an influenza vaccine, might ultimately lead to the conquest of the flu. "We may expect to find the lamps installed in barrack rooms, mess halls, classrooms, motion picture theatres, and other confined spaces wherever men congregate," he predicted.

With help from Hitchens, Wells started planning an experiment. He traveled one hundred twenty miles west from Philadelphia to Carlisle Barracks, a center for training military medical personnel, where he drew up plans for installing enough lamps to protect thousands of soldiers. "The larger the experiment the more significant will the experience become," Wells said. "It might be that a reasonably large share of our time and effort could well be devoted to national defense."

But the Carlisle Barracks experiment got lost in red tape. Meanwhile, the government launched a large-scale program of experiments on airborne infections and put someone else in charge: the University of Chicago professor Oswald Robertson. He showed little interest in ultraviolet lamps, preferring the possibility of spraying chemicals such as ethylene glycol in barracks to kill airborne bacteria.

Wells once more was left seething at being overlooked. While Robertson had an impressive résumé—he had invented the first blood bank in World War I—he had done hardly any research on airborne infection before the war. His hunch about ethylene glycol was based on a few small-scale experiments on mice. Wells tried to argue his case to other government advisors, but he would get lost in rambling diatribes that left other scientists annoyed rather than swayed. By the time Japan bombed Pearl Harbor on December 7, 1941, Robertson was doling out enormous sums of money for research, and not a penny went to the Wellses.

"We wasted much time trying to obtain the opportunity to try out radiant disinfection of air in Army and Navy barracks, but we have failed," William and Mildred later wrote.

Hugely disappointed, the Wellses returned to their work in Philadelphia. William worried that his opponents would realize he was right only when 1918 replayed itself. "An epidemic of respiratory contagion, as in the last War, will arouse an immediate demand for preventive measures," he said.

° ° ° °

Now ejected from the war effort, the Wellses turned their attention to a project they had been thinking about for a few years: a book on airborne infection. Back in 1939, they had gotten the idea to write a short manual, but the university and the Commonwealth Fund both discouraged them, pushing the couple to stick to their research. The Wellses couldn't resist the idea, though, and went ahead with writing a brief outline for a book. Three years later, having failed to launch an experiment on a military base, they returned to the project. The sting of rejection transformed their plan. Instead of a short manual, they wanted to write a manifesto. In October 1942 they asked Geddes Smith, the assistant director of publications at the Commonwealth Fund, to look over crude drafts of a few sections. The Commonwealth staff had been dealing with the Wellses for five years, and they had survived many meandering monologues by then. Smith now dreaded the prospect of getting mired in a book. "Frankly I'm afraid to get into it, as I can see it as taking practically endless time," he said to his colleagues.

When William and Mildred weren't working on their book, they were tackling a new puzzle about airborne infection. In October 1942, an explosive outbreak of mumps took place in Swarthmore, sweeping through the College Avenue school where the Wellses had installed their lamps. Two months later, another of their Swarthmore schools, on Rutgers Avenue, also experienced cases of mumps, but there the outbreak never got big. Why had their lamps apparently failed in one school and succeeded in another against the same disease?

The answer might have been lurking in the calendar. When the mumps came to College Avenue, the autumn air was relatively humid. By the time it got to Rutgers Avenue, the cold winter temperatures dried it out. Ultraviolet light might work differently at different times of year. Humid air might block the UV rays from reaching droplet nuclei. William tinkered with the Infection Machine to test the possibility, figuring out a way to adjust the humidity inside the device. When the humidity inside the Infection Machine was below 40 percent, he found, ultraviolet light was three times deadlier for airborne bacteria than at 70 percent humidity.

The experience with the mumps got Mildred thinking about other ways in which the seasons could affect airborne infections. In the winter, people spent more time together in unventilated indoor spaces where they might share droplet nuclei more easily. And if a disease had a seasonal cycle, that in itself might serve as a clue that it could easily spread by air. To look for those patterns, Mildred began collecting public health records from across the United States. She created two hundred tables of data on diseases such as measles, mumps, scarlet fever, and whooping cough, breaking down the data month by month for twenty years. In the tables, she began spotting seasonal ebbs and flows.

Mildred conferred with the Harvard epidemiologist Edwin Wilson, worried that she might be misreading the data. "My own feeling is that the conclusion to be drawn, if any, is that the seasonal behavior of measles and chickenpox is consistent with a theory of spread by air, but does not prove it," she told him. "I am, of course, fearful lest there are statistical fallacies or mathematical atrocities which I do not see." Wilson assured Mildred that he had confidence in her work.

While she and William might express doubts in private, they publicly treated their colleagues at the University of Pennsylvania more haughtily than ever. Max Lurie, the tuberculosis researcher, asked the Wellses if he could use the Infection Machine for an experiment of his own. They agreed at first, but when the time came for Lurie to carry out the studies, they refused to let their assistant help him. Then the Wellses went further, barring Lurie from even touching their equipment. Finally, in a fit of pique, they packed up the Infection Machine and moved it out of the Phipps Institute, to a tower in the medical school blocks away.

Even after treating Lurie so badly, William thought he could ask him for a favor. He had designed an upgraded Infection Machine that could test airborne tuberculosis on six rabbits at a time by putting their heads in the same bacteria-infused chamber. William wanted Lurie to help him run an experiment on the new device.

Lurie refused. When Lurie complained about his bad treatment, Heffron could only shake his head. "It is a rather sad commentary perhaps on the Wells to recognize that they can't even get along with their best friends," Heffron later said.

By 1943, things were getting so bad at Penn that William and Mildred started to wonder about the security of their jobs. Along with their failure at wartime research and their alienation of their colleagues, their flagship study at Germantown Friends School had started to run into trouble. The lamps were burning out, but a new principal knew little about the experiment and cared even less. Theodore Wilder, the school's physician, had been skeptical about the experiment from the start, and now he offered little help as the protection of ultraviolet light faded away.

At the Commonwealth Fund, Heffron heard gossip about those woes, and he began to wonder whether he should keep supporting the Wellses. It would be a catastrophic decision for Willian and Mildred: if the Commonwealth Fund money disappeared, the University of Pennsylvania was prepared to drop the Wellses with only two weeks' notice. The war was stretching American universities thin, and Penn was no exception. The administrators looked at the Wellses, who could not teach, as a luxury they might not be able to afford any longer.

Before taking such a drastic step, Heffron checked in with his networks of experts. For the time being, they told him to continue supporting William and Mildred. Edwin Wilson thought it was a good idea to let them keep the theory of airborne infection alive until it became a part of the public consciousness. "It may take quite a while to sell this thing," he advised Heffron. "It took a long time to sell sanitation of water supplies and Pasteurization of milk."

The instability ground down Mildred and William's marriage. They started to argue about the wisest next step in their work. Mildred was back to working on the book, and she felt that her tables of epidemiological data made it even more compelling. She was compiling a tower of index cards, each listing a paper by other scientists that backed up their theory. She wanted William to join her on the book so that they could finish it quickly. But he could not break away from his lab work. He kept thinking of new experiments he needed to run in order to offer a definitive account of airborne infection. Every study he came up with seemed to promise another essential revelation.

Along with those pressures, there was the ever-present matter of Bud. Now a twenty-five-year-old man, he still lived with his parents, and his care fell largely on Mildred. A glowing *Saturday Evening Post* feature from that difficult time glossed over the stress. It cast William as the heroic patriarch of a happy American family, a scientist "who enjoys the capable assistance of his wife, Dr. Mildred W. Wells, a physician and epidemiologist—expert on epidemics—who continues to manage a home and family as well as her duties on this research team."

In fact, around the end of 1943, the Wellses suffered a "domestic crisis," as Heffron later described it. They had a fight over epidemiology.

○ ○ ○ ○

By 1943, Mildred had been overseeing the school experiments for six years, and it was now clear to her that they were too small for her ambitions. Many childhood diseases did not depend solely on classrooms for their spread, and so she wanted to create a bigger shield with ultraviolet lamps. "Rightly or wrongly, I believe that an expansion of the epidemiological studies is in order," she told Edwin Wilson.

William, however, did not see the point of a huge new experiment. Putting lights in schools might not provide a perfect defense, but the lamps could still tamp down outbreaks. Their fight left Mildred wondering if her work was valued any longer—not only by William, but by the university and the Commonwealth Fund. "The studies have, of course, been pretty crude, for I have been groping my way, but I do not see how the importance of such studies can be doubted," she said.

The conflict reached the breaking point in early 1944 when Mildred lost her maid, who likely also helped with Bud's care. Mildred abruptly announced she would cut her work at the Laboratories for the Study of Air-borne Infection to half time, "in order to take care of her home," as Heffron later recalled.

Mildred used her half time with the lab to write up her analysis of seasonal diseases. When she showed Wilson her hundreds of statistical tables, he was dazzled. "We have a better set of case material than I think has ever been collected before," he declared. Wilson encouraged Mildred to publish the data, but in wartime there was little interest in her seeming abstractions. No one would see the full scope of Mildred's careful work. Unable to get her tables published, Mildred worked instead on a paper summarizing her work on measles and chicken pox.

When it appeared in the *American Journal of Hygiene*, Charles-Edward Amory Winslow, one of the greatest figures in American public health, wrote a letter of praise to her. "I think it is one of the most interesting approaches which has been made to a study of the relation of season and climate to disease," he declared.

Mildred put her name alone at the top of the paper. At the end she thanked Edwin Wilson for helping her make sense of her data, but she made no mention of her husband. "Will and I are working together less harmoniously (an understatement)," she told Wilson. By the middle of 1944, Mildred decided to leave the Laboratories for the Study of Airborne Infection for good. The decision robbed the Wells family of a quarter of its income. It also meant abandoning Mildred's years of research in Philadelphia schools. She ended her time at Penn by writing the final analysis of her work in Germantown Friends School and Swarthmore, and then she resigned.

Mildred planned to mark this milestone by going to New York in October 1944 for the annual meeting of the American Public Health Association. She hoped to run into old colleagues like Roderick Heffron and bid them farewell. But Bud fell ill as the meeting started, and Mildred could not find a nurse. She managed to spend all of one hour at the conference, to hear Theodore Wilder, the Germantown Friends School doctor, talk about the ultraviolet lights there. Then she had to head back to Philadelphia.

She headed home in a fury. Wilder, who had provided so little help to the Wellses, had the temerity to stand up in front of an audience of experts in public health and recount the research as if it were his own. Wilder ended up the hero of a story that Waldemar Kaempffert wrote for the *New York Times*.

"He found that ultra-violet lamps in classrooms protected pupils markedly against cross-infection in cases of measles, mumps, chicken-pox, and colds," Kaempffert wrote. His article included only a brief nod to the Wellses: "Dr. Wilder's studies were based on those of Drs. William F. and Mildred W. Wells, which showed that the air we breathe in a room may be a major means of infection."

No one would have guessed from the story that the Wellses had designed the experiment, brought the lamps to the school, installed them, maintained them, and pored over the data for years. But what seems to have bothered Mildred most was that Wilder stole one of her ideas.

Wilder mused in his talk about why the ultraviolet lamps were not as effective against colds as other diseases, like measles. He argued that colds must be able to spread easily in the air at home, away from the neutralizing rays. Kaempffert shared the idea with his readers as if it had been Wilder's own.

Mildred called Wilder a few days later to complain. She reminded him that the idea about colds was hers, and that she was including it in the final analysis she was still writing. After the call, she sat down and sent Wilder a letter as a reminder.

"The question is more than one of 'claiming credit' for scientific work," Mildred wrote. She did not want to be remembered as a purveyor of hype. "There was, as you know, quite a bit of publicizing of the measles curves from the first school paper, and considerably less attention given to the negative results on colds. If now you publish the colds analysis without giving due credit, I very much fear that what I actually will get credit for will be the soft pedaling of these data."

There's no record of how the conflict ended, if it did. Perhaps Mildred and Wilder reached a truce. In any case, Wilder never published his New York talk, and Mildred made no mention of colds in her final report. Or of William Firth Wells.

○ ○ ○ ○

William went with Mildred to the 1944 New York conference, but he did not return home with her to care for Bud. He stayed on to listen to younger scientists talk about airborne infection. The experience made him feel as if he were the author of a novel that had been turned into a movie. Now he was sitting in a theater watching it with people who had never read the book. "Many of the points which irritated me did not seem to bother those I talked to and they seemed impressed with points that I considered jaded," William later recalled.

Some of the scientists who spoke at the meeting presented the first results from the wartime experiments that William had wanted to run. Oswald Robertson paid for scientists at Northwestern University to spray ethylene glycol on the walls of barracks. That experiment went badly, as the gas condensed into flammable stains. So Robertson doled out more money to test a safer variant, triethylene glycol. The Northwestern team came to the New York meeting to report how they sprayed barracks night and day for six weeks, in the hopes of protecting the thousand soldiers who slept there.

Wells had been skeptical that they would succeed, because so little was known about how the chemical behaved in the air. The Northwestern researchers announced that triethylene glycol appeared to lower the hospitalization rate among the soldiers by 11 percent. The results, they acknowledged, were so modest as to be inconclusive.

Wells listened to another talk that vexed him even more. Alexander Hollaender described what happened when his team installed ultraviolet lights at a navy base.

Hollaender, an expert on radiation damage, put the lamps in eleven barracks at the U.S. Naval Training Center in Sampson, New York, in December 1943. His team then tracked the infections in the buildings, comparing them with those in eleven unprotected barracks. When Wells had learned of Hollaender's experiment, he predicted it would fail because the sailors who slept in protected barracks spent time during the day with unprotected sailors. "The mingling of sailors from the irradiated and control barracks in common unirradiated meeting places should serve to lessen the difference in incidence of disease between the groups," Wells warned. "This is unfortunate, for the military camps provided an unusual and rare opportunity for almost complete coverage of groups."

Now, in October 1944, Hollaender came to the New York meeting to share the first results from Sampson. It was Hollaender, not Wells, who got to be the subject of a newspaper account about testing ultraviolet light on thousands of soldiers. "Germ-killing ultraviolet light has previously been used to check the spread of disease in hospital wards and school rooms," the reporter wrote. "This, however, was its first test in military barracks."

Wells had been right to be skeptical. Hollaender had tried out two kinds of lamps, one 121 watts, the other 235 watts. The low-watt lights failed to slow the spread of respiratory diseases. The more powerful ones cut them by as much as 25 percent, but only at the start of the study. That result was promising, Hollaender said, but hardly a vindication.

"In view of the fact that reduction in morbidity rates was marked in the high intensity irradiated group only in the first months of the study period, the results of the experiment of 1943–1944 should be interpreted with caution," he and his colleagues warned. Hollaender said he would not render a final judgment until he turned the lights back on at Sampson the next winter.

For the US military, these inconclusive results were evidence enough to abandon any attempt to stop airborne infections carried in droplet nuclei. Instead, it set out to fight dust. In a 1944 study, scientists found bacteria growing on the dust coating blankets and floors in military hospitals. The discovery raised the prospect that sweeping or bed making could spread diseases to people nearby. The army responded to the research by applying mineral oil to blankets and floors.

That decision was particularly maddening to Wells. He suspected that dust settled so quickly out of the air that any bacteria it contained were unlikely to cause outbreaks. Not only would anti-dust measures be a waste of time, but they would distract the scientific community from his own theory. "Such a cloud of infected dust, raised by war effort, obscures the true meaning of the droplet nuclei hypothesis," Wells later said.

An official military history acknowledged that the army took the easy way out: "Oiling of floors and bedclothes received better acceptance since it was a simple, inexpensive procedure." Ultraviolet light would have required a huge investment with uncertain returns, the authors observed, "not only of sleeping quarters in barracks but of other confined spaces in which men congregate, such as mess halls and drill halls, ship's service buildings, post exchanges, dispensaries, and washrooms."

Wells was left with a mix of vindication and yearning. "I envy your great opportunity to work on such a large canvas," he confessed to Robertson in December 1944.

o o o o

Japan surrendered to the United States in August 1945, bringing the war to a close. By some estimates, 15 million soldiers had died in battle, while 45 million civilians were killed. Unlike the First World War, the second failed to unleash a global pandemic. But that merciful turn of events was probably due more to luck than scientific prevention.

Over the next few years, the US government shared some of the fruits of wartime medical research. Penicillin, originally restricted to soldiers, became a cheap cure for bacterial infections. The world's first effective influenza vaccine, developed with wartime support, went into civilian arms. And the researchers who studied the spread of diseases in army camps looked back at their experiments to see what lessons they offered for peacetime public health. The scientists began holding conferences, publishing their results, and forming committees to make recommendations.

The end of the war also meant that civilian scientists could start running new tests of their own. One of them was James Perkins, New York's deputy commissioner of public health. Perkins considered William Firth Wells a visionary, and was so impressed by the Germantown experiments that he wanted to replicate them. "Only the Wells have conducted what appears to be a well-controlled experiment, and it is well not to assume too much on the basis of experience by only one group of investigators," he said.

Perkins chose three elementary schools in rural upstate New York for his experiment. They were fairly new buildings with identical ventilation systems. Perkins had ultraviolet lamps installed throughout the

Cato-Meridian school. In Port Byron, he had half the rooms fitted with the lamps. A third school, in the village of Mexico, got no lamps. It would serve as a control.

Wells waited eagerly for the results. "I think it would go a long way to crystallizing the thinking in the field and clarifying the situation," he said. And in the winter of 1945, a measles epidemic swept through New York State. With the lamps switched on at Cato-Meridian and Port Byron, Perkins was ready.

The measles epidemic washed over all three schools in waves of three different shapes. At Mexico, without UV lamps, measles hit fast and hard. Most of its 146 cases occurred between late January and late February. The half-irradiated Port Byron school saw two hundred twenty cases of measles, but they did not arrive until early March, in a wave that stretched out to May. At Cato-Meridian, where all the children studied under ultraviolet lamps, 144 got measles. But the cases were stretched out along an even gentler curve that lasted three months. The ultraviolet lamps slowed down the speed of the epidemic, Perkins concluded, but they did not cut the percentage of students who ultimately got sick.

As for why he failed to replicate the Philadelphia results, Perkins blamed the buses that the students took to school. In Philadelphia, the test subjects either walked home or were picked up in cars by their wealthy parents. But while riding on school buses, students were easy prey for measles.

"It is not to be construed," Perkins and his colleagues cautioned, "that upon the basis of these findings, the authors are recommending routine installation of ultra-violet lamps in classrooms."

The disappointing results from New York strengthened the growing skepticism about airborne infection. More doubts arose when Alexander Hollaender finished off the navy barracks experiments and delivered his final verdict: ultraviolet lights only provided modest, inconsistent protection. Scientific conferences filled with the buzz of disappointment. "The striking success of ultra-violet light in classrooms in the Swarthmore School, Philadelphia, in preventing the spread of measles does not seem to have been repeated elsewhere," the *British Medical Journal* concluded.

Ronald Hare, the Canadian microbiologist who had published an experiment in 1940 that cast doubt on airborne infection, took an even harsher view after the war. To him, the entire theory was useless. "Unless extremely violent and impolite methods are employed," Hare said in 1946, "it is probably very difficult for a carrier to infect the atmosphere in his vicinity to any great extent directly from his mouth or nose."

Wells moaned about these "ravages of misinterpretation and exploitation." The invitations to give lectures or write commentaries dried up. One of the few venues where Wells still had a chance to defend his legacy was a new committee created by the American Public Health Association to review wartime research. But for the other members of the committee, dealing with Wells was akin to torture. Oswald Robertson later remembered how Wells would hijack meetings to drone on about his own research, leaving the rest of the scientists confused as to what he had actually done. When they tried to get Wells to explain himself, Robertson said, he would lash out in anger. Another committee member, the Johns Hopkins University epidemiologist Alexander Langmuir, had similar memories. Wells scrambled his data as he delivered his harangues, Langmuir recalled, so that in the end no one believed a word he had to say.

For the few friends William Firth Wells had left, the reason for his decline was clear: Mildred was gone. She did not divorce him, but a wide gulf had opened up between them nevertheless. "He has now fallen out with her," Heffron said.

Edwin Wilson, who had known the couple for fifteen years, found the falling-out a mystery. "I don't understand exactly what the relationship is between Mr. and Mrs. Wells," he said. "So far as I can see neither of these persons is a particularly easy person to get on with as is often the case with people of a good deal of ability and people who keep themselves tired trying to do things. Whether they get on at all well with one another is more than I can make out. Sometimes I think they don't and sometimes I think they do."

The conflict left William alone, at least as a scientist and perhaps as a husband. "The split-up between Wells and his wife," Heffron said, "has removed from the scene some sort of balance wheel."

○ ○ ○ ○

Even as Wells alienated his colleagues and his own wife, he still managed to carry out experiments that would be recognized decades later as some of his most important research. He found a new scientist to take Max Lurie's place and help him run the Infection Machine. Herbert Ratcliffe, a veterinary scientist, studied tuberculosis and other diseases in the animals at the Philadelphia Zoo. He was game to help Wells study them in more intimate detail.

Ratcliffe and Wells embarked on experiments to trace the path of tuberculosis from the air into the lungs. Researchers had long known that the disease gets its start when *Mycobacterium tuberculosis* reaches the finest branches of the lungs and invades the alveoli. It's in these sacs that oxygen crosses into the bloodstream. Each alveolus houses its own squad of immune cells known as alveolar macrophages, which prowl the sac and gobble up anything that doesn't belong there—be it bacteria or inhaled bits of ash from a fire. But *Mycobacterium tuberculosis* turns the tables: when a macrophage eats it, it doesn't die. Instead, it starts feeding on the macrophage from within and multiplying. The new bacteria burst out and feed on other alveolar macrophages. The feasting produces a tubercle of scar tissue.

From his previous experiments with Lurie, Wells knew that droplet nuclei could deliver the bacteria into the airway of a rabbit and cause tuberculosis. Now he wondered if their size mattered to their success. Wells figured out how to control the production of droplet nuclei in the Infection Machine so that he could create them in two sizes, which he called coarse and fine. Only the fine droplet nuclei could infect the rabbits.

From that result, Wells and Ratcliffe concluded that tuberculosis occurred only when fine droplet nuclei managed to float into the lungs, through narrowing passageways, until they reached an alveolus, where they could lure a macrophage to its doom. Wells estimated that droplets larger than five microns were too big to make the full journey. And dust grains, which could get up to a hundred microns in size, had no chance of success. "Most dust-born bacteria would not reach the lung," Wells predicted.

Those experiments brought Wells visits from officials from the Veterans Administration. The Department of Defense was no longer worried about outbreaks of diseases like influenza in overcrowded wartime barracks, but tuberculosis continued to take a heavy toll on soldiers. It wasn't clear exactly how they were getting sick: some seemed to fall ill right after getting infected, while others might harbor the bacteria for years before developing a bloody cough. Without any drug to cure the disease, the Veterans Administration built dedicated TB hospitals where bedridden veterans could only hope they might recover. The hospital staff did their best to keep dust down, in the belief that it carried the disease to new victims. And yet, despite those precautions, all too many VA nurses and doctors came down with tuberculosis as well.

Wells's research led the Veterans Administration to consider the possibility that tuberculosis was airborne. Some officials were so persuaded that they made plans to install ultraviolet lamps in some VA hospitals. But Wells worried that his new admirers were jumping the gun. Before they installed the lights, they should put them to a proper test to confirm that they actually helped people.

Despite those reservations, Wells was grateful that at least someone was still taking his ideas seriously. The epidemiologists rising to prominence after World War II were giving up on his entire theory of airborne infection, erasing over fifteen years of work. "The relative importance of the air-borne route of infection compared to direct contact and droplets is far from established in any particular environment," Alexander Langmuir wrote to Wells. "It seems to me that your concepts are still in the stage of development."

Langmuir made his opinions public as well, attacking Wells in reviews he published in scientific journals. New studies on the epidemiology of diseases pointed away from the air as the deciding factor in how diseases spread, and toward what Langmuir called "close personal association." It seemed to him that Charles Chapin had been right all along: "Thus, the challenge to contact spread of infection failed."

The negative opinions came at a bad time for Wells. Roderick Heffron was talking once more to experts about whether to renew support for his work. Heffron consulted with Langmuir, who said it would be foolish to do so. Langmuir warned that the Commonwealth Fund would be betting on someone in whom other people in his field had no confidence, and whose results they could no longer trust.

Heffron also sat down with Wilson Smillie, who had originally hired Wells to teach at Harvard in 1930 and later served with Wells on committees in World War II. Smillie told Heffron that Wells deserved credit for opening the field of airborne infection, but he had become impossible to work with.

On June 12, 1947, Heffron delivered his verdict to the Commonwealth board: "Mr. Wells is his own worst enemy." The board voted not to renew its support. Wells would receive an extra payment of three thousand dollars so that he could finish up his research and look for other opportunities. And then his work would be at an end.

"I cannot say that the final action was altogether a surprise though I was of course greatly disappointed," Wells wrote to Heffron.

Wells, incapable of teaching, could not ask the University of Pennsylvania for help. At sixty, he witnessed the Laboratories for the Study of Air-borne Infection get taken apart piece by piece. But as his career collapsed and his scientific legacy disappeared, Wells got wind of some disturbing news. During the war, as he had tried to fight airborne infections, the military had been secretly running its own experiments on how to create them. And in its quest to invent a weapon that could unleash an epidemic, it ended up supporting Wells's theory in a way he could never have imagined.

eight

WINGS FOR DEATH

In September 1941, Elvin Stakman traveled to Chicago to lead the world's first aerobiology meeting. In his lecture at the symposium, he took his audience on a sweeping tour of life in the air. Aerobiology, he declared, "is almost literally as broad as the world and as high as the sky." Soon after his return to Minnesota, Stakman got an ominous request from a secret government committee. They asked Stakman if he thought aerobiology could be brought onto the battlefield. They wanted to turn the spores he and Fred Meier had hunted in the clouds into weapons of war.

The idea of biological warfare was not new. For centuries, armies had tried to sow diseases among their enemies. In 1346, for example, the Mongols attempted to spread the plague inside the city of Caffa on the coast of the Black Sea.

The Mongols had spent three years besieging the city when the Black Death arrived and started decimating their ranks. As the Mongol army prepared to retreat, it launched one last assault. "They ordered corpses to be placed in catapults and lobbed into the city in the hope that the intolerable stench would kill everyone inside," the Italian notary Gabriele de' Mussi recorded soon afterward.

The flying bodies terrified the residents of Caffa, who dumped them into the sea as fast as they could. But they started to die of the plague anyway. "Soon the rotting corpses tainted the air and poisoned the water supply, and the stench was so overwhelming that hardly one in several thousand was in a position to flee the remains of the Tartar army," de' Mussi wrote. The outbreak was so brutal that the people of Caffa thought the Last Judgment had come.

The Mongols believed they were carrying out biological warfare, but they might not have been the ones who actually delivered plague into Caffa. Contrary to what de' Mussi wrote, the disease does not spread as a poison in the water or a stench in the air. No one in the fourteenth century knew that fleas bearing *Yersinia pestis* bacteria were responsible. It's unlikely that any fleas were on the corpses that the Mongols loaded into their catapults. Fleas jump off a dead body as soon as its temperature drops. It's more likely that the plague came to Caffa when flea-infested rats wriggled their way through cracks in the city's walls.

In the centuries that followed, history records a few more examples of soldiers trying to spread diseases among their enemies. In 1763, British soldiers in Pennsylvania tried to kill Delaware Indians with blankets and a handkerchief taken from a smallpox hospital. In the Civil War, Confederate doctors used a similar tactic: they tried using bedding and clothing to spread yellow fever around Washington. There's no firm evidence that any of these attempts killed anyone. It certainly didn't help that the would-be killers didn't understand how diseases spread.

The rise of the germ theory in the late nineteenth century made it easy for people to imagine death coming from a Petri dish. When anarchists set off bombs in England in the 1890s, fears grew that they were brewing Koch's comma bacillus to unleash cholera in London. No such plot existed, but the writer H. G. Wells knew a good storyline when he saw one. He wrote a tale in which an anarchist dreaming of terrorist glory steals a vial of the bacteria from a microbiologist: "All those people who had sneered at him, neglected him, preferred other people to him, found his company undesirable, should consider him at last. Death, death, death!"

With the start of World War I, those nightmares grew more intense. "How Deadly New Bacilli May Be Bred to Take the Place of Bullets" was the headline of a 1915 *San Francisco Examiner* article. "Distribution from airships would at present seem the most feasible," the newspaper predicted. The idea seemed to become real when German airships dropped bombs on London. A strikingly large fraction of the blast victims developed gangrene. "The extraordinarily high proportion of deaths following injuries caused by Zeppelin bombs has led to strong suspicions among surgeons who treated these cases that the bombs have been impregnated with deadly disease germs," a *Daily Mail* correspondent wrote.

While there's no evidence that Germany really rained gangrene-causing bacteria on London, it did send agents to the United States and several other countries to carry out biological warfare. The agents took deadly brews of bacteria and injected them into horses and mules, hoping to wipe out the animals that the Allies needed to haul their cannons and wagons. The German agents succeeded in killing a few thousand animals, but the sabotage made little difference to the course of World War I—the United States alone deployed more than a million horses.

Still, the end of World War I did nothing to dim fears of germ warfare. In 1925, they hung like a pall over a meeting of the League of Nations in Geneva, where diplomats were working out an agreement to ban chemical weapons. The Polish general Kazimierz Sosnkowski urged the conference to also ban biological weapons, such as bacteria-packed bombs dropped from airplanes. "Great masses of men, animals and plants would be exterminated," Sosnkowski warned. The representatives from other countries agreed; in the end, the Geneva Protocol of 1925 included a prohibition against "bacteriological methods of warfare."

At the time, the United States military didn't worry much about its enemies using bacteriological methods. If the German attacks on horses and mules in World War I were as bad as germ warfare could get, it seemed like a minor concern. But during the 1930s, American generals had a change of heart. It got easier for them to imagine a catastrophic biological assault. Reports surfaced that Germany was going to unleash clouds of bacteria against their enemies. Nazi spies were supposedly sneaking into London and Paris, where they were reenacting Carl Flügge's experiments with *Serratia marcescens*. Instead of having people talk after a microbial mouthwash, the spies were allegedly releasing a *Serratia* mist into subway stations and then using Petri dishes to measure how far the bacteria floated.

From Asia, more reports of germ warfare emerged. Japan invaded Manchuria in 1931 and began terrorizing the populace. American spies heard that among their atrocities, Japanese soldiers were intentionally spreading diseases. It was hard to know how many of the stories about Germany and Japan were true. But the American military was now open to the prospect of strange new weapons. Physicists were warning that quantum theory could give rise to a bomb that could erase cities. Perhaps microbiology could unleash another kind of annihilation.

○ ○ ○ ○

In October 1941, Secretary of War Henry L. Stimson invited nine prominent scientists to meet secretly in order to determine the threat of germ warfare. To help conceal their mission, the government gave the committee a bland name: the War Bureau of Consultants. The War Bureau contacted dozens of scientists to gather their thoughts. Elvin Stakman was among them.

Stakman was intimately acquainted with how food could steer the course of battle. In World War I, late potato blight had brutally undermined the German war effort, while the abundant harvest from American farms had kept Allied soldiers well fed. In response to the War Bureau's inquiry, Stakman told them it was entirely possible for America's enemies to threaten the country's food supply by dropping rust spores on farms. The fungi might wipe out millions of acres of crops and starve the nation. To prepare for this possibility, Stakman recommended the creation of a corps of plant pathologists. They would monitor American fields for signs of trouble and respond to attacks.

But Stakman went further. He believed that the United States should not just defend itself from biological weapons. It should create its own arsenal of deadly spores that might be used against its enemies.

"Indeed, to fail to make the attempt would seem gross negligence," Stakman warned.

Stakman laid out a plan for starving Axis forces. "In general, the best method of disseminating infective material would probably be from airplanes," he wrote. His plan was a reverse image of the flights he and Meier had taken before the war. They had caught airborne spores to learn how to protect the fields below. Now Stakman proposed releasing spores to destroy enemy farms.

As for what the planes should release, Stakman offered a catalog of terrors, much of it drawn from recent research by plant pathologists. The planes could drop late potato blight spores. Or they might try a newly discovered type of rust from Peru called Strain 189 that killed every variety of wheat scientists tried it on. The military shouldn't limit itself to pathogens that caused famines, Stakman added. Japan's fleets of trucks rolled on tires made from rubber grown on Southeast Asian plantations. Stakman proposed wiping out rubber trees with South American leaf blight.

To make sure that its bombs wiped out crops, Stakman recommended that the military search for as many pathogens as could do the job, and then pack them together. Each pathogen might work only under certain conditions—one in wet soil, another in dry, for example—but a mixture of germs could guarantee destruction. "A SHOT/GUN ATTACK SHOULD BE MADE," Stakman wrote.

After sending off his gruesome ideas, Stakman kept the inquiry a secret. When he published his Chicago lecture in 1942, he made no mention of his ideas for turning aerobiology into a weapon. By then, Stakman's advice to the government was being put into deadly practice. And for decades to come, the rise of germ warfare would throw the science of aerobiology off the course Fred Meier had dreamed of.

○ ○ ○ ○

The War Bureau of Consultants distilled everything it learned from experts like Stakman into a report that it sent to Stimson in February 1942. Biological warfare, the experts concluded, was "distinctly feasible." They conjured a vast panorama of destruction, including everything from typhoid-laced milk to subway strap handles smeared with diphtheria.

Among all the ways to deliver disease, their favorite route was by air. "Serious consideration should be given to the possibility of the dissemination, as mists or dusts widely distributed by airplanes, of diseases which may be contracted through the respiratory tract," they wrote. "Fine dusts might be very effective against crops as well as against animals and man."

The War Bureau of Consultants recommended preparing for germ warfare against the United States by building up stockpiles of vaccines.

They also called for the nation to build its own arsenal of biological weapons. "The best defense for the United States is to be fully prepared to start a wholesale offensive whenever it becomes necessary to retaliate," the consultants told Stimson. They advocated for brewing pathogens on an industrial scale and putting them in bombs, shells, and spray tanks.

Stimson approved. "Biological warfare is, of course, 'dirty business,'" he told President Roosevelt. "But in the light of the committee's report, I think we must be prepared."

Roosevelt ordered the development of biological weapons in order to learn how to defend against them as well as to build an arsenal. The effort was given its own anodyne name—the War Research Service—and put under the leadership of the pharmaceutical executive George Merck. For its headquarters, the army selected an old airfield in Maryland called Camp Detrick. A town's worth of buildings went up behind a high wall, and Merck filled them with four thousand workers, all sworn to secrecy.

Among the recruits who came to Camp Detrick was Theodor Rosebury, a thirty-nine-year-old professor from Columbia University. Like the other scientists who joined the effort, Rosebury did so with a clear conscience. He was not troubled by the fact that he was about to violate the Geneva Protocol. "The world was like a house on fire, and scientists were willing to dirty their hands to put the fire out," Rosebury recalled in a 1968 speech. "We could clean up later."

○ ○ ○ ○

Before Pearl Harbor, it would have been impossible to imagine that newspapers would call Theodor Rosebury "Mr. Biological Warfare." Until then, he was simply a left-wing dentist.

Rosebury was born in London in 1904 to Jewish immigrants from Russia. His family moved to New York when he was six, and his father worked as an editor, writer, and translator for garment unions. Aaron Rosebury was an ardent anti-Communist, his son later said, and he quit the unions when the party tried recruiting their members. He took a string of jobs—running a movie theater, opening a summer boardinghouse in upstate New York—and found success in none.

Theodor was the youngest of the four Rosebury children, a small, bookish boy who suffered from poor health. At fourteen, he came down with violent coughs and ended up in the tuberculosis ward at Bellevue Hospital in Manhattan. As he lay alone on a cot, he tried passing the time reading *Martin Eden*, Jack London's 1909 book about a self-taught proletarian who becomes a successful novelist. All around him, TB-ridden men groaned in pain. Every day, a few of them died. Rosebury spent over a week in the ward before his doctors informed him that he did not have tuberculosis after all. He was free to leave.

Rosebury went on to earn a spot in Manhattan's Stuyvesant High School, where he excelled in history and learned photography. After graduation, he spent three years kicking around New York, trying to become a professional photographer, but it gradually became clear he would not become Martin Eden with a camera. He didn't know what else he could do. When a friend from grade school decided to become a dentist, Rosebury tagged along to his dental classes at New York University. A year later, he followed his friend again, this time to Philadelphia. There they enrolled in dentistry school at the University of Pennsylvania.

To earn tuition money, Rosebury worked in a Philadelphia lunchroom and spent the summers back in New York, working as a photographer. He met Lily Aaronson, whom he married in 1925. Theodor and Lily went on to have two daughters, Joan and Celia. Rosebury could have supported his new family in middle-class comfort if he had opened a dentist's office after earning his degree. But by the time he graduated, he had become seduced by the biology of teeth.

In his school notebook, Rosebury drew elaborate sketches of the enamel, pulp, blood vessels, and nerves that add up to a tooth. The furrowed surfaces of teeth fascinated him even more. Germs could drill holes into teeth, but not all were enemies. Every person's mouth was slathered in an assortment of harmless species, such as alpha streptococci, the bacteria that Wells would later draw from his students with his sneezing powder at Harvard. Why some species could live peacefully on our teeth while others wreaked havoc, no one could say. Rosebury accepted a fellowship at Columbia to explore such oral mysteries. The meager twelve-hundred-dollar stipend, he later recalled, "looked to my dental classmates like insanity."

After the crash of 1929, Rosebury tried to supplement his salary by pulling teeth for a few hours a week. He didn't last a year. Rosebury gave himself up entirely to science and investigated the origin of cavities. He fed rats different diets in a lab at the base of the George Washington Bridge. He observed how certain kinds of food fostered tooth-attacking bacteria. He found that corn made rats more prone to cavities—perhaps by spurring the growth of bacteria that spewed acid on the enamel.

In 1936 the Bureau of Indian Affairs offered Rosebury a rare chance to study cavities in people. He traveled to Alaska to spend months inspecting the teeth of Eskimos. At remote fishing stations on the Bering Sea, he ended up in high demand as a dentist rather than as a scientist. When a young man needed a lower molar pulled out, Rosebury used a pair of carpenter's pliers, "with a fallen log as a dental chair," as he later recalled. "At one point later, faced with a good-sized batch of toothaches, I enjoyed the help of a competent nurse and was able to seat my patients luxuriously on a barrel."

When his survey was over, Rosebury took a six-day boat ride from Alaska to Seattle. He passed the hours looking back through his field notebooks. In the settlements where Rosebury had worked, some had a high rate of cavities, while others had no cavities at all. The diet of the Eskimos seemed to explain the differences. Those who ate traditional foods like fish did not get cavities, while those who usually ate pilot bread—a kind of hardtack that settlers brought to Alaska—did. "White man's food and decaying teeth have appeared at the same time" was how one newspaper later summed up Rosebury's findings.

On his way home to New York, Rosebury also listened to the radio to catch up on the news he had missed in the wilderness. He was shocked to learn that Spain had fallen into civil war and that General Francisco Franco was backed by Nazi Germany. "I began to read the foreign news more carefully," Rosebury later recalled.

Only then, in his early thirties, did Rosebury become politically active. He joined the American Association of Scientific Workers, a progressive organization that believed scientists must take political action to improve society. They threw themselves into battles against everything from astrology to fascism. Opposed to Nazism, the AASW launched a boycott of German scientific equipment.

Rosebury also resumed his research back at Columbia. He figured out how to keep some of the bacteria usually found only in the mouth growing in his laboratory. He planned to write a book about the "normal flora," as he called it. He taught microbiology to medical and dental students, livening up his lectures with gruesome stories about the Black Death and other epidemics. And in the newspapers, he came across stories raising the possibility that Hitler might start epidemics of his own with biological weapons. In one article, British authorities issued vague warnings that Hitler was preparing to conduct germ warfare. In another, an American microbiologist scoffed at the idea. No one would try such a thing, he insisted, because germs "neither recognize nor stop at the front line."

When Rosebury read these reports, they seemed to fall halfway between fantasy and reality. "At most this was a subject for light conversation with colleagues at luncheon," he later wrote.

○ ○ ○ ○

The attack on Pearl Harbor put an end to the light conversation at luncheon. "I shall never forget the spirit that was abroad in the land after Pearl Harbor," Rosebury would later say. "I had it; nearly everybody shared it, and there was plenty of evidence that my own circle was a fair sample of the whole country. We were not free from doubts as to whether the war might have been avoided, but war had been thrust upon us; we were convinced that the enemy was indeed totalitarian and utterly ruthless; and we rose to meet the challenge."

Rosebury believed that American scientists could—and should—help the country defeat Hitler. As he talked with fellow AASW members about what they could do, Rosebury thought back to the newspaper speculations about germ warfare. He might have been a dentist by training, but Rosebury believed his wide-ranging knowledge about pathogens qualified him to judge their true threat to the country.

He brought up the idea of investigating biological warfare with his department chairman, Alphonse Dochez, who had just joined a new secret government committee overseeing wartime medical research. Dochez thought it was a good idea and promised to pass along any report

Rosebury put together to the proper authorities—as long as Rosebury kept it a secret.

Rosebury embarked on a search through the scientific literature, but there was far too much information for him to sift through alone. He recruited the Columbia immunologist Elvin Kabat as a research partner, along with a medical student named Martin Boldt as an assistant. The three of them settled down in Columbia's library, paged through journals, and took notes.

They examined every infectious disease known to science, judging which might work well as a weapon and which would likely fail. The pathogens most likely to cause devastating wartime outbreaks were the ones that traveled by air, they concluded. "The air-borne route of infection seems to be the most practicable for bacterial warfare," Rosebury and Kabat later wrote in their report.

To explain why, they surveyed the history of research on germs, starting with Pasteur, moving on to Flügge's experiments on sneezes and coughs, and then describing Chapin's dogma that droplets could spread diseases over short distances. "This concept of droplet infection as the sole means of transmission of most infections of respiratory origin was generally accepted until recently," Rosebury and Kabat wrote.

But then William and Mildred Wells showed how droplet nuclei could spread diseases over long distances. Even as public health experts were turning against the Wellses, Rosebury and Kabat found their work persuasive. "The bearing of these findings on bacterial warfare is far-reaching," they wrote.

Before the Wellses did their research, it was reasonable to be skeptical that germ warfare in subways and other public spaces posed a real threat. But now, Rosebury and Kabat declared, "such justification no longer exists." Thanks to the Wellses, it was clear that airborne pathogens could be potent weapons. "By the air-borne route, they would meet no effective man-made sanitary safeguards," Rosebury later said. "The air is free in evil ways as well as in good ones."

Certain pathogens looked to Rosebury and Kabat as being naturally suited for use as airborne weapons. At the top of their list was parrot fever.

Psittacosis, as the disease is formally known, is caused by a species of bacteria now called *Chlamydia psittaci*. Parrots and other birds pass it between one another as a gut bug, but in 1929 hundreds of pet owners across the United States got infected by their birds, likely through the air. A doctor in Annapolis, Maryland, uncovered the outbreak on a house call to a patient. When he stepped into the living room, he found a man sprawled in a chair, muttering in a feverish sleep. The man's wife wandered in from the bedroom, shouting gibberish. From the kitchen, her mother emerged unsteadily, a rattling cough rising up from deep in her lungs. Lying on the bottom of a bird cage was a dead parrot, claws up.

All told, about eight hundred people contracted parrot fever over the course of six months, thirty-three of whom died. While a lot of the infections were caused by germ-carrying birds, the victims also included a few scientists who got sick in their labs as they tried to isolate the germ. Their cases hinted that *Chlamydia psittaci* could fly up out of its Petri dishes.

Epidemiologists also discovered four infections that formed a chain of contact: it started with a bird dealer who passed the germ on to his doctor, who passed it on to an intern, who passed it on to a nurse, who passed it on to another nurse. It looked as if parrot fever could spread between people as well as birds, at least under the right conditions.

Parrot fever did not return to the United States after the 1929 pandemic, thanks to the invention of a test that made it possible to block the spread of the disease. While it might no longer be a threat to American public health, Rosebury and Kabat thought the germ could become a powerful weapon. It could be grown to huge numbers and then released into the air to infect a vast number of people. Vaccinated soldiers could march into a zone where the disease was raging without fear of falling ill. All in all, Rosebury and Kabat concluded, the parrot fever germ "may be one of the most useful agents of bacterial warfare."

Other germs didn't look very promising in their natural state, but a little tinkering might have turned them into good weapons too. Anthrax spores would be especially worth investigating, Rosebury and Kabat believed.

Normally, anthrax spores lurked in the soil, where they struck grazing cattle and sheep. People occasionally became sick from handling contaminated hides or eating tainted meat, but they couldn't pass the bacteria on to other people. And yet, under the right conditions, anthrax could also become airborne.

The first clues to this lethal route emerged in the 1800s, when English doctors noticed that workers in wool factories sometimes suddenly died. They would struggle to breathe, cough up blood, and expire in a matter of hours. The mysterious condition came to be known as woolsorter's disease. "Scarcely a month passes without the record of some death in the district from this disease," John Bell, the medical officer of the Bradford Infirmary, wrote in 1880.

When Bell read about Robert Koch's experiments on anthrax, he wondered if it was the cause of woolsorter's disease as well. Perhaps, Bell thought, woolsorter's disease was what happened when the anthrax spores became airborne and invaded the lungs, instead entering the stomach or the skin.

Bell tested this idea by following Koch's example. He drew blood from a factory worker who had died of woolsorter's disease and injected it into a rabbit, a mouse, and a guinea pig. All three animals died within three days. Bell injected some of their blood into another rabbit, which died even faster. "The fluids from this animal were found to be crowded with the *bacillus anthracis*," Bell later wrote, "and the disease was recognised to be anthrax."

Rosebury and Kabat were impressed by how quickly inhaled anthrax spores could damage people's lungs and then spread throughout their bloodstream, releasing toxins along the way. "Its properties make it an obvious possibility as an agent in warfare," they concluded. They envisioned packing spores into a bomb, perhaps along with mustard gas to ravage the lungs, giving the spores better odds of establishing an infection. An army could unleash an outbreak of woolsorter's disease across a battlefield.

Rosebury and Kabat finished writing their report in June 1942. Dochez passed on the secret analysis to the National Research Council. The two scientists believed they had simply explored a world of deadly possibilities. They had no idea that the army was already starting to make those possibilities real.

○ ○ ○ ○

Rosebury got a quick letter of thanks for the report from E. H. Cushing, the commander of the division of medical sciences. "The immense amount of labor that has been spent upon this excellent report is evident, and we are very glad to have it," Cushing wrote. After that, no further word came from Washington. Having done his part for the war, Rosebury turned back to teaching at Columbia and investigating the bacteria of the mouth. But as American forces started fighting in the Pacific and Europe, Rosebury got restless.

"I have no doubt that in peace time this is the most important work for which my talents and training fit me," he wrote to a government assignment agency, "but in war time I believe I could be placed in more essential work." A colonel in the army's Dental Corps assured Rosebury that he was serving his country well by producing well-educated dentists, but Rosebury would not be put off. He learned of an opening for an army dental pathologist in January 1943 and put himself forward yet again. This time, he was told he was unqualified. "This is so ridiculous that it would be funny if it weren't outrageous," Rosebury complained to a friend.

He stopped pestering the army and resigned himself to watching the fight against Hitler from Manhattan. "I don't doubt that the Army will manage to win the war without me."

A few months later, the army came to him. Officials informed Rosebury that an entire military complex now existed in Maryland where thousands of workers were carrying out the research that he and Kabat had envisioned. Now they wanted the two Columbia scientists to help run it. The army might not have needed Rosebury's skills as a dentist, but it admired his deadly imagination.

Kabat agreed to go to Camp Detrick as a consultant for a few days each month. But Rosebury left his family and moved down to Maryland to work on germ warfare full-time. "It was a cause to which I gave myself completely," he later said.

Before letting him go to Camp Detrick, the government did a thorough investigation of Rosebury's background, including his membership in groups such as the American Association of Scientific Workers. The association had gained some notoriety early in the war. It included an anti-Nazi faction and a pro-Soviet one, and when Stalin and Hitler had formed a pact, the association fought over whether to condemn it or celebrate it. Some scientists, including the physicist Robert Oppenheimer, dropped out of the AASW for fear it might make them look like subversives. Rosebury remained an active member, but the government's investigation turned up nothing of concern. After getting a security clearance, he arrived at Camp Detrick in December 1943, ready to get to work.

○ ○ ○ ○

When George Merck's team got their hands on Rosebury and Kabat's report in June 1942, they were startled by how much it resembled their own plans. Even though the two scientists had created it from nothing more than library research, the government immediately classified the document. Each new recruit who came to Camp Detrick was handed a copy. Their work became the camp's bible.

The Detrick researchers had independently decided that airborne pathogens would be the best kind of biological weapon. And among the pathogens they started investigating were some of Rosebury and Kabat's favorites, such as parrot fever and anthrax. "These germs may have constituted the hottest batch of bugs ever handled at one time by a single group of men," Rosebury later recalled. Some of the staff were assigned to grow stocks of bacteria and run experiments to make them deadlier. Rosebury was appointed the chief of the Airborne Infection Project. He helped design a building where his team of thirty scientists would test the newest weapons. It was called the Aerobiology Building.

The lab was a far cry from the government aerobiology research center Meier had dreamed of five years before. Inside Camp Detrick's Aerobiology Building, a deadly inversion of William Wells's research took place. Rosebury developed military-grade versions of his Infection Machine called cloud chambers, in which the Detrick team could kill mice with germs. Wells had created clouds of droplet nuclei to figure out how to protect people from airborne infections. Now Rosebury worked on germ-laden clouds that might someday spread over army camps or even cities.

The lion's share of work at Camp Detrick was on anthrax. The army set out to create the woolsorter's disaster that Rosebury and Kabat had imagined. Detrick scientists learned how to grow the spores at an industrial scale and convert them into a powder that could be loaded into bombs.

Rosebury believed that at Camp Detrick he was not just helping fight fascism, but also making important fundamental discoveries about life. In the Aerobiology Building, his team was proving beyond a doubt that germs could spread by air under the right conditions. They even figured out the minimum number of germs necessary for a cloud to become lethal. Yet the findings always had a military application. Thanks in part to Rosebury's research, Camp Detrick researchers tripled the deadliness of their anthrax.

o o o o

On May 25, 1945, Rosebury went to Camp Detrick's officers' club to meet two fellow lab workers for drinks. One of his companions that night was a lab technician, Jerry Aaron. The other was Martin Boldt, Rosebury's former library assistant. After Boldt finished medical school, Rosebury had persuaded him to come to Camp Detrick to help make their report a reality. The three men finished their drinks, headed out into the beautiful spring night, and boarded an army bus headed to the nearby town of Frederick. They went to the Buffalo, a favorite restaurant, where they slid into a dark pine booth and all ordered lobster.

Rosebury had been looking forward to the meal, but he struggled to finish it. "I had been working almost without letup and considered that a good enough reason for feeling pretty tired," he later recalled. After taking the bus back to Camp Detrick, Rosebury said goodbye to Aaron and Boldt and headed for bed. His friends didn't think much of his departure as they headed off for a nightcap at the officers' club. "It was not unusual for me to be a bit moody," Rosebury said.

At two in the morning, Rosebury woke up feeling as if the lobster were trying to claw its way out of his stomach. He tossed around till morning. When he slid out of bed, he managed to have a little toast and coffee before heading to the Aerobiology Building. But by lunch he was shaking with chills and retreated to bed again. Boldt dropped by to examine him and diagnosed nothing more than an upset stomach. All Rosebury needed was sleep.

Rosebury dozed for four hours. But in the late afternoon, he woke again with a jolt. He feared that it wasn't the lobster that was making him sick. Now he was sure he had parrot fever.

A technician from the parrot fever team would regularly come to the Aerobiology Building with a sample of the germs to test. The germs were mixed in a dollop of frozen egg yolk sealed in a glass capsule. Rosebury and his team would carefully thaw the capsule, remove the germs, and run experiments to see how well they could kill mice by traveling through the air.

On May 14, eleven days before Rosebury got ill at the Buffalo, a technician had arrived carrying a pasteboard carton. Inside the carton, a glass capsule rested on a bed of cotton wool. Rosebury removed the capsule from the carton and placed it in the palm of his left hand so he could read the label. In the brief time he held it, the warmth of his skin thawed the frozen egg yolk in the capsule. A fine, gentle stream of liquid shot out of its base, forming a tiny pool in his cupped palm.

"I remembered saying something at the instant of the accident, something inane like 'I'm a dead duck!'" Rosebury later recalled. When he saw the fluid escape, he immediately put the capsule in a sealed chamber and bathed it in a sterilizing steam. He scrubbed his hands and washed down the area with a disinfectant.

That seemed like precaution enough. A drop of germ-laden fluid could not seep into his skin in a few seconds. And it was hard to imagine how it could have gotten into the air where he could breathe it. Rosebury and his team put the whole incident out of their minds and got back to work.

Lying in bed eleven days later, Rosebury realized he might have been wrong. Maybe the parrot fever had gone airborne. Eleven days was within the window of time it took for parrot fever's symptoms to emerge. His symptoms matched up nicely with what he had read about the disease. It often started out so mild that it fooled doctors. What felt like an upset stomach might actually have been pathogens multiplying in the lower reaches of his lungs.

If Rosebury did have parrot fever, he knew he had a one-in-four chance of dying. Sulfa drugs had shown some promise in infected lab animals, but no one knew if they'd work in people. He had seen reports of two people successfully treated with penicillin, but the antibiotic was still a rare novelty. Rosebury headed for Camp Detrick's small hospital, where he tried to persuade two medical corps officers that he had come down with a rare bird disease.

The skeptical doctors left Rosebury standing in agony while they examined his X-rays. The pictures convinced them to take him seriously and put him to bed. The next day, Rosebury was visited by the virologist Gordon Meiklejohn, the head of the parrot fever team. Meiklejohn collected a sample of Rosebury's spit to test.

As Rosebury suffered through his fever and aches, a soldier brought a document to his bedside for him to sign. Rosebury had to agree to keep the germ warfare program a secret, no matter what happened. If he died in the next few days, no one could ask for an authorized autopsy, not even his wife. His corpse would be placed in a lead casket that would be permanently sealed. It would be disposed of as the authorities saw fit, never to be opened.

Rosebury didn't hesitate to sign the paper. "Of course this business struck me as macabre, but less so than you might think," he later said. "Our whole operation was macabre. The war inured us to such things."

While Rosebury's doctors waited for Meiklejohn's test results, they gave him sulfa drugs and penicillin. Several days later Meiklejohn announced that the parrot fever pathogens were indeed growing from Rosebury's spit. By then, Rosebury was recovering, thanks either to the drugs or his own immune system.

It took weeks for Rosebury to recover fully. But even after he returned to work, he could not stop thinking about what had happened. He learned that Meiklejohn's team regularly grew three strains of parrot fever bacteria. On the morning of May 14, Rosebury determined, the technician had brought a capsule containing a strain called 6BC, which had been isolated from a parakeet in 1941. Rosebury now tested 6BC on mice in his cloud chambers. He found that it could kill the animals but only if he gave them a relatively large dose. Another strain, called Gleason, proved to be fifteen times deadlier than 6BC. And the third strain, known as the Borg virus, was more than ninety times deadlier. Rosebury suspected he was alive thanks to sheer luck, having been infected by the mildest strain of the three.

To figure out how 6BC had gotten into his lungs, Rosebury retrieved its capsule, now empty and sterilized. Peering through a hand lens, he noticed what looked like a tiny bubble floating inside the capsule's base. He discovered it was actually a hole barely wide enough for him to slide a hair through. Rosebury suspected that the hole had popped open as the frozen glass quickly warmed in his hand. The warmth had also increased the pressure in the capsule, forcing the fluid to shoot out of the hole in a hair-thin stream.

Rosebury tested the idea by replicating his experience. This time, he captured it on film. The inveterate photographer set up a camera with a flash that worked much like the stroboscope used by Marshall Jennison a few years earlier to capture droplet nuclei coming out of people's mouths. Rosebury had no intention of putting anyone in his lab at risk of getting parrot fever for his experiment, so he prepared a batch of harmless *Serratia marcescens*. He and his lab mates cut off the top of a capsule and poured in five milliliters of *Serratia*-infused fluid. They attached a rubber tube to pump in a little air and force the fluid out onto Rosebury's hand. The camera flashed.

When Rosebury developed the film, his palm looked like a ridged landscape of hills. A slender *Serratia* waterfall was tumbling into his hand and forming a clear pond. High above Rosebury's skin, floating like stars in a dark gray sky, were droplets. The impact of the fluid on his hand had hurled them into the air. They were so small, Rosebury concluded, that gravity could not pull them down. Instead, they kept rising and mixed into the air that he inhaled. Rosebury had become his own proof of the evil ways of the air.

nine

A FINE FROZEN DAIQUIRI

In the spring of 1944, British officials were in a secret panic. Their spies were hearing that Germany had created biological weapons and was preparing to send them across the English Channel, either inside a rocket or dropped from an unmanned plane. Britain, like the United States, had been working on biological weapons, but no nation had yet openly used them on the battlefield. If Hitler was on the verge of crossing that line, Winston Churchill wanted to deliver a similar response. Britain didn't have any weapons ready to use, but Churchill knew that Theodor Rosebury and his colleagues at Camp Detrick were making huge strides. So Churchill placed an order with the United States government. He asked for half a million anthrax bombs "for use if this mode of warfare is employed against us," he said.

To meet the request, army officials set up a plant in Indiana. While they were at it, they planned to make another half million anthrax bombs for America's own use. The factory would need months of safety tests before it could start production. But now that the weapons were so close to creation, officials began to talk seriously about what rules should govern the first use of germ warfare. Historians have suggested that when President Harry Truman was making plans for dropping atomic bombs on Japan in 1945, he considered using biological weapons if Japan still refused to surrender.

It never came to that. The intelligence reports about Hitler's microbial arsenal turned out to be wrong, and Germany's V-1 missiles delivered conventional explosives to England. When Hitler's army later retreated to Germany, it did not use germ warfare to slow down the advance of the Allies. Dropping atomic bombs on Hiroshima and Nagasaki was enough to force Japan's surrender. The war ended without any of the research at Camp Detrick making it onto a battlefield.

With peace restored, the Allies began investigating the wartime horrors that had taken place in Germany and Japan. Interrogations of Nazi scientists and doctors revealed that they had used prisoners at Buchenwald and other concentration camps as lab animals. They tested experimental vaccines for diseases like malaria, tuberculosis, and yellow fever on prisoners. Mrugowsky led the effort at Buchenwald to create a vaccine for typhus. As part of the experiments, the Nazi doctors intentionally infected the prisoners and watched them die in agony.

Friedrich Weyrauch, the sneeze photographer, might have witnessed some of those experiments. He was summoned to Buchenwald in 1938 to deal with a paratyphoid outbreak, and while it isn't clear if he actually participated in the experiments at Buchenwald, he did suffer a nervous breakdown while at the camp. Two years later, on November 16, 1940, he shot himself.

After the war, Mrugowsky was put on trial at Nuremberg. More than fifty-six thousand prisoners had died at Buchenwald alone, through medical experiments, torture, and outbreaks. Mrugowsky tried to distance himself from the carnage. He portrayed himself as a noble doctor developing vaccines for the good of humanity, a claim betrayed by all the people who had died in their development. He also claimed that he had saved countless lives by killing typhus-spreading lice with Zyklon gas. The trial revealed that Mrugowsky also supplied the gas for the slaughter of Jewish prisoners. For his genocidal acts, Mrugowsky was hanged.

Given all the horrors the Nazi scientists unleashed, it's surprising that they never created a place like Camp Detrick. German officials got as far as discussing the possibility of using biological weapons, but Hitler ordered that they limit their work to defensive measures. While American officials overestimated the threat of germ warfare from Nazi Germany, they wildly underestimated Japan.

After the invasion of Manchuria, the Japanese army built installations there to develop biological weapons. Researchers started testing them on animals and then on Chinese prisoners. They infected thousands of victims with diseases such as scarlet fever, smallpox, and whooping cough. Afterward, Japanese doctors investigated the effect of the weapons by performing autopsies on the prisoners—sometimes while they were still alive.

The Japanese military then began using their weapons on the Chinese people. They dumped plague-infected fleas from planes. They spiked village wells with cholera and typhoid. They spread anthrax in rice fields. It's believed that tens of thousands of Chinese people died during Japan's rampage. But the pathogens did not respect the nationality of their victims. They killed thousands of Japanese soldiers as well.

After World War II, the United States tried more than five thousand Japanese for war crimes. The leaders of the biological weapons program were not among them. American officials chose to keep those interrogations secret in order to keep the Japanese officers talking. They hoped to use what they learned from the interrogations for their continuing efforts. The war might have been over, but the Americans had no plans to stop building their own biological weapons.

○ ○ ○ ○

On November 1, 1945, two months after the surrender of Japan, Theodor Rosebury officially resigned from Camp Detrick. He returned to New York still sworn to secrecy about his wartime work, but his service to the government left him in good stead. Rosebury won generous support to restart his investigation of the harmful and harmless bacteria in our mouths. The army wanted to give Rosebury even more money for his research at Columbia, but he turned down the offer. He didn't think the military had any business sponsoring civilian science.

Rosebury had been back in New York for two months when his wartime work officially came to light. George Merck released a six-page statement acknowledging that the United States had conducted biological weapons research. It was long on abstractions and short on detail. Merck did not name Rosebury or any of the other scientists who worked at Camp Detrick, nor did he list the pathogens they studied. Instead, Merck wrote cryptically about the program's achievements: "significant contributions to knowledge of the control of airborne disease-producing agents."

The biggest thing missing from Merck's announcement was the fact that biological warfare research had not actually stopped with the end of the war. Although Rosebury had resigned from the army, he was still working for them as a consultant, making regular trips back to Camp Detrick to help upgrade his cloud chambers. "I hope that I have an opportunity to go on with it," he told Willard Rappleye, the dean of Columbia's medical school. Camp Detrick officials even dropped hints to Rosebury that they might offer him a full-time job. Rosebury later speculated that if he had gone to work there, he might have risen through the ranks to become Camp Detrick's technical director.

With the war over and the biological warfare program now a matter of public record, Rosebury wanted to share some of his experiences. Many of his colleagues had the same wish. They considered their work in the same league as research into antibiotics and vaccines during the war: its significance endured into peacetime. The government agreed to loosen its control of wartime work, although it kept ongoing research classified. Officials gave their approval, paper by paper, for scientists to publish a selection of results.

Rosebury was allowed to write up his bout of parrot fever. His account appeared in 1947 in the *Journal of Infectious Diseases*. Nowhere did he explain that he had gotten infected while testing the pathogen as a weapon. At the end of the report, Rosebury merely said that his experience might "throw light on the general problem of airborne laboratory infections due to highly infective agents."

He then wrote about his cloud chambers. This time, the army did not let his paper slide so easily into print. "It still smacks considerably of classified motivations which are almost impossible to cover up, it seems to me," Oram Woolpert, the technical director at Camp Detrick, complained to Rosebury in June 1946.

Hoping to placate the army, Rosebury made changes to the manuscript and sent it back to Woolpert. "I think they have a considerable deodorant effect on the classified motivation which, as you know, worried me, too," Rosebury said. The final version of *Experimental Air-borne Infection* ran to 222 pages. Not a single page offered a clue about its classified motivation.

In 1947 the army awarded Rosebury the Decoration for Exceptional Civilian Service in recognition of the dangers he had faced in his wartime work. He accepted the medal with a mix of pride and misgiving. His conscience had been clear while he tested biological weapons, because he could tell himself he was working at an extraordinary moment in history. "We were in a crisis that was expected to pass in a limited time, with a return to normal values," he later said. Now the crisis was over, yet Rosebury was still helping Camp Detrick test biological weapons. He could see that the United States was entering a Cold War, with the Soviet Union stepping into Germany's role as the enemy. Rosebury kept waiting for normal values to come back.

○ ○ ○ ○

Rosebury was not alone. Albert Einstein and some of the other physicists who had helped usher in the nuclear era began protesting the construction of bigger bombs now in peacetime. They appealed to the public, warning that a full-scale nuclear arms race would not protect the United States. It would only end in annihilation. Rosebury decided to launch a similar campaign "to bring the subject of bacterial warfare to the attention of the public so that concrete action may be taken to prevent its use."

But bacterial warfare was much harder to bring to the attention of the public than nuclear weapons. Einstein could point to photographs of mushroom clouds. There was no mushroom cloud for parrot fever. The public knew next to nothing about what biological weapons could do or how hard the United States was working to build new ones. Rosebury knew, but much of his knowledge remained classified. If he shared the wrong piece of information, he would end up in prison.

Looking for a loophole, Rosebury thought back to the report he had written with Kabat in 1942. Even after the war, it remained classified. Rosebury asked Woolpert for permission to publish it. After all, he and Kabat had written it based solely on what they could read in the Columbia University library.

Woolpert saw no harm in the idea. "Frankly, I believe that it is good propaganda for us at this time," he told Rosebury.

Rosebury got a much colder reception from Columbia when he raised the idea. As the Cold War picked up, the university didn't relish his left-wing activism. Rappleye tried to scare Rosebury away from releasing his report. "Publication of this paper while you are Professor in the University is in the judgment of the administration not in the public interest so far as the University is concerned," Rappleye told Rosebury. "We do not wish it published while you are a member of this staff."

Rosebury and Kabat decided to ignore Columbia's wishes. They submitted their paper to the *Journal of Immunology*, which promptly accepted it. The journal was a technical publication that usually published little-read reports with titles like "Allergenic and Anaphylactogenic Properties of Vaccines Prepared from Embryonic Tissues of Developing Chicks." But in May 1947, it became the eye of a political hurricane.

The American Association of Scientific Workers held a press conference to announce the publication of Rosebury and Kabat's paper. The two scientists had originally written their report to coax the government to investigate biological weapons—both to defend against them and to prepare to use them in war. Five years later, they used it as part of a campaign to end germ warfare once and for all. "It is now presented by the American Association of Scientific Workers as an argument for peace," Waldemar Kaempffert reported in the *New York Times*.

The report came as a shock; few people knew at the time how readily germs could be turned into weapons or how many there were to choose from. The nightmare scenarios that had been conjured by the likes of H. G. Wells and General Sosnkowski seemed closer than ever. "Set down with a kind of desperate, scientific calm," *Time* reported, "the report would make as pleasantly alarming reading as any outrageous fictional chiller—except for the fact that it might all come true."

○ ○ ○ ○

As the press seized on Rosebury and Kabat's report, the government realized that it was not good propaganda after all. And they meted out swift reprisals against the two scientists for their betrayal.

Rosebury and Kabat lost their clearance at Camp Detrick. The FBI opened investigations on them both. Rumors circulated that Kabat was a Communist sympathizer, leading to his forced resignation from the

Bronx Veterans Administration Hospital. Meanwhile at Columbia, Rosebury sensed that his superiors were growing even more hostile toward him. "It was clear that this sort of thing was not taken kindly at the medical school," he later said.

At home, Rosebury found no refuge. His time away at Camp Detrick had done irrevocable damage to his family. "I had returned to find my home life in shambles," he later recalled. Theodor and Lily fought bitterly. Their daughter Joan spiraled down into depression. After she attempted suicide, she underwent electroshock therapy. Their other daughter, Celia, sided with Lily and became estranged from Theodor. After two years, he moved out and filed for separation.

Rosebury soon fell in love with a psychotherapist named Amy Loeb and they decided to get married. Rosebury took a year's leave from his research at Columbia so that he could move with Loeb to Florida to take advantage of the state's loose divorce laws. "Leaving my lab wasn't easy, but leaving the kids was amputation," he told Kabat.

Without any research or teaching to take up his time, Rosebury became "an ardent peace activist," as he put it, determined to fight America's drift into a militarized existence. Hitler might have lost the war, but fascism still seemed to be winning. Rosebury was appalled when President Truman created a federal loyalty program, which he feared would lead to government-supported scientists being forced to take an oath against Communism. Rosebury and the physicist Melba Phillips condemned the oaths in the pages of *Science*. "The integrity of science and scientific education in this country is seriously jeopardized," they warned. "Unless we resist completely this latest invasion of the personal beliefs of students and their teachers, we are inviting the kind of atmosphere which pervaded Germany in the thirties."

Rosebury also continued to fight the escalation of biological warfare by writing a book that would give the public a sense of its true threat. He mined his own published work, along with other papers and news articles, while gingerly avoiding even a mention of classified work. In 1949, Rosebury published *Peace or Pestilence: Biological Warfare and How to Avoid It*, the first book-length treatment of the new threat.

"We need not doubt that BW is capable of taking its place beside the atomic bomb and other major weapons adaptable to mass destruction,"

Rosebury warned. He urged the United States not to get locked into an endless race with the Russians to build more weapons—either atomic or biological. "It seems to me that the so-called 'menace' of Soviet Communism is vastly overrated," he wrote. "If the United States is really strong—as I believe it is—and if our strength resides, not in military power and aggressiveness, but in the character and the way of life of our people—as I believe it does—then I am sure we have nothing to fear from the Russians or from any other nation."

Rosebury chose a bad time to make the case for calm. After four years of uncontested postwar supremacy, the United States watched with dread as the Soviet Union became a nuclear power. Only a few months after the publication of *Peace or Pestilence*, a mushroom cloud rose over the steppes of Kazakhstan in August 1949. Suddenly the United States was in a race, and it had to keep ahead.

It also had to prepare itself for a surprise attack. John Kennedy, then a young congressman, called for the country to build defenses against the new nuclear threat lest it suffer an "atomic Pearl Harbor." A biological Pearl Harbor seemed possible too. "There are indications that the Soviet Union is prepared in the BW field and will not hesitate to use BW if she deems it to her advantage to do so," a Department of Defense team warned in a top secret 1950 report.

Now the US government needed to prepare the public for both kinds of attacks. The Federal Civil Defense Administration urged Americans to build bomb shelters and offered tips on how to duck and cover in classrooms. In 1951 it also released a pamphlet on the danger of a biological attack. Drawing on the work of scientists like Wells and Rosebury, it described how airborne germs might be spread from airplanes or submarines. Or a saboteur might spray a germ-laden mist into a factory's ventilation system. "The big danger to people is in new ways of spreading diseases already known," the administration warned.

○ ○ ○ ○

The American government also prepared for biological attacks in a way that would influence the country for decades to come: it made its public health system part of its military defenses. This Cold War transformation was led by none other than Alexander Langmuir, the Johns Hopkins epidemiologist who had clashed with William Firth Wells, disputed his theory of airborne infection, and helped to end his career.

Even as Langmuir publicly cast doubt on Wells's work, he was secretly visiting Camp Detrick during the war to consult with the government on its biological warfare program. As an epidemiologist, Langmuir had no advice to offer about how to load spores into bomb casings, but he could make predictions about the devastation a biological weapon would cause in an American city. After the end of the war, Langmuir joined the Committee on Biological Warfare to assess the risk of a Cold War attack.

Those secret experiences had a paradoxical effect on Langmuir's thinking. He dismissed the risk of natural airborne infections, but he warned that airborne biological weapons were a genuine threat. "Specially designed bombs, shells, or other types of disseminating devices discharged from enemy aircraft or from warships offshore could create large clouds," Langmuir later said. They might, "like smog, hang over a city for many hours."

In 1949, Langmuir gave up his job at Johns Hopkins to join a new government agency known then as the Communicable Disease Center—what today is known as the Centers for Disease Control and Prevention. The CDC made Langmuir their top epidemiologist, a post he relished after his long stint in academia. "Epidemiology means getting out to the scene to see what's happening," he later said. "We ring doorbells. We pound the pavements. We go to the patients instead of waiting for the patients to come to us."

In his new job, Langmuir proved irascible and unpredictable. A reporter once asked Langmuir's secretary at the CDC what it was like to work for him. She pointed to a vent pipe in the roof of their building. "When we see that smoking we hardly move because we know he's blowing off steam," she said. But unlike Wells, Langmuir was shrewd about how to reach his goal: build a modern team of government epidemiologists who would track a host of infectious diseases, from parrot fever to typhus to dysentery.

Langmuir believed he needed to protect the country not just from natural outbreaks, but also from biological attacks. He later estimated that he had spent about a fifth of his time as CDC's top epidemiologist working on germ warfare. He worried that a biological weapons attack would come without an explosion's flash or the thunder of a shock wave to announce itself. Bomblets might break open silently high over a city and rain down anthrax spores. Agents might spray subway stations with fine mists of bacteria. It might take days before a surge of sick people showing up in emergency rooms would reveal the attack. To uncover the earliest signs of a biological weapons assault, Langmuir wanted to conduct an epidemiological version of military intelligence. Instead of gathering clues about tank attacks or missile strikes, his epidemiologists would track local medical reports for odd clues of disaster.

Langmuir chose the right moment to build his team. In June 1950, the Korean War brought the United States into armed conflict with proxies for the Soviet Union for the first time. North Korea accused the United States of deploying biological weapons on the battlefield (a claim that was never substantiated). Worried that the Soviet Union would respond to the accusation with germ warfare attacks on American soil, Congress handed Langmuir the money he asked for. He used it to create the CDC's Epidemic Intelligence Service.

Langmuir also helped the American public imagine the invisible threat his detectives would be looking for. He did so by taking to the new medium of television. Langmuir was the star of a nationally broadcast show called "What You Should Know About Biological Warfare."

The balding forty-one-year-old epidemiologist arrived in the studio dressed in a three-piece suit and a polka-dotted tie. He leaned against a desk and began to explain in clinical detail what it would be like if germ warfare came to the United States.

"Although we can't predict the exact form that some enemy attack might take in the future, we can use our best judgment from our present knowledge to try to forecast as well as we possibly can," he said. "I and many others who have studied this problem feel that from the evidence the chances seem best that an attack might take the form of contamination of the air."

Langmuir stood up from the desk and walked over to an easel with a blown-up photograph. It was one of Jennison's photographs: the profile of a jowly man blasting out a sneeze. Langmuir pointed out the droplets spewing from the man's mouth. He then moved from the photograph to a table strewn with props to show how easy it was to artificially create germ-rich mists. He motioned to a Waring blender. "It makes a fine frozen daiquiri, and yet it is used in the laboratory to grind highly infectious material," Langmuir said. He poured dry ice in the blender and switched it on. After letting it spin for a few seconds, he removed the lid, and a gassy cascade spilled over its rim. Langmuir warned that if he had added germs to his cocktail, the blender would have created a cloud that could infect most of the people in the studio.

The blender showed how readily germs could accidentally escape from a laboratory. To demonstrate how someone could intentionally spread them, Langmuir held up an insecticide bomb from a hardware store. He invited his audience to imagine it filled with anthrax. He turned to a small mock-up of an ivy-covered university lab and sprayed the insecticide into its side. "See how the cloud permeates throughout the rooms," Langmuir said.

He moved back to the desk and glanced nervously at index cards for guidance. From buildings, Langmuir turned to attacking whole cities. He described a ship or submarine putting up a cloud offshore. Airplanes could drop bombs loaded with germs. Langmuir injected ink into a network of tubes to show how a secret agent could contaminate a water supply.

The United States needed to prepare as a nation for such attacks, Langmuir warned. "We can examine our existing protective networks and build them into a strong defense system," he said. It was vital to catch attacks early to trace outbreaks and to launch vaccination campaigns to stop their spread. "These together form an epidemic intelligence service," Langmuir told his viewers.

Off camera, Langmuir was already building that service. In their first full year of work in 1952, the CDC disease detectives responded to more than two hundred calls from local authorities to investigate outbreaks, including polio and rabies. In later decades, the team would become legendary for jetting off at a moment's notice to a distant country to assess a medical crisis. Some of the epidemiologists Langmuir trained would go on to help eradicate smallpox.

But along the way, they never discovered a Soviet germ attack of the sort Langmuir had gone on television to warn the country about. And yet Langmuir's nightmare, in one form or another, would endure in the American psyche for generations.

○ ○ ○ ○

While Langmuir built his team of detectives, Rosebury fell under more scrutiny as a subversive. Anti-Communist politicians began to treat left-wing scientists as a dangerous threat. "The ranks of American scientists have been infiltrated to an alarming degree by the Communist enemies of the United States," Senator Joseph McCarthy declared.

The government took note of Rosebury's membership in the American Association of Scientific Workers, the American Labor Party, and other suspicious organizations. Rosebury had a vague hunch that he had gained enemies in high places. But he didn't realize how much trouble he was in until the summer of 1950.

It was then that the United Nations hired Rosebury as a consultant to the Atomic Energy Commission. The commission was looking into hosting a conference about biological weapons, and Trygve Lie, the Secretary-General of the UN, wanted a list of essential readings on the subject. In June 1950, Rosebury started creating that list, a project he expected to take several months.

Anti-Communist senators considered the United Nations a haven for Soviet spies, and they started pressuring the State Department to investigate the loyalty of American citizens working for the organization. In August 1950, Rosebury got a call from the United Nations letting him know that someone in the American government had sent a letter complaining about his work for the Atomic Energy Commission. The mysterious official warned that Rosebury might divulge military secrets. The United Nations decided to abandon the project, and canceled Rosebury's contract in September 1950.

Rosebury kept the incident quiet, but it might have spurred him to declare his loyalty in public. In a speech soon afterward, he emphasized that an American scientist could be a patriot and a skeptic about the Soviet threat at the same time. "It may be well for me to assert—and this I do without hesitation—that I am a loyal American, as loyal as Mark Twain's Connecticut Yankee," Rosebury said. "And, indeed, it is as much my loyalty as my scientific training which prompts me to suggest certain doubts."

At Columbia, Rosebury's superiors did not agree. "I came to feel increasingly less welcome there," he later recalled. "I had tenure, but it had little meaning for me. I do not stay when it seems clear that I am not wanted."

In 1951, at age forty-seven, Rosebury found an escape route: an old acquaintance had become the dean of Washington University in St. Louis and now invited him to join the school of dentistry as a full professor. Theodor moved out west with Amy. He would later call it one of the three biggest mistakes of his life. (The second was deciding to become a dentist in the first place. The third was marrying Lily.) Rosebury found St. Louis much more conservative than New York. The dental school in particular felt like a parochial, small-minded place. "If we could tolerate it at its dubious best, we found it ludicrous, irksome, or even abominable at its more frequent worst," he said.

Life in St. Louis got even more irksome, thanks to the anti-Communist senators. In January 1953, the Senate Internal Security Subcommittee released the names of thirty-eight past and present government employees "believed to be Communists or under Communist discipline." The roster was printed in newspapers across the country. It included an entry describing Rosebury's aborted service at the United Nations: "Theodor Rosebury, hired in June, 1950; adverse comment Aug. 23, 1950; terminated Sept. 18, 1950."

On top of that old embarrassment, a new scandal broke out when Rosebury was invited to give a series of lectures at the University of California San Diego. Under California law, Rosebury had to sign a loyalty oath. When it arrived in the mail. Rosebury wrote back that he refused to sign it, and his lectures were canceled. "Some of my colleagues applauded this action," Rosebury later said, "but it tied another knot in my reputation."

Soon Rosebury was struggling to find money for his research. When he had returned from Camp Detrick a few years before, he had been flush with cash. Now the National Institutes of Health cut him off. He got enigmatic letters that informed him his grants were approved but could not be funded.

It dawned on Rosebury that he had been blacklisted.

"It is ironic that during the period I am speaking of now, just as in my war period at Detrick, I was too busy with the immediate job in hand to do anything that could be called political or in any conceivable sense subversive," he later wrote. "Except refusing to sign a loyalty oath, which I think I would do with my last breath."

○ ○ ○ ○

While Rosebury was being persecuted for speaking out against germ warfare, the government was expanding the research at Camp Detrick far beyond the experiments Rosebury had worked on in World War II. His cloud chambers were now too small for its ambitions. Instead of misting animals with pathogens, Detrick scientists invented a way to expose them to germs released by exploding bombs. For these new experiments, engineers constructed an enormous sphere with metal walls thick enough to withstand explosions. They nicknamed it Eight Ball. One observer later wrote that Eight Ball "looks like a spaceship hovering three stories above the ground."

Inside Eight Ball, a new chapter of scientific research took place. "We were pioneers in the science of aerobiology," one of the scientists later declared. They started by setting off small blasts for animal experiments, and in 1954 they began exposing human volunteers to the clouds produced in Eight Ball. More than twenty-three hundred military draftees signed up for the research, mostly Seventh-day Adventists who wanted to serve their country but refused to go into battle. In some trials, the volunteers were injected with experimental vaccines, after which they sat down in front of the Eight Ball and had their head strapped in place in front of a sealed window. They would breathe air laced with pathogens that caused diseases like tularemia and Q fever to see if the vaccines protected them. Unvaccinated volunteers inhaled some of the germs released inside Eight Ball, after which they got experimental treatments to see if they could be cured. While none of the volunteers died, some got violently ill.

In addition to the research at Eight Ball, the army started carrying out secret germ warfare research far outside the walls of Camp Detrick. One of the earliest clues of their tests came from San Francisco in 1950, when a seventy-five-year-old retired pipe fitter named Edward Nevin died of a strange infection.

After a minor surgical procedure at Stanford University Hospital, Nevin returned with chills and a fever. He had an infection that spread to his heart valves and could not be stopped with antibiotics, and he died after three weeks. It was not surprising for a frail man to pick up a deadly infection, especially after surgery. What was surprising was the species that infected Nevin.

It was *Serratia marcescens*, the seemingly harmless bacteria that lived in the soil and that microbiologists from Carl Flügge to Theodor Rosebury had used in their experiments. The Stanford doctors grew more suspicious when they discovered ten other patients who had come down with urinary infections of *Serratia* at the same time.

In 1951, Nevin's doctors published a report on the outbreak: "a curious clinical observation," as they called it. It showed that what seemed like a harmless species of bacteria could be dangerous under certain conditions, especially for people with frail immune systems. The researchers had no idea how the microbe, which had hardly ever been reported to cause infections, had attacked so many people at once.

Unknown to the Stanford doctors, their paper sparked a panic in the army. After World War II, germ warfare researchers had wondered if an enemy could infect an American city from a ship miles offshore. If the ship sprayed biological weapons into the air, could germs reach land, waft down streets and alleys, and kill people who inhaled them? In September 1950, a navy mine-laying vessel headed for San Francisco to find out. The army offered the city no warning of the experiment it was about to run.

The crew loaded the ship with *Serratia marcescens*, along with *Bacillus globigii*, another supposedly harmless microbe. When the weather turned in their favor, the sailors opened up giant hoses and sprayed a cocktail of the two microbes into the air. They sprayed bacteria from the hoses for an hour a day for six days in a row, and then sailed away.

Later, investigators secretly fanned out through cities and towns along the coast, swabbed samples, and smeared them into Petri dishes. Colored colonies grew in the dishes—red dots of *Serratia marcescens*, black dots of *Bacillus globigii*. Army researchers concluded that the cloud had successfully exposed hundreds of thousands of people to the microbes, not just in San Francisco but as far away as Sausalito and Daly City.

Seven months later, when Alexander Langmuir appeared on television, he cryptically referred to enemy ships releasing clouds of bacteria off American shores. His audience had no idea that American ships had done just that. But when the report on the *Serratia* outbreak came out in October 1951, biological warfare officials worried that their San Francisco exercise was to blame. Camp Detrick's safety officer secretly ordered no further tests be carried out near hospitals or other facilities where people might be endangered.

Eventually, however, the army came to believe spraying bacteria did not put Americans at risk. After the Stanford report, more *Serratia* outbreaks came to light, and they occurred in places where the army was not conducting tests. It might have been pure coincidence that the Stanford cluster of cases occurred right after the secret operation in 1950. The army went on spraying *Serratia*, blowing clouds into New York subways and other locations into the 1960s.

By 1956, biological weapons work at Camp Detrick had progressed so far that the government made the camp a permanent installation, changing its name to Fort Detrick. It also began building facilities in Utah and elsewhere so that it could start testing the weapons outdoors. In one trial, a plane streaked over the Pacific while releasing plumes of microbes that caused Q fever and tularemia. The clouds descended on cages of monkeys hauled by tugboats. While no one is known to have died in those exercises, lab accidents at Fort Detrick took a toll. More than four hundred workers became infected between 1943 and 1969, and three died.

A network of factories also opened up to manufacture the pathogens. By the mid-1960s, the air force had enough weapons to unleash germ warfare on a staggering scale. A single B-52 bomber sortie could shower anthrax bomblets over ten thousand square miles. Alternatively, the planes could be fitted with spray tanks to unleash microbes over fifty thousand square miles.

Along with making human pathogens, the US military began mass-producing crop-killing spores. Elvin Stakman continued helping the biological warfare efforts after World War II, developing a standardized weapons-grade form of wheat rust. The army also built up stockpiles of other fungi that could attack oats, rye, and rice.

Detrick scientists maximized the destructive force of the spores by ensuring that they fell together, instead of dispersing in the wind. Inspired by some of Japan's World War II experiments, the scientists sprinkled the spores onto turkey feathers, which they then dropped on farms in New York and California. In some tests, the feathers were loaded into balloons; in others, they were dropped from planes in bombs originally designed for showering leaflets on enemy territory. These tests and others like them led the air force to estimate that a single sortie could trigger 100,000 rust outbreaks over a fifty-square-mile stretch of wheat farms. Officials drew up plans for dropping the spores across the Soviet Union's grain fields and China's rice paddies. All told, the United States stockpiled enough rust to kill all the wheat on Earth.

○ ○ ○ ○

On a blustery afternoon in March 1958, Rosebury suffered a massive heart attack. At the hospital, a cardiologist visited Rosebury's bed and told him he would have to stop working in order to recover. "Spare me your sermon," Rosebury barked. "If I feel the need for a priest, I'll ask for one."

Rosebury got over his stubbornness, took a year's leave from Washington University, and headed to Cape Cod with Amy to recover. There he painted watercolors, read Shakespeare aloud with friends, and played clarinet next to his woodpile. After six months, Rosebury felt strong enough to resume work. Still blacklisted, he couldn't launch any major new research. Instead, he turned back to the mystery that had consumed him as a young dentist in the 1920s: the microbes that live peacefully on and inside us, known today as the microbiome.

From birth, each of us becomes home to trillions of microbes—not just in our mouths, but in our guts, on our skin, and throughout much of our body. Microbiologists traditionally paid little attention to them. They investigated the pathogens that invaded people's bodies and made them sick. But Rosebury suspected the microbiome was not made up of opportunistic passersby, species as likely to grow on an elbow as on a rock. He believed that the species in our microbiome were adapted to humans in particular—"indigenous to man," as he liked to put it. Rosebury also suspected that they did not survive in isolation. They depended on one another, created what he called "microbic ecology." Rosebury spent the last six months of his leave gathering his research into a book.

Rosebury's spirits lifted further when he got a letter from the philosopher Bertrand Russell, who invited him to a conference in the remote Canadian village of Pugwash. Since 1957, the Pugwash Conferences had brought together some of the world's leading scientists to discuss the dangers of nuclear war. Now they wanted to turn their attention to biological weapons, and they asked Rosebury to guide them. He was the obvious choice: the only scientist on Earth with a deep knowledge of biological weapons and the willingness to talk about them in public.

"I was back in the real world again, recovered," Rosebury said.

At the conference, Rosebury emerged once more as Mr. Biological Warfare. He provided a detailed history of the field up to 1959 and ended his speech by quoting Einstein's warning about nuclear weapons: "Science has brought forth this danger, but the real problem is in the minds and hearts of men."

The meeting, which brought Soviet and Western scientists together, drew international attention; Rosebury and the other Pugwash scientists seized the opportunity to release an open letter calling for the United States and the Soviet Union to stop hiding their chemical and biological warfare programs, "to dispel the miasma of secrecy that fosters international suspicion and tension."

As the 1950s came to a close, American public opinion slowly began to turn Rosebury's way. No one had any idea of the full scope of the American germ warfare program, but they knew enough to start protesting. Church groups stood vigil at the gates to Fort Detrick for months. Rosebury began receiving invitations from across the United States and

Canada to give talks about the dangers of biological warfare. As the Red Scare subsided, he got support from the National Institutes of Health again. He used his new grants to study the microbiome, as well as the bacteria that cause syphilis, hoping his research might lead to a vaccine. In 1962, he published *Microorganisms Indigenous to Man,* the first book ever written on the microbiome.

Yet a fear of arrest still gnawed at Rosebury. In April 1964, he traveled to Portland to give a scientific talk at the University of Oregon Dental School about the microbiome. A reporter from the *Oregonian* named Jack Ostergren buttonholed Rosebury afterward for an interview. Once they got through some friendly palaver about the billions of bacteria that live in the human body, Ostergren changed the subject.

"Is it true that you are sometimes called 'the father of biological warfare'?" he asked.

Rosebury's twinkle disappeared. "I do not wish to talk on that subject," he said.

"Oh," Ostergren replied in a tone that sounded as if he were changing topics. "Well, I understand you're the author of two books on airborne infections."

"That's the same subject," Rosebury declared coldly. "Let's put it this way: Because of my bacteriological research I became involved in BW during the war, and I have been in it since, but with the idea of getting it stopped. Unless you're prepared to go into the subject, exhaustively, you could get us both in trouble by going into it at all."

○ ○ ○ ○

Rosebury soon got himself into trouble anyway. As the war in Vietnam expanded, he spoke out against it. He was particularly concerned about the army's use of plant-killing herbicides that had been developed at Camp Detrick during his time there. American planes dumped Agent Orange on the Vietnamese countryside to wipe out crops and to rob Vietcong soldiers of their hiding places. As students began organizing anti-war demonstrations, Rosebury joined the cause, giving speeches and running teach-ins.

FBI director J. Edgar Hoover considered the Vietnam protests essentially a Communist plot. He saw the student activists as dupes and punks misled by older subversives like Rosebury. The McCarthy era might have been over, but Hoover deployed FBI agents to sabotage the fresh Communist assault.

On February 13, 1965, Rosebury drove to Lambert Field, St. Louis's airport, to meet a middle-aged Marxist named Herbert Aptheker. Aptheker was a popular speaker on the university lecture circuit, addressing thousands of students at a time about racial injustice and the war. He had flown to St. Louis to give a talk about why Marxism should be treated like any other political theory rather than as a crime. "Although I had never met him (Amy had done so briefly on one occasion), we elected to meet him at the airport and have him stay with us for the few days," Rosebury later recalled.

The FBI, which was tracing Aptheker's movements, tipped off reporters at the *St. Louis Globe-Democrat* that he was on his way to their city. The agents suggested that they go to the airport as well. "One of their men met us both at the airport, and our pictures and a typical piece of character assassination was front page news the next morning," Rosebury later wrote.

Over the next few months, Rosebury had a few "cardiac rumbles." He blamed them on the fallout from the Aptheker affair. He grew paranoid. An old Camp Detrick colleague whom he had hired as an assistant professor seemed to be acting oddly. "I got him into my inner office and he crumpled under a grilling and admitted that he had been asked by the Navy to keep an eye on me," Rosebury later claimed. "He was, in short, a spy."

Rosebury was certainly right to suspect that he was being watched. The FBI was keeping a file on him that would ultimately grow to some twelve hundred pages. A few details in the file were included in an October 1965 Senate report about the anti-war movement. Hoover had encouraged Connecticut senator Thomas Dodd, the chairman of the Senate Internal Security Subcommittee, to investigate Rosebury and other opponents of the Vietnam War. Dodd—himself a former FBI agent—was happy to comply.

"The control of the anti-Vietnam movement has clearly passed from the hands of the moderate elements who may have controlled it at one time, into the hands of Communists and extremist elements who are openly sympathetic to the Vietcong and openly hostile to the United States," the subcommittee announced. The subcommittee's report singled out twenty-six subversives, including artists, philosophers, and scientists. Rosebury was on the list. "Theodor Rosebury has for a number of years held posts in the Communist-controlled American Association of Scientific Workers," the subcommittee explained, and then listed a number of other suspicious ties he had formed since 1940.

The *St. Louis Globe-Democrat*, which had embarrassed Rosebury at the airport eight months before, jumped on the report. It attacked Rosebury not only in a news article, but also in a scathing editorial.

> *If Rosebury is not a Communist, he at least has been undeviating in his Communist sympathies and activities over the years. Fired by the United Nations as one "believed to be Communists or under Communist discipline," he found refuge here in St. Louis in a great university.*
>
> *He has betrayed this trust. He has brought disgrace and shame to the university which, in 1951, foolishly took him in. He has done harm which is difficult to estimate.*
>
> *We hope that Rosebury is decent enough to resign his position at Washington University and get out of our city—which does not need him and does not want him. We doubt that he is man enough to do this.*

The university responded to the scandal by stripping Rosebury of most of his responsibilities and privileges. "My position at the school had become insupportable, tenure or not," he later said.

With backing from sympathetic professors, he negotiated a graceful exit. The university announced in April 1966 that Rosebury would take a second year's leave of absence on account of his ill health. He was sixty-one.

The staff of the *St. Louis Post-Dispatch* looked more kindly on Rosebury than their *Globe-Democrat* rivals, and they published an editorial praising him as a scientist and condemning "the intermittent attack by local vigilantes." They wished him a speedy recovery and looked forward to his return. But the leave turned quietly to retirement, and Rosebury never went back.

○ ○ ○ ○

Theodor and Amy left St. Louis for Chicago, followed by another move to rural Massachusetts. Theodor let his gray hair grow long and added a goatee. He could no longer carry out his laboratory work, but playing the clarinet was not enough. He refashioned himself into a literary writer, turning his scientific knowledge of the microbiome into a stylish new book.

In *Life on Man*, published in 1969, he argued that people should give up their fetish with cleanliness and accept that they were home to countless microbes. He wove Rabelais, Shakespeare, and Lenny Bruce into his science. In the *New York Times*, John Leonard praised *Life on Man* as "sane, elegant, and informative, a joy to read, making elegant ripples that go on widening long after the words have dropped in the mind." Rosebury became famous in his new role as a public intellectual. He chatted on the radio with Studs Terkel about how the microbiome was essential for good health. On television, he cracked jokes with Johnny Carson about syphilis.

Rosebury also kept working against biological weapons, warning of their potential use in Vietnam. "Since the United States started using so-called 'non-lethal' chemical warfare and crop-destroying agents in the course of escalating the war in Vietnam, people have been wondering how far this sort of thing might go," Rosebury wrote in a newspaper editorial in 1967. "The United States is going forward with research on airborne BW, and the question that needs an answer at this point is, what do the hawks have in mind for it?" Rosebury wouldn't have put it past the military to unleash pneumonic plague or some other disease on Hanoi. Agent Orange might have served as "the preparation of world opinion for the use of greater ones."

By the late 1960s, political pressure began to crack the shield that had protected the biological warfare program for a generation. The journalist Seymour Hersh published an explosive series of stories on the secret project, including experiments that had gone awry. Hersh got his scoops from confidential sources who had once been in the program and grown disillusioned. Rosebury was one of them. Congress responded to

Hersh's stories by summoning army officials to divulge the staggering scope of their program.

Rosebury was now in his mid-sixties, but when he spoke, young antiwar activists listened. He told them he was "someone over thirty who understands." Rosebury also now enjoyed the help of a new generation of scientists who were opposed to biological warfare. In 1969, a team led by the Harvard biologist Matthew Meselson sent Richard Nixon's incoming administration a report in which they argued that biological weapons did not aid in the defense of the United States at all. They believed that germ warfare actually made the country less safe by providing a deadly model for other nations to mimic.

On November 25, 1969, Nixon responded to the report by canceling the entire biological warfare program. "Biological weapons have massive, unpredictable and potentially uncontrollable consequences," he said in his announcement. "Mankind already carries in its own hands too many of the seeds of its own destruction."

On Nixon's order, the United States gave up its vast stockpile of germs. At the Pine Bluff Arsenal in Arkansas, the Department of Defense destroyed 220 pounds of dried anthrax spores, 804 pounds of dried tularemia bacteria, and 334 pounds of dried Venezuelan equine encephalitis virus. Meanwhile, the Rocky Mountain Arsenal in Colorado destroyed 158,684 pounds of wheat rust spores and 1,865 pounds of rice blast spores.

Nixon's renunciation led to negotiations with the Soviet Union and a treaty to prohibit biological weapons altogether. After Nixon resigned, it fell to Gerald Ford to sign the Biological Weapons Convention on January 22, 1975. Never before had nations agreed to eliminate a class of weapons. "Our entire stockpile of biological and toxin agents and weapons has been destroyed, and our biological warfare facilities have been converted to peaceful uses," Ford declared. "It is my earnest hope that all nations will find it in their interest to join in this prohibition against biological weapons."

Whatever relief those historical events gave Rosebury was soured by cynicism. He wondered why Nixon hadn't taken the opportunity to ban chemical weapons such as Agent Orange or push for a reduction in nuclear bombs. Rosebury suspected that Nixon simply saw biological weapons as a political embarrassment that he didn't need to support in order to fight wars.

"He knew the stuff was no good; and besides, we had 'better' weapons," Rosebury said near the end of his life.

Rosebury died at age seventy-two, almost two years after the signing of the Biological Weapons Convention. He may have reached the end of his life believing that he had seen the entire story of biological warfare, from its genesis to its end. If he did, he died under an illusion. The Soviet Union—which Rosebury had assured Americans was not the great menace that cold warriors claimed—had turned itself into the greatest biological warfare threat in the entire world. And it largely ignored the treaty it had signed shortly before Rosebury's death.

○ ○ ○ ○

Russia was among the countries into which Germany sent spies bearing biological weapons during World War I. Its agents wedged glass capsules full of anthrax and glanders into sugar cubes, which they dropped into feed troughs for Russian horses and cattle. When the animals bit down on the sugar, the cracked glass cut open their gums and allowed the microbes to slip into their blood.

After the war, the leaders of the new Soviet Union saw many lessons to be learned from the attack. It needed to defend not only its livestock from germ warfare, but its citizens. Soviet scientists developed new vaccines against anthrax and other potential biological weapons. They also tested those same pathogens as weapons, to provide the Soviet Union with a secret new arsenal to use against the enemies of Communism.

In the 1930s, the Soviets ramped up these efforts after receiving intelligence—later proven false—that the Nazis were planning biological warfare against them. They built a network of biological warfare labs across the Soviet Union and set aside entire islands for open-air tests of weapons that sowed diseases such as plague and leprosy. In 1938, Kliment Y. Voroshilov, a Soviet marshal, declared that his country was obeying the Geneva Protocol's ban on biological weapons. "But if our enemies use such methods against us," he warned, "I tell you that we are prepared—fully prepared—to use them also and to use them against aggressors on their own soil."

The Soviet Union has been accused of using biological weapons in World War II against Hitler's forces. But historians have not been able to confirm the allegations. The history of Russian germ warfare grows even murkier in the years immediately after World War II. Once the Soviets defeated Japanese forces in China, they seized many of Japan's germ warfare units. They put twelve Japanese servicemen on trial for their work on biological weapons, but the men were punished lightly, serving out their sentences in a manor house near Moscow. Historians suspect that the Japanese were providing information to Soviet scientists who, like the Americans, were continuing to develop biological weapons.

Rosebury himself might have inadvertently helped fuel a germ warfare arms race. When he began publishing details about biological weapons, Soviet leaders read them. His reports helped persuade them that pathogens could indeed perform as powerful weapons. And Rosebury left them believing that the Americans were building a vast arsenal. The Soviets felt they had no choice but to keep up.

And keep up they did. The Soviet germ warfare program employed tens of thousands of workers who built an arsenal to rival the American one. Along with pathogens to kill people, they created weapons for livestock, using pathogens such as the rinderpest virus. They made an assortment of fungal bombs to kill off crops. In the 1960s, the Soviet Union broke new ground by creating smallpox bombs. "Annual quotas of smallpox were required as it decayed over time," the Soviet biological warfare expert Ken Alibek wrote in his book *Biohazard*. "We never wanted to be caught short."

When the United States renounced germ warfare, the Soviet military assumed the move was pure deception. "We didn't believe a word of Nixon's announcement," Alibek later said. The United States would continue making biological weapons, the Soviets believed, and so their researchers worked even harder on their biological weapons after the treaty was signed. They stepped up their production of anthrax and created new weapons, such as tularemia that could resist antibiotics. They designed long-range missiles that could deliver death to American cities. "A single SS-18 intercontinental ballistic missile equipped with multiple warheads filled with a strategic biological agent would be sufficient to cover a city the size of New York, killing at least 50 percent of the population," said Alibek.

Alibek, who would become one of the West's key sources of information on Soviet biological weapons, started out in the 1970s as a young microbiologist working at a Siberian biological warfare lab where he grew cattle-killing *Brucella*. His superiors seemed indifferent to his work, and he worried that his career was stalling. One afternoon, as he entertained a visiting colonel with vodka on a riverbank, he complained about idling in limbo.

The colonel told him not to be an ass. "You know about Sverdlovsk, don't you?" he asked.

All Alibek knew was that the city of Sverdlovsk was home to one of the Soviet Union's top anthrax-producing plants. The colonel informed him it had just experienced an accident. "I only brought it up to show you how lucky you are not to be doing the kind of work you want to do, the 'important work,'" the colonel said.

○ ○ ○ ○

It took months for word of what had happened in Sverdlovsk to reach the West. In Germany, a Russian-language newspaper published a brief report that an explosion had occurred at the plant. In early 1980, another German tabloid offered a lurid story of its own that claimed that a thousand people had died. In March, the US State Department declared that it had information suggesting "inadvertent exposure of large numbers of people to some sort of lethal biological agent." The United States accused the Soviet Union of betraying the Biological Warfare Convention by building up an arsenal that included anthrax and perhaps other pathogens.

Moscow brushed off the accusations as "anti-Soviet hysteria," denying that it was making any biological weapons. A few people in Sverdlovsk had indeed gotten infected with anthrax, but it was due to a natural outbreak, and they all had been successfully treated for mild infections. As years passed, the picture of what actually happened in Siberia did not get any clearer. In 1986, the CIA asked Matthew Meselson to look over their intelligence about the incident. He told them he doubted the accounts from both the United States and the Soviet Union.

Only after the fall of the Soviet Union in 1991 did the truth begin to emerge. Russia's new president, Boris Yeltsin, revealed that the country had indeed run a biological weapons program for decades. He invited a joint team of Russian and American scientists to travel to Sverdlovsk in 1992 to figure out what had happened thirteen years before. Meselson led the team. They reviewed a list from the KGB of sixty-eight people who had died, supposedly due to anthrax picked up from contaminated animals. Meselson's team inspected pathology reports, looked at autopsy slides, and mapped the location of each victim.

The evidence definitively ruled out contaminated meat as the trigger of the outbreak. Instead, Meselson and his colleagues concluded the incident was a biological weapons disaster. Subsequent research has filled in more of the gaps in the story, although some mysteries endure.

In 1979, a team of forty workers grew anthrax in a four-story building. They reared a special strain prepared by Soviet scientists to be especially deadly when inhaled. On the building's second and third stories, they brewed the bacteria in fermentation tanks as large as fifty cubic meters. The liquid was sent down to the first floor, where the bacteria were pulled out and delivered to the basement. There they were mixed with chemicals and sprayed dry. The workers milled the dried anthrax mixture to a dustlike consistency, the particles measuring five microns or less in diameter. The basement air was then filtered and released from the fourth floor, from which it drifted out across the city of 1.2 million people.

It's not clear to this day exactly how the anthrax dust got out of the building. Two arms control experts, Milton Leitenberg and Raymond Zilinskas, talked to Russian sources and developed their own hypothesis. On April 2 the crew on the day shift removed some of the facility's filters to inspect them. They warned the operation center to stop spray-drying until the filters were back in place. But the night crew didn't get the message and went on spraying the anthrax for hours, letting the dust fly out the chimney.

That story doesn't square with Meselson's research. After his team looked at weather reports, diaries, and other evidence from the disaster, they concluded the anthrax must have been released from the factory the following day.

What happened next is clearer. The wind on April 3 was blowing southeast over the plant's chimney. Meselson's team pinned down the locations of sixty-six victims of the disaster and found that sixty had been in a narrow slice stretching four kilometers precisely downwind from the factory. They did not die by swallowing bad meat or touching contaminated animals. They died in a cloud wrought by aerobiology.

o o o o

From its origins in World War I, modern biological warfare may have caused tens of thousands of deaths in total. Most of its victims died during Japan's criminal rampage in Manchuria, while a tiny fraction perished in accidents like the one at Sverdlovsk. When you consider the size of the Soviet and American arsenals—big enough in theory to kill every human and stalk of wheat on Earth—biological weapons have, so far, caused remarkably little harm.

And yet germ warfare has left profound marks on the modern world—on politics, medicine, and science. Thanks to the efforts of people like Alexander Langmuir, clouds of pestilence became familiar in our collective imagination. In later years, the fear they provoked would help push the United States into a pointless war in Iraq and would lead the government to spend huge sums defending against terrorist attacks that never came.

In the United States, germ warfare also altered public health. Nineteenth-century sanitarians launched it as a social crusade to fight poverty and other systemic causes of disease. But in postwar America, diseases were transformed into external threats to be defended against like incoming nuclear missiles. Instead of reform, public health became an endless surveillance for attacks.

The race for biological weapons also had a profound influence on scientific research. It produced some valuable advances, including vaccines for diseases such as Q fever. After Nixon banned the creation of biological weapons, the military continued to support important research, transforming the scientists working at Eight Ball into the United

States Army Medical Research Institute of Infectious Diseases. In that new incarnation, the researchers shifted their focus to what came to be known as biodefense, investigating diseases that could wreak havoc either as biological weapons or as natural epidemics.

But potential biological weapons always remained their top priority. In a world that didn't suffer from self-inflicted terror of biological warfare, the billions of dollars that were spent on obscure diseases like Q fever might have been used to fight diseases that claimed millions of lives every year, such as tuberculosis, malaria, and HIV.

Germ warfare also took a heavy toll on aerobiology, the science that made it possible in the first place. Many branches of American science became part of the war effort in World War II—physics for making bombs and radar, meteorology for planning the invasion of Normandy—but the young field of aerobiology was entirely engulfed by the military. The American government built the Aerobiology Building not for the objective study of life in the air as Fred Meier had hoped, but to carry out classified research on how to release pathogens into the air to destroy an enemy. After the end of the war, the military complexes of the United States and the Soviet Union continued to employ tens of thousands of scientists to study aerobiology, but only in secret. Much of their work remains classified today.

What might a scientist like Theodor Rosebury have discovered about aerobiology if he had not taken a long detour to Camp Detrick, if he had not been hounded for decades after the war? He might have helped unite aerobiology's split personality: the indoor and outdoor worlds of living air. Or he might have recognized connections between the microbiome inside our bodies and the air that surrounds us. Ever since life began, the atmosphere has not been sterile. For billions of years, it has welcomed a vast diversity of life that extends far beyond the pathogens that make us sick or attack our crops, a sphere of life that scientists have only recently come to appreciate and to give a name: the aerobiome.

But perhaps no scientist was more vexed by germ warfare than William Firth Wells. He struggled in World War II to persuade the army to take the threat of airborne infection seriously, only to get pushed by Langmuir and others to the margins of public health. After the war, as his career was being chopped short, Wells discovered that the army had secretly built a vast facility where swarms of scientists ran experiments profoundly influenced by his work. The army might not have considered airborne infections a serious threat in their barracks, but it did consider them a serious threat when dumped out of a plane.

It's not clear why Wells was not summoned along with Rosebury to Camp Detrick. But when he finally learned about what had happened there during the war, he was relieved he never got the call. "Luckily, I was not drafted for biological warfare," he later said. "I know only what I read in the papers about the ghastly business." While aerobiology boomed at places like Camp Detrick after the war, Wells would struggle for the rest of his life to carry out one last experiment that he hoped would vindicate him. It would do exactly that, and yet Wells would die soon afterward in obscurity. One force that drove him to his tragic ending was, as he once called it, "the suicide of bacteriology by biological warfare."

And yet, when Wells first learned about America's biological warfare program, he took some grim satisfaction from it. Biological weapons, in their own ghastly way, confirmed his theory. And Wells believed that ultraviolet light could offer defense against an enemy's biological attacks. "I only wish that we had as simple and effective a means of civilian defense against atomic bombs," he said.

PART 3

Afterlife

ten

LOCH RAVEN

By 1948, William Firth Wells's life was in a shambles. Four years had passed since his wife had left him, at least as a scientific partner, and at age sixty, he watched the Laboratories for the Study of Air-borne Infection get dismantled. In a letter to the engineer Cyril Tasker, he described how the University of Pennsylvania let him haunt his old office to take care of unfinished business before he left for good. "This is a rather quiet corner in the Tower of the Medical School and I have little contact with what is going on in the field outside," he wrote.

In his quiet corner, William took up the book that he and Mildred had started nine years earlier. Mildred had long since abandoned the work, along with the experiments she had run with William. Without his Infection Machine, without fresh data to analyze from the ultraviolet lights in schools, William turned to the manuscript, hoping it would become his salvation.

"I am now engaged in a rather serious review of the whole subject ultimately I hope to appear in book form," he told Tasker. William was binding together all the work he and Mildred had carried out over the previous eighteen years while also pulling in other research that would bolster his theory. When Theodor Rosebury publicly revealed details of the germ warfare work at Camp Detrick, William copied out the doses of different biological weapons required to kill animals. He hoped the sheer weight of information would show doctors and public health experts that airborne infection had the same inescapable logic as Newton's Laws of Motion.

In September 1948, William decided he was done and sent the manuscript to Roderick Heffron at the Commonwealth Fund. Although Heffron had shut down William's lab the year before, he was still willing to consider helping William publish a book. "I submit for your perusal our manuscript on an Essay on Air-Borne Infection," William announced. "I believe the material now is ready for judgment."

Heffron struggled to read it. It might well have contained hugely important ideas, but William—without Mildred's guidance—could not make a readable story out of them. "The sheer multitude and weight of the words in some of his statements gets in the way of the sense of them, and this often makes some of his written material nearly unintelligible to the uninformed reader," Heffron told a colleague.

Perhaps Heffron expected William to look back at his career as he moved gently into retirement. "I am not too sure how many more active years he has ahead of him," Heffron said. "Neither am I sure he is going to find himself a setting where he can continue his air-borne studies." But even in his quiet tower corner, William was still in the fight. He was searching for a way to run the biggest experiment of his life.

° ° ° °

The experiment Wells had in mind would demonstrate that people infected with tuberculosis could spread the disease by droplet nuclei through the air. He believed that many diseases were airborne, but he chose tuberculosis as the one that would give him the best chance of success. In the late 1940s, tuberculosis experts still considered its transmission a matter of conjecture. But for his own part, Wells judged the matter settled, thanks to his experiments with the Infection Machine. Only the finest droplet nuclei could make rabbits sick, by delivering bacteria to the alveoli in their lungs.

TB was also a good choice for an experiment because so many people still suffered from it. In 1947 alone, it claimed more than forty-eight thousand American lives. No effective drug could yet cure tuberculosis, and so patients had to fight against the bacteria on their own. Many of them lost the fight after years of agonizing decline. During World War II, the army worried about explosions of influenza and other infectious diseases. Now, in peacetime, those risks faded back, but tuberculosis continued to kill soldiers. The army could still do little more than give sick veterans a bed where they might recover. The nurses and doctors attending them sometimes got infected as well and faced the same bleak prospects.

Wells believed a Veterans Administration hospital would be the perfect place to run his experiment. The sick soldiers were living Infection Machines, spraying out bacteria-packed droplets day and night. The bigger droplets plummeted to the floor, but others—the most dangerous ones—floated in the air. Wells wanted to shunt the air they breathed out of their wards and into cages of guinea pigs. He expected that the droplet nuclei would, over time, make the animals sick. Any given breath from a tuberculosis patient would not contain many bacteria, so it might take weeks for a new infection to develop in the animals. Once the experiment started, Wells would have to wait for months for clear results to emerge.

VA doctors were among the few medical experts who took much interest in Wells's work. His experiments with ultraviolet light had prompted some of them to put the lamps in their tuberculosis wards, even if there was no evidence beyond lab rabbits that they worked. And Joseph Stokes, Wells's old ally, had become the chairman of the Veterans Administration's Committee on Air Hygiene.

"We must feel our way along both on the installation and its follow up," he told Stokes. "Our greatest hazard is I believe to make a false step rather than to step too slowly. As it is, we are already far ahead of anything hitherto attempted and we have only to hold this leadership by being sure of our ground."

In July 1948, Stokes delivered a proposal from Wells to John Barnwell, the chief of the Tuberculosis Division at the Veterans Administration. "He is a true pioneer in this field," Stokes told Barnwell. It's clear from Stokes's letter that the two doctors were well aware of how cantankerous and wearying Wells could be. Stokes politely acknowledged Wells's "personality difficulties." But he was certain that Wells could continue to do important work "if given a proper chance."

° ° ° °

While William plotted in his quiet tower corner, Mildred was running a new experiment of her own. The idea had come to her in late 1944, after she and William had their falling-out and she resigned from the Laboratories for the Study of Air-borne Infection. The experiments she and William ran in Philadelphia schools could have produced better results, she believed, if ultraviolet light had protected the children beyond their classrooms. To show that was true, she needed to find a town where she could install lamps not only in schools, but in other public spaces.

Mildred shared her idea with Edwin Wilson to gauge his opinion. "When I get the outline whipped into shape, I shall start on the wholly distasteful task of trying to sell it to someone," she told him.

"There is one opportunity which has interested me a little bit," Wilson replied. He knew that the health commissioner of Westchester County, just north of New York City, was looking for a new statistician.

William Holla seemed to Wilson like a good patron for Mildred Wells. He had turned Westchester's health department into one of the best in the United States. He enjoyed the support of the county's wealthiest residents. Nelson Rockefeller, who would go on to become governor and then vice president, sat on Holla's board of health. Holla built a staff of two hundred workers and led them into battles with landlords who shut off heat in apartments, butchers who sold black-market meat, and doctors who failed to vaccinate enough children. He wiped out rabies in Westchester and drove maternal deaths almost to zero.

Holla waged his public health campaigns with a swagger. He made grand statements to the press. When he wrote up the health department's annual reports, he would sometimes break into verse. One year he resorted to poetry to explain why the county needed to keep testing for tuberculosis, even as rates were falling.

"One can well cite the words of Sir Walter Scott," Holla wrote.

If the pilot slumber at the helm
The very wind that wafts us towards the port
May dash us on the shelves.

Edwin Wilson told Wells that working for Holla would give her a chance to pore over Westchester's impeccable statistics in her free time. Wells traveled north to meet with Holla in November 1944, but she did not ask him for a job. Instead, she proposed that the two of them run her experiment.

Holla was intrigued. He and Wells agreed to work on drumming up the money for the project once she produced a detailed proposal. She began to think about what she could do in Westchester, and by early 1945 the study had morphed into an even more ambitious undertaking. In addition to installing UV lamps throughout an entire Westchester town, they would monitor a second one as a control.

Wells struggled to write the proposal, fretting that she wouldn't convey the importance of her findings in Philadelphia and how they promised even greater results in Westchester. "I feel wholly inadequate to sum up the experience here in a way to please everyone, probably I should say anyone," she confided to Wilson. She sent the proposal to the Commonwealth Fund and was unsurprised when they rejected it. Holla then turned to his own powerful network. He got some money from the county government and additional funds from the Milbank Memorial Fund, a New York philanthropy. The budget totaled a generous thirty-two thousand dollars. Wells got to work.

On October 9, 1945, Holla announced the launch of the Study on Community Prevention of Air-borne Infection. Ultraviolet lamps would be installed in Pleasantville, an affluent village of 4,357 souls, many of whom took the hour-long train to Manhattan for work. Getting carried away with the publicity, Holla declared that the experiment would test UV lamps "in the control of air-borne diseases, such as influenza and colds." In fact, Wells designed the experiment to examine only measles and chicken pox.

In early 1946, Mildred began spending weeks at a time in Westchester, away from William, working with Holla on installing ultraviolet lights around Pleasantville. No one had ever undertaken such an experiment before, and in May, Holla complained to the Westchester Board of Health about "difficulties encountered in the application of ultra violet light in the Pleasantville area." When Holla and Mildred tried to install lights in the town movie theater, for instance, they couldn't figure out how to extinguish the glare when films played. Holla and Mildred pushed on, gradually bringing lights to schools, churches, a local factory, a restaurant, the town library, the train station, and what Holla and Wells later described as "the most patronized of the soda fountains."

As the lamps started glowing overhead, six county health workers embarked on a door-knocking campaign. They visited the 987 children who were under the protection of ultraviolet light in Pleasantville. They then traveled to Mount Kisco, another village seven miles away that did not enjoy the protection. There, they kept track of another 1,161 children. The workers filled out a medical form for each child, noting if they had previously had measles or chicken pox, which would make them completely immune to new infections. Every four weeks, the workers checked up on every child. When they encountered children sick with measles or chicken pox, they traced their contacts to see how the viruses spread.

As the health workers traveled through the villages, Holla continued to promote the study. He wrote a full-page essay about it—without once mentioning Wells—in the *American Weekly*, a Sunday magazine included in many US newspapers. "In the next three years we hope to make the little town of Pleasantville, N.Y., the most unpleasant place on earth for diseases which prey on children," Holla declared. "Our only weapon is the 'community ray'—a death ray for germs."

Other public health experts scolded Holla for his chest puffing, but he did not stop boasting. Just a couple months later, he was crowing about a new victory. The spring of 1946 saw a bloom of measles in Westchester County. The susceptible children in Pleasantville fared better than the unprotected ones in Mount Kisco. Holla jumped on the preliminary results, telling reporters that the experiment was already working as he had hoped.

"These germs," he explained, "are contained in tiny droplets which are coughed and sneezed into the air, and which, after evaporation, drift about on air currents much as do particles of tobacco smoke." Holla assured the reporters that the ultraviolet lights were sanitizing the air and slowing the spread of measles and chicken pox. The fact that the Westchester experiment was only five months into a three-year plan did not hold him back.

Mildred remained committed to the crusade for clean air, but she would not jump ahead of the evidence. Going through the epidemiological data, she recognized that some children in Pleasantville were getting sick despite the lights. She suspected they were picking up infections in parts of town where she had yet to reach. Wells knew she was now in a treacherous position. She hoped that Westchester would deliver results so clear that skeptics about airborne infection would see the error of their ways. But as ambitious as her study might be, it might not be big enough to work. And if it failed, the theory of airborne infection might suffer a mortal blow.

When a correspondent named Peter Huntington wrote from Chicago about putting up ultraviolet lights in a school there, Wells replied frankly about her frustration. "We are dubious that our coverage is adequate," she wrote. "We have some 600 commuters going into New York daily, and bringing back respiratory infections from the city or the train itself. The big question in our minds is whether or not protection in the homes will be necessary."

Rather than boost Huntington's interest, Wells scolded him for thinking he could cut down disease simply by installing a few schoolroom lamps. "While I do not like to be discouraging, I think the only way to sum up your plan is to say bluntly that you are setting up a beautifully controlled experiment, fore-doomed to failure." If her admirers carried out a string of failed experiments, they might destroy the theory of airborne infection for good.

○ ○ ○ ○

William Firth Wells waited for the Veterans Administration to consider his proposal. He filled the time with his book, traveling regularly by train to New York to harangue the staff of the Commonwealth Fund for hours at a time about all the ideas he wanted to cram into it. From time to time, they would ask him to turn his manuscript into something comprehensible. "I think you have a big story to tell and I hope you will tell it just as plainly and directly as possible," Geddes Smith told Wells.

The prospect of telling a plain story panicked Wells. He drafted new outlines. He rewrote great swaths. He proposed that the Commonwealth

Fund send him to Europe for three months to tour great labs and talk to other scientists who studied airborne infection so that he could add even more information to the manuscript. Wells's anxiety might have been fueled by his looming termination at the University of Pennsylvania in June 1949—and with it, the termination of his salary. Mildred remained busy in Westchester on the Pleasantville study, unable or unwilling to donate her time to what might have looked like a doomed effort. "Until now she has been so tied up with her own problems that I get little sympathy—and can ask for little," Wells told Stokes. He was finally rescued from his crisis by his old assistant Richard Riley.

After Riley finished medical school in 1937, he left Boston for New York. But he remained in touch with Wells as he started work on infectious diseases at Saint Luke's Hospital and then moved downtown to Bellevue to work with its chest service, where he studied the workings of the heart and lungs in patients who came to the emergency room. During World War II, Riley joined the navy and was assigned to the Aviation Medical Research Laboratory in Pensacola, Florida. He studied what happened when navy pilots were shot out of the air and had to breathe the scarce oxygen in the upper atmosphere. In the course of his research, he invented a way to measure the oxygen and carbon dioxide in blood, which allowed him to learn about how the gases passed between the blood and the lungs. After the war, Riley returned to Bellevue to continue his research, and in 1948, he was offered a position at New York University's Institute of Industrial Medicine.

Shortly after getting the offer, Riley was diagnosed with tuberculosis. Normally he would have been sent to a VA sanatorium, but he had a tiny lesion in his lung and a doctor for his wife. Polly Riley got permission to care for Richard in their Manhattan apartment. And as he lay in bed, Riley received a request to help Wells with his book. He agreed without hesitation. "He was an aristocrat with impeccable manners," Riley's student Solbert Permutt later said.

As a medical student, Riley had ghostwritten some of Wells's papers, and now he helped the old wizard again. Lying in bed, Riley worked his way through Wells's jumble of pages. He would send Wells comments, Wells would refuse to accept any of them, the two men would argue for a while, and then Wells would relent. When he wasn't fixing the book,

Riley was trying to get both William and Mildred jobs. He inquired about spots for them at New York University, where he was about to start himself. But as a new junior professor, Riley didn't have enough pull. And so, as of June 1949, William was officially out of work.

○ ○ ○ ○

June 1949 also marked the end of Mildred's experiment in Westchester. She spent the next few months analyzing the records. Patterns emerged from the data, some that encouraged her and others that disappointed. In May 1948 Pleasantville experienced an explosion of measles despite its ultraviolet lights. Mildred blamed heavy rain, which might have made the air too humid for the ultraviolet rays to do their work. On the other hand, a wave of chicken pox that hit Westchester later that year largely spared Pleasantville.

Holla was impressed by the final results. He talked of extending the ultraviolet shield all the way across Westchester—from Ossining on the west side of the county to Armonk on the east—and creating a barrier of immunity that could stop waves of airborne outbreaks. Holla's excitement might have given Mildred hope for a comfortable future for her family. She registered the Wells family on the 1950 census as residing in Somers, a small Westchester town. With William fired from Penn and having no luck with his tuberculosis experiment, the Wellses were running out of money. "Do you suppose the Fund might advance a few hundred dollars to help get the book in final shape?" William asked Heffron.

Mildred and Holla wrote up the results of the Pleasantville experiment. "Ventilation in the Flow of Measles and Chickenpox Through a Community—Progress Report, Jan. 1, 1946 to June 15, 1949, Airborne Infection Study, Westchester County Department of Health" appeared in the *Journal of the American Medical Association* on April 29, 1950. Holla guaranteed that it got lavish attention. "Ultra-violet radiation disinfects dry indoor air and prevents 'explosive outbreaks' of measles and chicken pox," the *New York Herald Tribune* reported. Holla also made sure to remind reporters that the State of New York paid counties for their schools based on the attendance rate of their children. The ultraviolet lights reduced so many sick days that the thirty-two-thousand-dollar experiment paid for itself.

Holla also informed reporters that he had asked the navy for money to put lights across Westchester to block not just chicken pox and measles, but the common cold and influenza as well. Holla was reaching far beyond what he and Mildred had actually discovered. A statistician with the Milbank Memorial Fund named Jean Downes confirmed this when she looked over all the records of the study from 1946 to 1949. While the lights might have worked against chicken pox and measles, she saw no difference whatsoever in the rates of colds and influenza between Pleasantville and Mount Kisco. "It is apparent that the introduction of ultraviolet lights in the Pleasantville schools and other places where children congregate did not affect the illness rates," she concluded.

Downes's sober analysis may have scared Holla away from airborne infection. He did not give Mildred any more work. He never followed up on his big talk about a curtain of black light across the county. After years of promises and boasts, he had nothing more to say about the entire airborne infection study until his retirement in 1956. Writing about his decades of achievements, Holla acknowledged a few disappointments. One was the ultraviolet experiment. "It proved inconclusive" was all Holla had to say.

In its ambition and rigor, the Westchester Airborne Infection Study was unlike anything ever attempted before in the field of epidemiology. But, just as Mildred had feared, it failed to convince public health authorities to take airborne infection seriously. Richard Riley would puzzle for years about why it did not make more of an impression. It's possible that many experts only skimmed the results. The small reduction of infections overall may have made the effort look like a failure. If readers had looked more closely, they would have noticed that the UV lamps did manage to block the spread of diseases in the particular places where they glowed. Unfortunately, the residents of Pleasantville could still get infected outside of the village, and pass diseases to one another where the air remained unprotected.

But they did not come away with that message from the study. Instead, Riley later said, "it contributed to the blackballing of UV air disinfection for years to come."

o o o o

In July 1950, Mildred bought a decrepit colonial farmhouse called Pleasanton Abbey a few miles outside the city of Dover, Delaware. She moved in with Bud soon afterward. William appears to have lived part-time there as well, but also continued to work in his quiet corner in Philadelphia.

With no progress on his tuberculosis experiment, Wells had nothing to do but continue working on the book. The more he worked on it, the more chaotic it became. Wells told Roger Crane, the director of publications at the Commonwealth Fund, that he had recast the whole manuscript in what he called "the Euclidean approach," presenting the case for airborne infection like a mathematical proof. Crane found the Euclidean version even harder to understand. "We can't make a really good book out of it," he confessed to his colleagues.

The Commonwealth Fund advanced William the hundred dollars he requested, and over the next few months he asked for more. He was, he explained, "without salary from any source." In November 1950 William upped his request to fifteen hundred dollars, to pay for three months of collaboration with Mildred. Now that her Westchester experiment was behind her, she agreed to help William "improve the form, simplify the diction and clarify the meaning," he wrote. She also agreed that he could list her as a collaborator, but only if everything in the book agreed with her views as a physician.

Crane told Wells that his division had no money for that kind of help. Mildred turned to the New York Academy of Medicine for a fellowship. The fund was overseen by Wilson Smillie, who had hired William at Harvard in 1930. Now, two decades later, Smillie turned Mildred down.

"I hope that I can find some way finally to clear up this manuscript and get back to earning a living," William complained in January 1951. "We have gained all by our objectives and dug in. But there is some mopping up before it is a book. I thought Mildred could help me on this as she has been doing but she also feels that she should be doing something remunerative."

Mildred also complained about their dire straits in March. "Will seems to have about completed his book," she told Heffron. "I was sorry that I could not get a grant to work on it—or that I could not work on it without a grant, but someone in this family has got to be realistic about paying the bills." It's not clear how the bills got paid, but it could not have been easy. In May, William reported to Heffron that his work was going slowly "as I had to get in a crop on the farm to carry us through the summer and possibly the winter until the book is finished."

Over the course of 1951, Wells kept sending new pages to the Commonwealth Fund. "The sad truth is that the book is ten years beyond the field," he declared, saying that he had reached "a goal beyond anything we could have hoped for." Each batch of changes would be the last, he promised. "The wedding guest may now depart—the ancient mariner has finished his story," he wrote in June, only to send another letter the next month with a new opening paragraph to the second chapter. "I hope this will not be annoying and promise not to mull over the manuscript any more."

Wells did not keep the promise. When he sent in another batch, he declared that "at last it has really crystalized and all that remains is to drain off the mother liquor and wash the crystals."

In July 1951, Harvard University Press sent out Wells's manuscript to a few scientists for their opinion. The replies were swift and brutal. John Snyder, a doctor at the Harvard School of Public Health, thought that the new threat of biological warfare made a book about airborne infection a timely one. But he found Wells's book to be partisan, contentious, and repetitious. Wells made unjustified jumps from his laboratory results on rabbits and mice to sweeping claims about human diseases.

Ronald Ferry, a professor at Harvard Medical School, was no less disappointed. "I have read it with interest and have tried to maintain sympathetic attention," he said. "It, nevertheless, leaves a tragic impression."

○ ○ ○ ○

Wells knew that skeptics like Snyder and Ferry wouldn't be satisfied until he had clear-cut results from human subjects. He dreamed of getting those results from his tuberculosis experiment, but one VA hospital after another tantalized him and Joseph Stokes with interest, only for red tape and personnel shuffles to send them back to square one. In February 1952, Richard Riley came to Wells's rescue once more.

After Riley recovered from tuberculosis, his wartime colleague Joseph Lilienthal poached him from New York University. Riley joined Lilienthal at Johns Hopkins University in Baltimore, where he continued his research on how blood in the lungs absorbs oxygen while releasing carbon dioxide. A few miles away, the Veterans Administration was just finishing up construction of a new hospital called Loch Raven. When Wells told Riley about his long-incubating experiment, Riley promptly persuaded Loch Raven officials to host it. He recognized that it might at last offer direct evidence that droplet nuclei released by people could spread pathogens.

Most scientists might have dreaded yoking themselves to Wells for such a risky undertaking. But Riley was delighted to work again with him. "Wells, in his cranky way, gave us an ingredient that scientists seldom mention: a mission to convince unbelievers," Riley later recalled.

In March, William had to step away from that mission when Mildred was rushed to a hospital for what he described as "a bilious attack." She then came down with the flu, which left her sick for weeks at home. Mildred, now sixty, began a steady decline, developing chronic heart and kidney trouble. And as William nursed Mildred, more bad news arrived.

Stokes had been tying up the last loose ends for the Loch Raven experiment. He had buttoned down the permissions from Johns Hopkins and the University of Pennsylvania and the Loch Raven staff. All that remained was the formality of an agreement from the Veterans Administration, which would pay for three hundred guinea pigs, air-conditioning equipment, and salaries. Stokes had no reason to expect anything but a rubber stamp.

But Congress was just launching a campaign to cut back on military spending. With severe budget cuts looming for the VA, Barnwell turned Stokes down. In desperation, Stokes called Roderick Heffron at the Commonwealth Fund a few days later. He hoped that Heffron could find the money to make up for the VA's rejection. After all, the Commonwealth Fund had supported Wells at Penn for over a decade. But Stokes might not have known just how badly Wells had been fraying Heffron's nerves with his book.

Heffron responded with a terse note. "I am sorry to have to say that there is no possibility of our being able to take this on," he wrote.

After four years of struggles to launch the TB experiment, Stokes finally lost his cool. He lashed out at Heffron, refusing to believe "that a man who has been THE PIONEER in the field" could no longer do his science.

○ ○ ○ ○

Harvard University Press looked at the harsh reviews of Wells's manuscript and declined to publish it. The Commonwealth Fund still held out hope and asked their editor, Beulah Chase, to turn the manuscript into something readable. "Let's hope Mrs. Chase has strength and endurance," Crane said.

Wells regularly traveled to New York to talk to Chase and ended up spending days pouring out his anxieties about the book. "Your patience served me as a psychoanalytic treatment," he later told her. Wells fretted that he now had one last chance to share all the new ideas that kept bubbling up in his mind. "As I recede toward posterity, I catch glimpses of the whole which were out of focus when intensively engaged in fabricating the parts," he wrote to Crane.

Thinking of approaching death, Wells ripped apart the manuscript once again. "The whole of 'Book One' must come down," he announced to Crane in October. In December, he described another round of changes that would, he promised, be the last: "This about exhausts my constructive ideas on the subject."

Crane underlined *exhausts* in red pencil. "Exhausts us too, doesn't it," he jotted.

○ ○ ○ ○

In the summer of 1953, William enjoyed a break from his anxieties. He and Mildred traveled to Rome to attend an international microbiology congress. William's river of letters to the Commonwealth Fund dried up, but only because he didn't know enough Italian to buy an envelope.

Both William and Mildred gave talks at the conference. William spoke about how an epidemic grew only if each infected person passed on germs to more than one new victim. When the germs ran out of vulnerable people to infect, the epidemic sputtered to a halt. If the germs spread like smoke in the air, it was possible to shift the threshold for epidemics with ventilation. Bringing in fresh air diluted indoor droplet nuclei, making it harder for germs to find their next victim.

Mildred talked about the checkered history of ultraviolet light as protection against diseases. She declared that their Philadelphia experiment had stopped a measles outbreak, but she also acknowledged that the Westchester experiment had been a disappointment. "Irradiation of a village in a metropolitan area did not stop infiltration from neighboring communities," she concluded. But Mildred still held out hope that air disinfection could show its worth, if only scientists could set up the right experiment.

On their way back home, William got some stationery in London and wrote to Chase about the thrill of presenting their research in the country of his hero, Girolamo Fracastoro, the Renaissance scholar who had also written about airborne contagion. "We added one more stone to the Eternal City in the land where modern science was born," William declared.

But when the Wellses returned home to Delaware, trouble flooded back into their marriage. William kept asking Mildred to help him with the book, and she kept refusing. She had to get back to paying the bills for Pleasanton Abbey, fixing up the decrepit house, and caring for Bud, who was now in his thirties.

Even if Mildred wouldn't help, William had long since decided to put her name on the book. He once said that her ideas suffused it, so "it would be nice to make her a co-author on that." But in September 1953, Mildred learned of William's plan and asked him to take her name off the title page. He responded with a peculiar formality, informing her that she would have to write to the Commonwealth Fund to that effect. Mildred let the matter rest.

Over the next six months, William continued to work on the manuscript with Chase. At last, she decided it was ready to send back to Harvard. "I cannot believe any unprejudiced hearer with any feeling for science can fail to be impressed," he told her giddily. He believed his book was a sequel to *Principles of Sanitary Science and the Public Health*, written fifty years earlier by his mentor, William Sedgwick. "Sanitary principles which have proved so effective in purifying the water we drink and the food we eat have not been applied to the air we breathe," Wells said. Now, with his new book, he believed that would change forever. "Prejudice dies but a feeling for science lives with our civilization, so I am confident that if we can get it published, posterity will take care of the rest," he said.

A few weeks later, Mildred had a free moment at Pleasanton Abbey and decided to look at last at William's finished manuscript. "I was shocked to find myself down as co-author," she recalled. Rather than speak to William again on the matter, she dispatched a note directly to the Commonwealth Fund. "Please take my name off entirely," she commanded. "I'm sure it's all Will hopes it is, but it is NOT MINE."

Chase could not understand the conflict, even after William tried to explain it to her a number of times. "It's hard to tell what to do because of the tangled skeins of responsibility and the ambivalent attitudes," she later complained to Crane. "Every time I see W.F.W. I get a new 'story behind the story.'"

She tried to smooth things over with a note to Mildred. "As you know, in many cases of joint authorship some of the persons whose names are listed have done little or none of the actual writing," Chase explained. "In the present instance, we all felt that full recognition should be given to your part in the experiments described in the book and further that your name would give added weight to our presentation of the case for 'airborne contagion.'"

Flattery did not work on Mildred. "Thank you for your nice letter which does not change my attitude in the least," she replied. "I am as aware as you are at the C.F. of the diversity of standards as to what constitutes authorship. The best anyone can hope for is to not do violence to his personal standards."

As forceful as Mildred could be, she did not display any hostility toward Chase or even William in her letter. After repeating her demand to be taken off the book, she shifted to describing the hot weather in Delaware that day. "Will had better compromise some of his Mass. ideas of when to take off his woolies or one of us will be writing a forward to a posthumous book. But the trees & spring flowers are lovely now—I hope when things are straightened out here you will visit Pleasanton Abbey & see why I have not co-authored Will's book."

When Roger Crane followed up with a call, Mildred remained cordial and firm. She explained that as far as she knew, there was not a statement in the book that she would not be willing to defend. But since she had not written any of it, her name should not appear on the title page.

Crane then checked in with William. "Mr. Wells told me that there have been ups and downs about the matter of collaboration but just at the moment they are in a down period," he later recalled.

○ ○ ○ ○

It had taken two years for Beulah Chase to work through Wells's manuscript, and in that time she performed an editorial miracle. In the summer of 1954, Harvard University Press decided to accept the revised book. "I think we broke through the sound barrier and I lost only one tooth," Wells told Chase afterward.

The summer brought more good news from Baltimore. Riley and his colleagues at Johns Hopkins managed to find the money for a tuberculosis study at Loch Raven. Wells, who had gone five years without regular income, became a paid consultant to the Veterans Administration, and his experiment had a formal title: the Air Hygiene Research Study Unit. At age sixty-seven, Wells was finally embarking on the biggest study of his life. Joseph Stokes was immensely relieved at the news. Wells, he declared, "has been somewhat of a martyr long enough."

Loch Raven had a wing with a layout perfectly suited to Wells's plan. The six-room unit, located on the top floor of the hospital, had been set aside for "psychiatric restraint." Its ventilation system, which pumped air out of the rooms and vented it out of the roof overhead, was also separate from the one that handled the rest of the hospital. A large empty penthouse was located above the ward. Riley and Wells converted it to a lab for their guinea pigs.

Wells and Riley prepped the rooms—which they called the Pilot Ward—by stretching plastic curtains over the windows and plugging holes with tape and cardboard. They rerouted the ventilation ducts to pump air from the rooms into the penthouse, where Wells built an expanded version of the Infection Machine. The guinea pigs lived in a cage where they breathed in only air shunted from the ward below. The air then flowed out of the building. "This arrangement prevented unsavory odors and also had the great advantage of insuring that infection could only be transmitted by droplet nuclei," Riley later wrote.

For the most part, they created an experiment that matched what Wells had imagined seven years before. But during his long wait, something important had changed: scientists had found a cure for tuberculosis. In the 1940s, a team of Rutgers University microbiologists led by Selman Waksman discovered a compound made by soil bacteria that killed *Mycobacterium tuberculosis*. Doctors tested the drug, called streptomycin, and found that it cured many patients. For the discovery, Waksman won the Nobel Prize in 1952. Wells and Riley decided they could still get good results from their experiment even if some patients were getting treated with streptomycin. They decided that any patient who was cured of TB would be moved out of the ward and replaced with someone with an active infection to replenish the air with bacteria.

The job of making sure all the parts of the experiment ran smoothly fell to a young woman named Cretyl Mills. Mills had worked in Wells's lab in Philadelphia, and now she came to Baltimore to work in the Loch Raven penthouse. She cared for the guinea pigs, tested them each month for tuberculosis, recorded all the data, and fended off complaints from the hospital staff.

Riley couldn't blame them for being angry. After all, the Air Hygiene Research Study Unit demanded that doctors and nurses rotate patients in and out of the Pilot Ward, that lab workers run a lot of extra tuberculosis tests, and that pathologists autopsy dozens of guinea pigs and inspect their lungs.

"In a word, a damn nuisance for everybody," Riley later said. "And this went on for years."

It wasn't a nuisance for Wells. In fact, it might have been a welcome relief from the troubles at Pleasanton Abbey that he alluded to in his letters at the time. Perhaps they were arguments with Mildred about his book, or perhaps difficulties with Bud, or perhaps some combination of the two. It's hard to tell exactly from William's cryptic notes, such as a

January 1955 letter to Chase asking that she send the book index to Pleasanton Abbey rather than Loch Raven. "Until I am thrown out of Dover I think that is my best address to send the Index—certainly during the next week," he explained. "Even though page proofs precipitated a second crisis there I believe it is my best landing field—depending of course upon weather changes."

∘ ∘ ∘ ∘

In April 1955, a package from Harvard University Press arrived at Wells's office in Baltimore. He opened it and pulled out a hardback copy of *Airborne Contagion and Air Hygiene: An Ecological Study of Droplet Infections.* On the cover, against a black background, were alternating Petri dishes and glass cylinders, all peppered with gleaming colonies of bacteria captured in his air centrifuge. Sixteen years after he and Mildred had first conceived of the book, it was now published. "After I looked into it I am dazzled beyond words," he wrote to Crane. "Everything is perfect."

Wells now believed that the book would become a sensation not just with doctors, but with architects, teachers, and historians, and that it would be recognized as a classic. "The book is not for here and now. It is from now on," he declared.

But *Airborne Contagion* attracted devastatingly little notice. And what little attention it did get was disappointing. John J. Phair, an epidemiologist at the University of Cincinnati, published a short review in the *American Journal of Public Health.* Phair praised Wells's inventions as "unique and ingenious," but he warned that anyone lacking a deep knowledge of aerobiology "will be quickly lost." Despite the years Wells had spent trying to tell a clear story, Phair warned that "this reviewer cannot believe that laymen without a professional understanding will be able to find a connected story of air hygiene and its place in the control of infections."

Theodor Rosebury, who had used William's work as a starting point for making biological weapons, reached the same conclusion. Writing in the *Quarterly Review of Biology,* he called Wells "an acknowledged pioneer" in aerobiology. Rosebury dove into the book expecting to read a definitive presentation of two decades of research. "While this seems to have been the author's objective, he has not achieved it," Rosebury wrote.

"The work suffers from poor organization, and fails to reach acceptable standards of coherence or clarity."

Airborne Contagion was not welcome at Pleasanton Abbey either. "The only fly in the ointment is the reaction at home. It was violent!" Wells told Chase. "My 'medical' pretensions have always been a powerful emetic but a 'book' under the imprint of the Commonwealth Fund & Harvard! It was worse than carbon monoxide poisoning."

None of the criticism, whether at home or in journals, seemed to bother Wells. The book remained "the deepest satisfaction of my life," he said.

His good mood was lifted further by news from Baltimore. In a preliminary test, the team sprayed droplets of *Mycobacterium tuberculosis* into the sealed rooms of the Pilot Ward. The vents pumped the air to the penthouse, where they housed a group of rabbits. Some of the animals came down with tuberculosis, confirming that the Pilot Ward might have acted as a giant Infection Machine.

The team got ready for the more realistic test. The six rooms were filled with veterans suffering from tuberculosis, and Mills put 156 guinea pigs into the penthouse test chamber. Wells could now only wait for the guinea pigs to breathe the air exhaled by the tuberculosis patients. Approaching seventy, Wells wondered if he would survive to see the experiment to its end. "How long, oh Lord, can I be spared?" he asked Stokes.

In late 1956, the first results emerged. "You will be pleased to know that we have finally begun to catch 'bugs' from the human patients in the pilot ward," Riley told Stokes. Riley and Wells dissected the guinea pigs and began finding some with tuberculosis. "Mr. Wells and I are considerably relieved to open these animals and find one solitary tubercle exactly according to theory," Riley wrote. A single droplet nucleus less than five microns must have made its way deep into the lungs of each animal, creating one nodule of scar tissue. The experiment was working as Wells had predicted.

Over the ensuing months they continued opening guinea pigs, and they found more tubercles. The numbers added up slowly but relentlessly. Between December 1956 and December 1958, seventy-one guinea pigs out of 156 exposed to the air from the Pilot Ward got tuberculosis.

By early 1957, Riley had seen enough. He started traveling to tuberculosis conferences to deliver lectures about the results. Riley still felt the need to preface them with a promise that the theory of airborne infection was not a relic of a superstitious past. "These ideas do not constitute a reversion to miasmic theory, according to which bad air from marshes or even hospitals was thought capable of transmitting disease over considerable distances outdoors," Riley said. He was instead arguing for droplet nuclei carrying germs out of infected bodies and traveling through the air.

Riley would then describe the Loch Raven experiment to his audiences, as well as Wells's predictions about how it would turn out. "These findings show a remarkable correspondence between theory and practice," Riley announced. "Air-borne infection may be the most important mechanism of spread."

The results brought particular pleasure to Joseph Stokes, who had struggled to help William Firth Wells for so long. "It always does me good to think of Will as a Phoenix rising from the ashes of his defeat," he said.

○ ○ ○ ○

As the guinea pigs fell ill, Wells shared the news with Roger Crane at the Commonwealth Fund. He hoped that Crane would consider bringing out a new edition of *Airborne Contagion*. "If and when you consider a reprint," he told Crane in October 1956, "I think we have another chapter—Air Hygiene in Tuberculosis."

And then Wells went quiet. For years, he had inundated the Commonwealth Fund with letters and calls—sometimes several in a single day—and so the silence might have felt as unsettling as an endless shriek. Eight months passed before Crane heard from Wells again. The letter Wells sent on July 1, 1957, left Crane both baffled and horrified.

"I have come back alive from another scientific adventure, and brought my dead out with me. I would like to add IN MEMORIAM to the dedication page of the book," Wells wrote.

> *Our trophy is a new chapter—INFECTIVITY OF DROPLET NUCLEI CONTAGION—which may hasten exhaustion of the first edition.*

Before this can happen, however, I may have embarked on another dangerous adventure. I shall feel much better to know this is in your hands—just in case I do not come back.

Along with the letter, Wells sent Crane a new chapter on Loch Raven. He also sent a separate sheet of paper on which Wells wrote by hand a list of allies—both in science and at the Commonwealth Fund—who had died in recent years.

IN MEMORIAM

Died in action
WADE HAMPTON FROST
Sir PATRICK LAIDLAW
BARBARA QUINN
GEDDES SMITH
JOSEPH LILIENTHAL
MILDRED WASHINGTON WEEKS WELLS

Crane was shocked to see Mildred's name at the end of the list. It was the first he had heard of her death.

Four months had passed since Mildred Weeks Wells had died at age sixty-five. The official cause was heart disease, although Richard Riley later said that she died of cancer. The only public notice of her death was a short obituary in a Dover newspaper. It briefly mentioned that Mildred Weeks Wells studied contagious diseases.

After years of poor health, Mildred's death could hardly have come as a surprise. Still, William must have suffered a brutal loss when she died. No letter by William from around this time survives. There is no account of his grief. When he started writing letters again a few months later, he never mentioned his family, although he was now left to care for Bud alone at Pleasanton Abbey. He often reached out to Riley for company. "Wells kept in touch by phone, invariably at dinner time and at length and much to the annoyance of my wife," Riley wrote.

The two men had a lot to talk about. Their experiment had met with skepticism. Critics said the results were not strong enough to make airborne infection the only possible conclusion. "Maybe the guinea pigs were infected by the feed, the water, the caretaker, who knows," Riley later wrote. To quell the skeptics, Riley and Wells planned a second experiment.

This time, they created two identical colonies of guinea pigs, both getting air from the Pilot Ward. The only difference between them was that the vent to one of the colonies had an ultraviolet lamp inside. If Mills was passing along TB on her hands to the guinea pigs, then the researchers expected that animals in both colonies would get sick. But if the bacteria could only spread through the atmosphere, then the guinea pigs breathing UV-treated air would stay healthy.

Wells hoped the new experiment at Loch Raven might at last change the course of medicine. "This may profoundly modify our attitude toward air hygiene in public health, opening a new chapter in the epidemiology of contagion," he promised.

○ ○ ○ ○

In 1957, a new opportunity emerged to test Wells's theory on a different disease. A vast wave of influenza was heading to the United States.

It first emerged in Hong Kong, where it quickly overwhelmed hospitals. "Throughout each day, thousands of sick persons have stood in long lines awaiting treatment in clinics," a correspondent wrote. "Many women carried glassy-eyed children tied to their backs."

Maurice Hilleman, a microbiologist at the Walter Reed Army Institute of Research, did not like the news from Hong Kong. "I said, 'My God, this is the pandemic. It's here,'" he recalled later.

Hilleman arranged for saliva from patients to be sent to him in Washington. In his lab he isolated the virus, confirmed it was a new strain, and started work on a vaccine for it. The strain would later be named H2N2, but Hilleman gave it a nickname that stuck: Asian flu.

Hospitals across the United States braced for the arrival of H2N2. It might come by ship or plane, but come it must. At a VA hospital in Livermore, California, doctors wondered if they might already have some protection in place. They had been inspired by Wells's research to put ultraviolet lights in a new tuberculosis ward that housed one hundred fifty patients. But Wells had also carried out experiments at Penn in which he protected mice from getting influenza with UV. If Wells was right, Livermore's patients would be shielded from H2N2.

Accompanied by Ross McLean, a Loch Raven doctor, Richard Riley traveled to California in July 1957 to help prepare the hospital for the study. He came home optimistic that it would deliver Wells another win. "The Livermore study may yet turn into an important thing," he told Stokes.

Wells waited eagerly for the results. He promised Crane that the experiment "will not be ended with a report in a scientific journal or in the book, but by the welfare of mankind."

○ ○ ○ ○

Maurice Hilleman was right. In the summer of 1957, the H2N2 strain of influenza reached the United States. By then, several million Americans had received Hilleman's vaccine, and it likely prevented hundreds of thousands of deaths. Yet the virus still managed to kill about 100,000 people in the United States, and it is believed to have killed more than a million worldwide.

At Pleasanton Abbey, Wells escaped infection. He spent the fall compulsively sketching new outlines for an updated edition of *Airborne Contagion*. He reflected on how his theory fit into the history of public health. "The mission of the sanitarian is to cut the parasitic life line," he wrote, "to break the circuit of enteric infection between the anus of a host and the mouth of a victim; to exterminate an insect carrying a parasite from the blood of a host to the blood of a victim; and before being breathed by a victim, to vent infective droplet nuclei expelled indoors by a host. Sanitation is scientific isolation."

Wells made no effort to turn these ideas into a new manuscript. He simply sent loose pages, batch after batch, to New York. "I hope these running notes are not too much a bother to you," he told Crane in December 1957. "In my present disordered existence, I have no way to keep them from getting lost."

By January 1958, Joseph Stokes was getting worried about Wells's safety. Now seventy, he was living on a farm with an outhouse, caring for Bud by himself. He survived primarily on oatmeal, Riley would later write.

"I have been wanting to talk to you sometime about the dangers that Will is continually skirting in his own family situation," Stokes wrote to Riley. "If his own health fails, even to the extent of only moderate illness, it might precipitate him into a sudden calamity of considerable magnitude." In March, Riley assured Stokes that Wells was fine, "having weathered the cold weather without a shiver."

Riley was wrong, and Wells's health soon spiraled downward. He suffered a nagging backache over the summer, and in September he collapsed on his farmhouse floor.

When Wells tried to stand up again, he discovered he could not feel his legs. He could not stand, nor could he reach his telephone. Bud could not help him either. The sun set on Pleasanton Abbey, then rose, then set again. Two days passed before Wells was discovered on the floor of his house and taken to the Johns Hopkins Hospital in Baltimore.

His doctors puzzled over his collapse. His oatmeal diet had left him badly malnourished, but they eventually determined it was not to blame. They discovered Wells had prostate cancer, and that a metastatic tumor had grown in his back, causing a vertebra to collapse and sever his spinal cord.

Thanks to his service in the 301st Water Tank Train forty years earlier, Wells was eligible for care at a Veterans Administration hospital. His doctors moved him to the nearest one: Loch Raven.

○ ○ ○ ○

For days, Wells raved in his hospital bed, shouting about a hidden enemy who was applying electric shocks to his legs. Eventually the psychosis began to ebb, but when Riley stopped by Wells's bed to talk about the experiment, he would sometimes find the old man raging again. Riley would wait until his mentor's outburst ran its course, and then they got back to work. "We continue to discuss problems of the Air Hygiene Project almost as if nothing had happened," Riley wrote in astonishment to Joseph Stokes.

Between his cancer and his paralysis, Wells had no choice now but to make Loch Raven his new home. Mildred's sister, Marion Weeks, took over Bud's care, but it proved too much for her. In February 1959, she brought William Jr. to the director of the Delaware State Hospital for a psychiatric evaluation. Stokes, who was helping Riley oversee Wells's care, received a copy of the doctor's report. "It was impossible to do a routine psychologic examination, but certainly his vocabulary and range of ideation precluded the diagnosis of mental retardation," the doctor wrote. "However he did definitely give the impression of being psychotic. I felt he was suffering from schizophrenia and showing evidence of considerable regression."

The doctor recommended that Bud, now forty, be institutionalized. Riley paid a visit to Wells at Loch Raven to share the news. "This came as no surprise to either of us," Riley told Stokes. They arranged for a trusteeship to pay for Bud's care, and Pleasanton Abbey was sold off in 1961.

Wells's world was now limited to Loch Raven. It was his home, his lab, and the place where his friends could keep him company. Cretyl Mills would sometimes descend from the guinea pig penthouse to update him on the latest developments. Francis O'Grady, a young British doctor who had recently joined the Air Hygiene Research Study Unit, visited Wells for an hour each morning to get a daily tutorial on airborne infections.

In April 1959, the ultraviolet light experiment finally began. Precisely as Wells had predicted, the guinea pigs in the untreated cage got sick—sixty-three animals over the course of two years. None of the animals breathing air zapped with ultraviolet light developed tuberculosis. The skeptics had been answered.

Thanks to Cretyl Mills's exacting recordkeeping, Riley also observed that only a few of the veterans were responsible for most of the infections of the guinea pigs. He surmised they released many more bacteria in their droplet nuclei. And when the veterans responded to streptomycin, the guinea pigs upstairs stopped getting sick. Once the tuberculosis patients were cured, their droplet nuclei were no longer an airborne threat.

Wells had been right through and through, Riley concluded. Hospital workers needed to reckon with the fact that floating droplet nuclei could spread bacteria through a building. Inhaling just one droplet nucleus was enough to doom a victim to tuberculosis. Riley championed Wells's cause in papers and lectures. Tuberculosis might have been the first disease to gain definitive proof of airborne spread, but Riley expected other diseases would prove to be airborne as well, if only the right experiments were carried out.

As for stopping airborne diseases, Riley believed masks might help. He considered it wise for tuberculosis patients to wear masks for a week or two while they were receiving streptomycin in order to block bacteria escaping from their mouths and noses. But Riley doubted doctors and nurses would be able to protect themselves by wearing masks. Droplet nuclei were so small that they could slip through a cotton barrier or sneak around the sides of poorly fitted masks.

Riley was even more skeptical about masks outside hospitals. The only strategy to protect people in bustling department stores or crowded buses would be for everyone—both the healthy and the unknowingly infected—to wear masks at all times. "To achieve the prerequisites requires a kind of benevolent despotism," Riley said.

He also objected to masks on political grounds. People should not have to take personal responsibility for breathing in clean air any more than for drinking clean water. Water was a shared good, so it required shared protection. And just as the public water supply was purified for one and all, so should the air be.

"The individual would be relieved of direct responsibility," Riley wrote. "This is preventive medicine at its best, but it can only be bought at the price of civic responsibility and vigilance."

Riley did not imagine scrubbing the entire atmosphere clean. Instead, he called for purifying indoor air to the point that epidemics could not take off. Opening windows could help by bringing in fresh breezes. But by the 1950s, a growing number of hospitals, schools, and other large buildings kept their windows closed and used air conditioners and ventilation systems to maintain their temperature. It would be hard to filter the air well enough to reduce the risk of infections. Riley preferred the Wellses' dream of ultraviolet lamps quietly killing off pathogens in the upper air.

"If the lessons of the past mean anything, sanitary science will some day accomplish this hygienic triumph," Riley said.

As Riley spread the word, Wells lay paralyzed in bed. His thoughts kept returning to his book, which was now painfully out of date. He sent a note from Loch Raven to direct his lawyer back in Dover to send a new preface to the Commonwealth Fund. When O'Grady visited, Wells talked about his plans to reorganize the whole book and add a new chapter about the aerodynamics of droplet nuclei. In November 1959, Wells wrote to the Commonwealth Fund to lay out his plan for the revision. "I make note of these changes because of my precarious situation," he warned Crane.

At last, Crane broke the bad news to Wells: there would be no revision. Most of the copies of *Airborne Contagion* still sat unsold in a warehouse. "It seems therefore that the question of a reprinting probably will not arise in the near future," Crane wrote.

○ ○ ○ ○

In February 1960, a group of influenza experts gathered at the National Institutes of Health in Bethesda, Maryland, to share their observations on the H2N2 pandemic. Among their ranks was Ross McLean, who had come to describe the ultraviolet experiment at the Livermore hospital.

"This study was designed to demonstrate whether the disinfection of droplet nuclei would block the transmission of influenza to a susceptible population during an epidemic period," McLean said. The H2N2 pandemic had arrived at Livermore in December 1957, and it circulated among the patients and staff until late January. In the main section of the hospital, thirty-nine patients got infected. In the tuberculosis wing—where one hundred fifty patients lay beneath ultraviolet lamps—not one got the flu.

"The radiated patient group alone remained virtually free of infection, suggesting that an important mechanism of transmission had been significantly blocked during an epidemic period by the radiant disinfection of droplet nuclei," McLean concluded.

McLean hoped his talk would make his audience as excited as he was. He believed that the Livermore experiment fulfilled the predictions that came out of Wells's studies of influenza in animals two decades before. It now offered hope that ultraviolet light could hold back respiratory diseases. "Their epidemic control may well be accomplished," McLean told a colleague.

But his talk did not sway the Bethesda audience. They apparently still embraced Chapin's dogma and would not accept evidence to the contrary. McLean was particularly rankled when the British influenza expert Christopher Andrewes "dismissed it very abruptly by saying something to the effect that he did not think much of the droplet nucleus theory."

Riley saw the Livermore flu experiment as adding more strength to the airborne theory. But he also recognized that skeptics were still left with a little wriggle room. "The question is: would an outbreak have occurred in the absence of air disinfection?" Riley and O'Grady later observed. "This question cannot be answered."

Perhaps Riley pulled up a chair to Wells's bed and told him about the results from Livermore. Perhaps not. No record survives. The letter Wells wrote to Crane about his book in November 1959 appears to be the last he ever sent. In April 1962, Crane received another letter from Loch Raven, but this time it came from Cretyl Mills.

Mills told Crane how she paid Wells bedside visits and listened to him talk about his unquenchable plans for a second edition of *Airborne Contagion*. And then she began to cough. Working with the guinea pigs, Mills had inhaled a droplet nucleus harboring *Mycobacterium tuberculosis*, and it invaded her lungs. "I was hospitalized myself (also at the VA) for a long period at this time, and lost track of some important events and papers I regret to say," she wrote to Crane.

Now that she had recovered from tuberculosis, Mills wanted to see if she could help Wells with his book. She asked Crane if he could share any news about a second edition. Crane had to break the news to her too. There would be none.

Later that year, Riley went to Loch Raven to visit Wells before a long trip to India. He wanted to see his cranky old guide one more time before he left. When Wells was feeling well, Riley found his scientific insights as keen as ever. They still had things to talk about. They hoped to carry out one more version of the experiment, one that would demonstrate just how effective ultraviolet light was in protecting people against tuberculosis. So far, they had irradiated the air only after it flowed into the ventilation system and was headed for the guinea pig colony. Now they wanted to install lamps in the rooms themselves. The lamps would disinfect the air around the patients. If the guinea pigs inhaling this irradiated air were protected from tuberculosis, it would mean the patients and medical staff working in the patient rooms were protected too.

But Wells had declined in recent months, slipping again into bouts of psychosis. On Riley's last visit, he found Wells lashed by his wrists to the sides of the bed. He bled as he struggled to free himself.

○ ○ ○ ○

William Firth Wells died on September 19, 1963, at age seventy-six. "The end came without pain," his sister Clara Masland said. His family gathered the following month on Maryland's Eastern Shore for a memorial at the church where Mildred was buried.

Cretyl Mills traveled to the memorial from Baltimore. When she sat down in her pew, she looked around for scientific colleagues. There were none. It dawned on her that she was the only person at the memorial who wasn't a relative.

Mills sent off letters to let people know that Wells was gone. Joseph Stokes replied to Mills with a celebration of Wells's work. It had ended, he wrote, with "the final proof of the importance of the air as a mode of transmission under natural conditions." But Stokes also told Mills how hard Wells had made it for his friends to help him. "He had managed sometime to raise so many antagonisms in so many people that it did take a small number of us who understood and respected him to hold the balance in his life between real accomplishment and complete destruction," Stokes wrote.

Other doctors did not remember Wells so kindly. In October 1963—the same month as Wells's memorial—the microbiologist Ronald Hare offered a harsh eulogy in London.

Twenty-three years had passed since Hare had tested droplet nuclei for himself at the University of Toronto. He had had volunteers gargle with bacteria, and then instructed them to talk and cough. Hare observed that only at close range could they seed a Petri dish with bacteria. Droplet nuclei did not deliver the microbes over longer distances. That failure persuaded Hare that airborne transmission didn't matter much.

He turned his attention to other matters, such as boosting the production of penicillin to an industrial scale.

Hare kept a close eye on new research into the spread of diseases. But nothing changed his mind. When the Canadian navy asked Hare how to reduce infections in the windowless sleeping quarters on ships, he did not see any reason to recommend ultraviolet light. Instead, he recommended spacing out the bunks. With fewer sailors in a room, there would be fewer pathogens spread on the beds and floors. When wartime experiments failed to show any benefit from ultraviolet lamps, Hare was not surprised.

By 1963, Hare had become an eminent figure in the world of infectious disease. Now sixty-four, he was a professor at St. Thomas's Hospital Medical School in London and the president of the pathology section of the Royal Society of Medicine. He was required to deliver the annual presidential lecture. He entitled it "The Transmission of Respiratory Infections."

Hare offered a sweeping look at seventy years of science, starting with Carl Flügge laying Petri dishes out on his lab floors in Breslau. It seemed to Hare that respiratory infections were probably spread mostly at close range by heavy droplets, either released in coughs or stirred up in dust from clothes and bedding. Toward the end of Hare's lecture, he turned his attention to Wells. He granted that people release droplet nuclei from their mouths and noses. But those tiny globes of water were mostly free of pathogens, Hare believed. The evidence did not support Wells's theory.

"Despite the contentions of Wells," Hare said, "it is extremely doubtful whether droplet nuclei play any important part in the transmission of respiratory infection."

The following year, Alexander Langmuir published his own eulogy of sorts. In an essay called "Airborne Infection: How Important for Public Health?" he damned Wells with faint praise. Wells deserved credit for reviving airborne infection as a serious hypothesis after the death of miasmas in the nineteenth century, Langmuir said. But to Langmuir, the failure of the World War II ultraviolet experiments suggested it didn't matter much.

Langmuir did not dismiss airborne infection altogether. The best evidence for it came not from Wells's lab at Harvard in the 1930s, he said, but from Fort Detrick. Theodor Rosebury and the other germ warfare scientists had demonstrated that diseases like parrot fever could spread through the air, and they could do so very effectively if they were engineered by scientists. Writing in 1964, Langmuir still believed in the Cold War threat of biological weapons. He warned public health workers to prepare for a Soviet attack by brushing up on the principles of airborne infections.

But beyond germ warfare, Langmuir considered only tuberculosis as a genuine airborne threat. And he credited Wells with establishing that fact. Langmuir declared that Wells's experiment on rabbits and the Loch Raven study that followed "confirm the theory." Yet he rejected the idea that Wells's theory applied to any other diseases.

"The particle size of microbial aerosols is a major factor in the occurrence of airborne infection, whether natural, accidental, or experimental," Langmuir explained. Pathogens that infected only the alveoli at the tips of the lungs were "intrinsically and exclusively airborne."

That was because only particles less than five microns in diameter were small enough to fit in the narrow passageways leading to the alveoli. Larger droplets could not make that journey, nor could dust grains coated with bacteria. The strict five-micron rule, Langmuir believed, meant that virtually no diseases could be airborne, aside from tuberculosis and biological weapons like parrot fever.

In making this argument, Langmuir was distorting Wells's ideas. Wells had never established a five-micron rule. While droplet nuclei might have to be smaller than five microns to deliver tuberculosis to the alveoli, he believed that bigger droplets could also cause infections in other parts of the airway. And by Wells's calculations, droplets as big as one hundred microns could potentially float for hours in an indoor space.

Langmuir's misreading would grow into a major misunderstanding in later years. Public health experts would maintain that only droplets smaller than five microns could cause any airborne disease—in effect, ruling out the route for any disease that did not infect the alveoli. They cited Wells as their authority, despite the fact that Wells had said no such thing. The confusion would prove deadly.

In the 1960s, there was little incentive to clear up the confusion, because doctors and microbiologists believed that infectious diseases were on the verge of being eliminated. Vaccines were already wiping out polio in many countries, and in 1963, researchers in Boston invented a powerful vaccine against measles. Meanwhile, the World Health Organization was launching a campaign to eradicate smallpox from the planet; the campaign would end in triumph less than two decades later. And for diseases that vaccines couldn't prevent, researchers were inventing antivirals and antibiotics.

In 1964, the year after Wells's death, Selman Waksman published a triumphant history called *The Conquest of Tuberculosis*. Waksman gave a nod to the Loch Raven experiments in a few quick sentences. He treated them as a digression on the road to victory. Antituberculosis drugs were so potent that it would have been a waste of time to design ultraviolet lamps for stopping bacteria in the air.

"The future appears bright indeed," Waksman predicted, "and the complete eradication of this disease is in sight."

o o o o

When Richard Riley returned home from India, he planned to carry out the last part of the tuberculosis experiment. He would install ultraviolet lights in the Pilot Ward rather than in the air ducts overhead. But when Riley came back to Loch Raven, he discovered that the Pilot Ward had been reclaimed.

The tuberculosis patients had been moved to other rooms. The hospital had cleared out the guinea pig penthouse. Cretyl Mills lost her job and ended up teaching biology at a Baltimore junior college. And Wells, of course, was gone. Wells had needed Riley to make his dream of an experiment real. Now Riley could not carry it forward alone.

eleven

THE BROTHERS RILEY

After Wells's death, Richard Riley returned to his research on how we breathe. But he still found time for small-scale experiments on airborne infections. He once sealed his office, hung an ultraviolet lamp from the ceiling, put on a germ warfare gas mask, and flooded the air with *Mycobacterium tuberculosis*. Riley found that the UV lamp wiped out all the bacteria from the air in under twenty minutes. Without the light, the bacteria lingered for hours.

At the end of each day, Riley would ride home from Johns Hopkins with his colleague Solbert Permutt. They stopped off to pick up Riley's son, Richard Jr., at school and then went on talking as they drove. They would argue about what kind of mathematical model could capture the air moving in a room where ultraviolet lamps were hanging. When I asked Richard Jr. about his father, he recalled sitting in the back of the car on winter evenings, watching his father and Permutt run their fingers over the foggy windshield to draw graphs.

Riley had little company in the study of airborne diseases aside from Permutt. Other scientists were not carrying out experiments of their own to see how droplet nuclei moved through buildings or to find the best way to render them harmless. Riley made up for his solitude by searching for natural experiments. In 1966, a seaman aboard the guided missile destroyer USS *Richard E. Byrd* gave tuberculosis to 139 crewmates, many of whom never came into close contact with him. In 1969, a twenty-year-old electrician diagnosed with smallpox managed to infect twenty people in a German hospital, including one patient in a room two floors above him. Three of them died.

But the outbreak that offered Riley the best evidence for airborne infection occurred in 1974. This time, the disease was measles. And this time, it was Riley's own brother who served as the medical detective.

◦ ◦ ◦ ◦

Ed Riley worked with Wells for four years all told. When Wells moved to Philadelphia, he tried to lure his young assistant with him to help run the Laboratories for the Study of Air-borne Infection. But Ed chose medical school instead, and he never worked professionally on airborne infection again. After serving in World War II, Ed moved to Rochester, New York, where he oversaw the health of the workers who produced photographic film at Eastman Kodak. Thanks to his training as an engineer, Ed could apply complex equations to the spread of chemicals in factory air. In 1970, he retired at age sixty-two.

With time now on his hands, his thoughts turned back to Wells, who had died almost a decade earlier. "I shared Wells' conviction that many contagious diseases were airborne and that sterilization of the ambient air should prevent airborne infections," Ed later said. He began spending his retirement perusing the reports William and Mildred had made from schools in Philadelphia in the 1930s and 1940s when they had fought measles with ultraviolet light.

The Wellses had argued that the results of their experiments showed that measles was an airborne disease that spread like smoke. But in 1970, that was hardly the consensus. Alexander Langmuir, who studied measles in the 1960s, concluded that it did not spread over long distances. He based that conclusion in part on the fact that the measles virus infected the nose and the upper airway before spreading to other parts of the body. Langmuir assumed this meant that the virus had to spread in large, short-range droplets, or by getting onto people's hands.

If measles only spread at close range, as Langmuir believed, that meant that it could be eradicated now that a powerful vaccine existed.

An infected child would be surrounded mostly by vaccinated children, and the virus would have a hard time finding a new host. If the United States launched a large-scale vaccination campaign, measles might vanish from the country in just six months. "I have rarely been more confident of a prediction," he later recalled. He had rarely been more wrong.

The US measles campaign bolted out of the gate in 1966. Charles Schulz helped rally children and their parents, dedicating a week of his *Peanuts* comic strip to the story of Linus getting vaccinated. Millions of children got immunized, and Langmuir was heartened to see measles cases in the United States drop from 261,904 in 1965 to just twenty-two thousand in 1968.

But measles did not vanish as Langmuir had predicted. Cases bottomed out over the next few years and then started to rise again. In 1971, seventy-five thousand people got infected. "Where did we get derailed?" Langmuir later wondered.

Ed Riley thought back to William and Mildred Wells for an answer. After thirty years, their research struck him as both crude and visionary. When the Wellses went into a school, they used simple arithmetic to figure out how to install their lamps. They merely took into account the volume of the classrooms and the number of children. They did not try to factor the droplets being exhaled or inhaled by the children, or the speed at which the air moved around them. If the children experienced fewer infections after the lamps were turned on, the Wellses declared victory. Ed Riley set out to translate their ideas into mathematical equations that contained the variables important to airborne infections. Those equations would reveal how many infections would occur in a building of a certain size filled with a certain number of people.

Ed carried out the work as a way to honor the memory of the Wellses. He borrowed formulas engineers used to simulate the flow of air through a building. He melded them with equations that epidemiologists had developed to describe the spread of infections through groups of people. The equations could take into account how many people were vulnerable to a pathogen and how many were immune thanks to a previous infection. The equations were so big that Ed could not solve them by hand. Computers were starting to become available to scientists in the early 1970s, and Ed persuaded one of his younger colleagues at Kodak to help him write a program to run the model. Now all he needed was an actual outbreak to test it against.

Ed thought at first he would find an outbreak by traveling back in time. He would reconstruct the measles outbreaks that the Wellses had fought. He traveled to Philadelphia to inspect the schools where they had installed their lamps, hoping to create an accurate model of what happened there in 1942. But the buildings had been renovated in the intervening decades, and Ed could not reconstruct their original physics.

But then an even better opportunity fell in his lap. In 1974, measles hit an elementary school near Ed's home in Rochester. Twenty-eight children scattered across fourteen classrooms got sick.

Ed got in touch with the Rochester school nurse, who provided him with details about the children who had gotten sick, along with the layout of the school. As Ed built a model of the Rochester outbreak, he also took into account that the children rode to school on buses, where the virus could move readily among them. He arranged for a driver to take an empty bus along its regular route, opening and closing the door at each stop. Along the way, Ed recorded how much air flowed in and how much flowed out.

Running his computer simulation, Ed produced measles outbreaks with the same explosive velocity as the one that had hit the Rochester school. A second-grade girl had been the index patient, and her virus-laden breath was swept into the school's ventilation system. Only 3 percent of the children in the school were unvaccinated and thus vulnerable. The girl's plume of measles must have spread like smoke throughout the school in order to find those vulnerable students scattered in separate classrooms. The model worked, based on airborne infection alone.

° ° ° °

As Ed Riley breathed new life into the work of the Wellses, his brother was winding down his own career. Richard tried warning his colleagues in the 1970s that tuberculosis was not, in fact, a conquered foe. Public health workers needed to be more vigilant about stopping outbreaks, and hospitals needed to defend against patients spreading the bacteria down their halls and ventilation systems. "In my opinion, air disinfection by ultraviolet irradiation, a method that is both feasible and effective, should be widely used," he wrote in 1976. The following year, at the age of sixty-seven, Richard retired from Johns Hopkins and moved with his wife to Petersham, Massachusetts.

Richard was delighted to learn that his brother had returned to studying airborne infection after a forty-year break. He helped Ed write up the results of the Rochester model and submitted the manuscript to the *American Journal of Epidemiology*. Richard likely expected an easy stroll to publication. After all, one of the journal's editors was Gardner Middlebrook, who had helped Richard turn his office into a TB test chamber. But when other scientists reviewed the paper for the journal, they judged that it wasn't significant enough to publish. They recommended rejection.

Richard pounded out a letter to one of the journal's editors on the 1928 Remington typewriter that he had gotten as a high school graduation present.

"This is an original contribution to epidemiology and has enormous public health significance," he declared. "From our point of view measles is simply a marker identifying a general phenomenon. Is no one interested in understanding the dynamics of airborne infection, probably the greatest single cause of morbidity in humans?"

At the bottom, Richard added,

Copies to:

Edward C. Riley
Gardner Middlebrook, Editor

And, by celestial mail,
William F. Wells, who also had trouble getting published in this journal.

R.I.P.

After a few months, the Riley brothers wore down their opposition, and the journal accepted their paper. "Airborne Spread of Measles in a Suburban Elementary School" was published in 1978. "This is something of a landmark, and I am very proud to be publishing it with you," Richard told Ed.

In the paper, the Riley brothers demonstrated how one highly infectious student filled a school's recirculated air with a cloud of measles viruses. They suggested that ultraviolet lights could have annihilated the viruses as they floated inside droplet nuclei. "The present epidemic would probably have been aborted," they wrote.

The Rochester study marked a turning point in the study of measles. Ed had discovered an important reason that Langmuir's vaccination campaign had failed to eradicate the disease. To Langmuir's credit, he acknowledged in 1980 that his prediction of a six-month eradication period had been a "rather egregious blunder." It failed because Langmuir had misjudged how the virus spread. By floating through the air, it could find vulnerable children far away. "Clearly we must revise our theory and recognize that these outbreaks must be airborne," Langmuir conceded.

In later years, Langmuir made more amends to Wells. He and his fellow epidemiologists had "sinned grievously," he said, in dismissing the airborne spread of measles. "In the 1930s, William Firth Wells argued with intensity that measles was primarily an airborne infection," Langmuir wrote. "We should have given him more attention and respect."

Ed Riley left behind another legacy with his work on measles. The equations he used to capture the invisible complexity of airborne germs became a new standard. Later generations of researchers came to call it the Wells-Riley model.

After publishing his model, Ed Riley gave up airborne infection and settled down in retirement in Florida, where he lived till his death in 2000. "I am happy to be remembered as part of the Wells-Riley team," Ed once told Richard. Even after life, Ed would be linked to his mentor by a hyphen.

o o o o

While Ed gave up airborne infection after the measles study, he enjoyed watching Richard "carry on the gospel according to Wells and Rileys." Richard stayed busy in his retirement, building a case for airborne infection and a parallel case for fighting it with ultraviolet light. He tinkered in his Petersham barn on devices to reduce the spray of droplets released during medical procedures. He tested ultraviolet lights to demonstrate their safety. "These are the grubby kinds of things that need doing," Richard told a colleague.

Richard also kept waging his letter-writing crusade on his old Remington. He sent out letters and wrote essays calling on his fellow doctors to recognize the threat of airborne infection. Tuberculosis, he warned, was not the only disease that could float into our airways. Influenza pandemics and chicken pox outbreaks also posed threats—and could also be countered with ultraviolet lamps installed in schools, hospitals, and other places where the air could fill with pathogens.

In his old age, Riley earned a following among architects and engineers who were becoming aware of how modern buildings could make indoor air dangerous to human health. New tight-sealed windows might conserve energy, but they also reduced ventilation. The air inside buildings filled with new chemicals off-gassing from carpets and plywood. Mold grew in ventilation ducts and cast off spores that irritated the lungs of residents. People started complaining of a new disorder: sick building syndrome. Riley raised the prospect that sick buildings could also harbor droplet nuclei that caused infectious diseases. "The construction of tighter buildings and the rising use of recirculation systems have exacerbated the problem of airborne infection and stepped up the need for control measures," he warned.

But Riley found few fans among doctors and public health experts. "The medical profession remains confused and, by and large, has not given its blessing to air disinfection in hospitals," he said in 1980. But then, in 1983, one doctor tracked down Riley to ask for his help. Ed Nardell, a tuberculosis expert in Boston, was struggling to rein in an outbreak.

Nardell had earned his medical degree in 1972 in Philadelphia and moved to Boston to become a pulmonologist. He spent a year at Boston University School of Medicine learning how to study the molecular biology of lung diseases. While he found the work intellectually stimulating, he couldn't see where it would take him. When Nardell was asked to serve as the TB control officer for Cambridge, Massachusetts, he took the job. Soon he was asked to take the job for the whole state.

"Gradually, TB was engulfing me," Nardell later told me. I paid him a visit in his own semiretirement. In his apartment overlooking Boston's Riverway, he was learning to play the recorder and had a stack of the instruments resting on his piano like a cord of wood. At seventy-six, Nardell was tall, broad shouldered, and heavyset. After serving me a lunch of salmon and salad, he cheerfully recalled the horror his superiors felt when he announced that he would be leaving the lab bench to make TB the center of his career.

"They were thinking, 'Where did we go wrong?'"

◦ ◦ ◦ ◦

Nardell's disappointed mentors had trained at the peak of postwar optimism, at a time when Selman Waksman was promising that antibiotics would soon eradicate tuberculosis. To them, it seemed foolish for a young pulmonologist to bother with a nearly vanquished disease. Nardell's advisors thought he should make his mark by searching for a cure to a lung disease that was still incurable.

He took the job as TB officer anyway. And it didn't take him long to recognize just how naive his mentors had been. Tuberculosis stopped declining, and before long it started to rise again.

The turnaround had a confluence of causes. *Mycobacterium tuberculosis* was becoming resistant to standard antibiotics, forcing doctors to prescribe backup drugs that had harsh side effects and had to be taken for months. Then those drugs started failing too.

Tuberculosis had always inflicted outsized damage on the poor and the marginalized, and now it did so again. Poor Black neighborhoods across the United States experienced a sharp rebound, even as the rate of infection continued to drop among white Americans. TB rates were particularly high among Haitians and other immigrants from poor countries where the disease was also raging. In Boston, Nardell's Haitian patients lived in crowded apartments where the airborne bacteria could spread readily from one person to another.

Tuberculosis also got help from a newly arrived virus. In the early 1980s, California doctors recognized that a growing number of gay men were suffering a collapse of their immune systems, a condition known as acquired immunodeficiency syndrome. A virus that was dubbed human immunodeficiency virus proved to be the cause. Spread through unprotected sex, shared needles, or contaminated blood transfusions, HIV attacked immune cells and made its victims vulnerable to other diseases that they might otherwise have fought off. For years, the Reagan administration did little to stop the spread of HIV, dismissing AIDS as "the gay plague."

Reagan-era policies also helped tuberculosis by spurring a crisis in homelessness in the United States. Thanks to cutbacks in support for low-income housing and a push to move the mentally ill out of institutions, hundreds of thousands of Americans ended up living on the streets. Losing their homes added an extra burden on people's immune systems, making them more vulnerable to TB.

As all those factors combined to drive up the number of tuberculosis patients in Massachusetts, Nardell scrambled to manage their complicated care. He started getting requests to visit New York, San Francisco, and other cities seeing surges of tuberculosis among the homeless. "It suddenly became very gratifying to be the expert on TB, when nobody else had cared about it," Nardell said.

When a new outbreak hit in late 1983, Nardell realized he needed an expert of his own. Guests in Boston's biggest homeless shelter, the Pine Street Inn, were falling ill at a worrying rate. It was conceivable that Nardell's patients had been carrying TB infections for years that only now were activating. But Nardell and his team figured out that the new cases were the result of an outbreak spreading from a single Pine Street Inn guest. He recalled a lecture he had heard Richard Riley give years before about airborne infection and ultraviolet light. He wondered if ultraviolet lights might slow the spread at the shelter.

Nardell tracked the retired doctor down in Petersham. "I suggested to Riley that this might be the perfect place to put UV, because you're either in the shelter or you're outside on the street," Nardell told me. When homeless people were outdoors, they were unlikely to encounter *Mycobacterium tuberculosis* floating on a breeze. At night, ultraviolet lamps could purify the air in the one place where they were getting heavily exposed to the bacteria.

Riley agreed with the plan and offered Nardell guidance. "He was thrilled to get back in the saddle," Nardell said. Along the way Nardell became, in effect, a scientific grandchild of the Wellses.

At the Pine Street Inn, Riley advised Nardell on how to install ultraviolet lights in the common areas and three floors of dormitories. Nardell was trying to stop an outbreak, not run an experiment, so he knew it would be hard to measure the effects precisely. But he still managed to see some promising signs. New cases got rarer after the lights were installed. Nardell found a cluster of three cases among staff workers who worked together in a room. They admitted to sometimes turning off the ultraviolet light.

Nardell and Riley deemed the Pine Street Inn project a success, even if they had prevented only a few new cases. Curing the disease could require months of antibiotics, and ensuring that homeless and poor patients complied with the treatment became a labor of its own. "If just one case among the homeless is prevented, the installation will have more than paid for itself," they wrote.

Riley hoped that the project might remove the stigma that ultraviolet light had carried for more than thirty years since Mildred Wells's disappointing effort in Westchester County. He persuaded Lemuel Shattuck Hospital in Boston to install UV lamps; he urged other hospitals to do the same. He lobbied the federal government to update its guidelines for controlling TB in hospitals and other health care spaces. Fighting the disease in the air would help doctors drive TB back down again. Riley, now in his seventies, was grateful that Nardell carried on the fight. "He is courageous, tenacious, and unflappable," Riley said.

Nardell and like-minded researchers managed to sway government regulators, at least a little. In 1997 the Centers for Disease Control issued new guidelines for tuberculosis that acknowledged that droplet nuclei spread the disease. They even cited William Firth Wells and *Airborne Contagion* as evidence. Part of their guidance addressed what sort of mask to wear around people with tuberculosis to avoid getting infected. The only way to block the spread of tuberculosis was to stop the droplet nuclei. And for that job, the CDC recognized, traditional surgical masks probably weren't enough.

○ ○ ○ ○

Masks had evolved a lot in the century since German surgeons started wearing Carl Flügge's cotton strips over their beards. While some masks were designed to stop the spread of microbes, others defended against lifeless threats to the lungs, such as coal dust, asbestos fibers, and fiberglass. The most elaborate masks looked more like astronaut helmets with hoses connected to filters that people wore on their belts. Those space-age masks were also expensive, cumbersome, and uncomfortable. Then, in the 1970s, the chemical company 3M introduced a lightweight alternative. It was fashioned after a woman's bra.

The idea for the new masks was the work of a former editor for *House Beautiful*. In 1958, Sara Little Turnbull left the magazine to start her own industrial design firm. One of her first clients was 3M, which initially hired her to give them ideas about gift wrapping. She became intrigued by how the company made ribbons. Rather than weaving cotton, they sprayed polyester fibers into a fabric. Turnbull designed the first premade bows for 3M and then told the company to think bigger. The fiber technology could, she said, "create new solutions for old unsolved problems."

Turnbull offered 3M a hundred potential products that could be made from the new fabric. One of them was a molded bra. The sprayed fibers would provide women with more comfortable support than old cloth models, she predicted. Turnbull was assigned to design the bra, which 3M patented in 1959.

In the years leading up to her work with 3M, Turnbull had endured the slow deaths of her sister and her parents. She spent a lot of that time in hospitals, where she would watch doctors and nurses tying on their surgical masks. It occurred to Turnbull that the blown-fiber cups of her new bra might, with a little tinkering, serve as snug-fitting masks.

Turnbull designed a mask that used elastic bands rather than string ties to keep it on people's heads. In 1961, 3M patented it, only to discover that the pores between the fibers were so big that germs sailed through them. But the company's engineers found that the mask worked well against dust and other larger particles, and the 3M dust mask hit the market in the 1970s. It became a common sight at hardware stores. It blocked everything from sawdust to volcanic ash.

Years later, 3M dramatically improved the masks thanks to a trick of physics. A materials scientist named Peter Tsai at the University of Tennessee figured out how to give fibers a static charge. When 3M applied his treatment to their masks, they could stop tiny particles even if they could physically fit in the space between the fibers. The particles stuck to the fibers like balloons on a sweater.

The masks came to be known as N95 respirators, the name meaning that they could block 95 percent of nonoily particles. The new tuberculosis guidelines that government agencies issued in the 1990s recommended that doctors and other medical personnel caring for TB patients wear N95s. If health care workers needed even more protection, they should wear astronaut masks, officially known as powered air purifying respirators.

The new advice stirred up a lot of controversy. N95s might very well have been superior because they formed a tight seal—but only if they were properly fitted. No clinical trials had yet demonstrated that they actually reduced the rate of tuberculosis infection. As for powered air purifying respirators, some doctors complained that their science fiction look scared their patients. And many even pushed back against the notion that tuberculosis spread like smoke.

Writing to Riley in 1993, a Berkeley mask expert named Mark Nicas recounted the trouble he had had persuading hospital workers to wear powered air purifying respirators. They didn't see the need for the hassle, since surgical masks surely were protection enough. After all, the hospital workers told Nicas, tuberculosis spread only in the big droplets that sprayed a few feet when people coughed. "Unfortunately, I've witnessed some, might I say, irrational behavior when it comes to discussions of respirator use against TB transmission," Nicas told Riley. "A CDC official once told me that TB aerosols are akin to 'soggy raisins.'"

○ ○ ○ ○

It was hard enough to raise awareness of airborne infections inside hospitals. Outside them, it was next to impossible. Nardell got a rare opportunity to make the case on a massive scale when Philip Brickner, a New York doctor at the forefront of homeless care, asked about running the Pine Street Inn project nationwide.

Brickner proposed that they pick two shelters in each city, install ultraviolet lights in one and fake lights in the other. The staff at the shelters would be regularly tested for tuberculosis infection, but the scientists analyzing the medical records wouldn't know where the working lights were. In other words, the study would be a blinded experiment, making it a gold-standard test.

"I said, 'Phil, that's going to be really hard,'" Nardell recalled. Riley had taught him the tortured history of airborne infection, which was littered with trials that failed to deliver clear-cut results. "There's a reason why there's only one good epidemiological study of UV, and that's the Wells study in the 1930s," Nardell said.

Brickner still wanted to try. He persuaded Nardell to set aside his qualms and join him on the Tuberculosis Ultraviolet Shelter Study. They traveled across the country to convince homeless shelters to participate. They eventually got twelve hundred lights installed in fourteen shelters in six cities. They asked the staff and guests at shelters about whether they suffered any irritation to their eyes or their skin and found no overall difference between the shelters with real lamps and those with fake ones. Heartened that the lamps were safe, the researchers finally started collecting data in 1997.

But time was working against them. Tuberculosis rates in the United States were falling thanks to aggressive treatment of the sick and the declining rate of HIV infections. While the decline of TB was good news for public health, it also meant that fewer people in the shelters that Brickner and Nardell were studying had the disease, making it harder to detect an effect from the lamps. When the study ended in 2004, the scientists could find no statistically significant benefit.

During their long friendship, Nardell and Riley would sometimes dream about the ideal experiment to test the effect of ultraviolet light on tuberculosis. They would carry out the study that Riley never got to run at Loch Raven. In 1995, Nardell and Riley told Solbert Permutt about their idea over lunch. They would find a TB ward where they could install lamps on the walls. They would then vent the air from treated and untreated rooms into cages full of guinea pigs.

Permutt had a better idea, which he drew for Nardell and Riley on a napkin. To make the experiment stronger, he suggested turning the lamps on and off in a single ward. On the days when the lights were off, the air would get vented to one colony of guinea pigs. When the lights were on, the air would go to another one. It was a clean, elegant design, as Nardell and Riley both recognized. But with the number of TB cases falling in the United States, they couldn't figure out how they could use it to run an experiment.

At the time, Riley was in his late eighties, and he accepted that he had reached the end of his campaign. He would sometimes talk to his son, Richard Jr., about his regret that the research he and Wells had carried out had yet to gain broad acceptance. The timing, for whatever reason, had never been right. Riley brought out his Remington one more time. He typed out his final published work—a one-page essay for the *American Journal of Respiratory and Critical Care Medicine* recounting the experiment at Loch Raven, which came to an end with Wells tied to his bed. "Thus ended the career of a truly 'mad genius' who gave us the droplet nucleus hypothesis and changed our thinking about aerial transmission of infection," Riley wrote. Within a few months of the homage, he died at ninety.

In its obituary for Riley, the *Baltimore Sun* observed that his work lived on. In the 1990s, Nardell had traveled the world to help fight tuberculosis in prisons in Siberia, slums in Peru, and shantytowns in South Africa. A South African colleague let him know that a tuberculosis hospital was being built in the province of Mpumalanga. Nardell persuaded the administrators there to run Permutt's updated version of the Loch Raven experiment.

In 2005, the Airborne Infection Research Facility opened in Mpumalanga; it was the only facility in the world dedicated to experiments in which humans infect lab animals by air. It consisted of a set of three patient rooms with two beds apiece, along with bathrooms and a dayroom. For months, the lamps burned for a day and then were shut off the next day. Nardell and his colleagues found that ultraviolet light reduced the risk of infection in the guinea pigs by about 80 percent. A similar study carried out in Peru by another team of scientists found a comparable level of protection.

Nardell and his colleagues published their results in 2015, fourteen years after Riley's death. Four years later, the World Health Organization released new guidelines for preventing TB, and for the first time, it published a recommendation to use ultraviolet light. But the agency cautioned that it was only a "conditional recommendation based on moderate certainty in the estimates of effects."

In the long struggle to draw attention to airborne infection, that recommendation counted as a victory. But it was offset by a grave misunderstanding that continued to distort public health measures for keeping people safe from infection.

By the time Alexander Langmuir died in 1993, he had taken back a lot of his early criticisms of William Firth Wells. Yet the claims he made in the 1960s about the rareness of airborne infections—including the five-micron rule—endured after his death. In 1996, the Centers for Disease Control put them on prominent display when they released guidelines for controlling infectious diseases. They recommended elaborate "airborne precautions" for diseases that could be spread over long distances. Patients with these diseases had to be put in private rooms with negative air pressure to ensure the pathogens didn't leak out. The agency also recommended ventilating the rooms with fresh air from outside.

The CDC singled out only three diseases for airborne precautions: measles, chicken pox, and tuberculosis. Those diseases, the agency claimed, were spread by droplets smaller than five microns. Droplets of that size or smaller, the CDC claimed, "may remain suspended in the air for long periods of time."

In a separate category, the CDC placed diseases such as influenza and mumps that supposedly spread in droplets bigger than five microns. Those droplets, the agency said, were so heavy that they "generally travel only short distances (up to 3 feet) from infected patients who are coughing or sneezing." And since they only traveled short distances, people could avoid them by following a less cumbersome set of "droplet precautions." If people had to get within three feet of infected patients, they could protect themselves with surgical masks. Beyond that distance, people could rest easy. They were safe from the soggy raisins.

If William Firth Wells had survived until the 1990s, he would not have been all that surprised to see his work distorted in this way. After all, he had spent much of his career bemoaning misunderstanding and misrepresentation. And if he had survived until the twenty-first century, Wells would have experienced even more déjà vu. He would have seen how his work continued to be used as the basis for biological weapons—and to stoke an enduring fear of manmade plagues wafting through the air.

twelve

SACKS OF SUGAR, VIALS OF POWDER

In January 1988, the Nobel Prize–winning microbiologist Joshua Lederberg gave a lecture in New York in which he conjured a nightmare. He asked his audience to imagine AIDS becoming airborne.

"It is hard to imagine a worse threat to humanity than an airborne variant of AIDS," he said. "No rule of nature contradicts such a possibility."

Lederberg had been contemplating scientific nightmares for decades. When satellites went into orbit in the late 1950s, he worried that they might pick up microbes in space and deliver them back to Earth when they fell out of orbit, starting an unstoppable plague. He designed a plan for NASA to sterilize returning spacecraft. In the 1960s, Lederberg turned his attention to what he called America's "suicidal policies of biological warfare." He joined Matthew Meselson in the fight to stop the country's production of anthrax bombs, which they considered useless and foolish. Nixon's ban of biological weapons did not make Lederberg's nightmares go away. Even if the world escaped biological warfare and alien pathogens, he worried that natural microbes might doom humanity. By the 1970s, bacteria were evolving resistance to antibiotics at a frightening pace.

In the 1980s, the discovery of HIV fueled fresh nightmares. The virus caused a disease that could be found in no medical textbook. As far as infectious disease experts could tell, it had only recently evolved, and now it was swiftly spreading among humans. Anthony Fauci, a forty-one-year-old immunologist at the National Institutes of Health, was flummoxed by the first reports of gay men suffering from AIDS. "The idea of a totally new infectious agent was beyond my comprehension," he later wrote. "I had not heard of such a thing in recent recorded history."

Fauci and other researchers set about studying HIV and searching for ways to fight it. But AIDS remained a death sentence for the time being, and its origins inspired rumors and conspiracy theories. The Soviet Union spread a story that HIV had been created in a Fort Detrick lab by scientists who merged two viruses into a hybrid. When they tested it on human volunteers, the virus escaped. It would take years for scientists to quash such rumors by conclusively demonstrating that HIV had arisen in chimpanzees and jumped to humans in the early 1900s.

More rumors circulated about how HIV spread from one person to another. At first, doctors didn't know if it spread through the air, and so they suited up in full bodysuits to treat HIV patients. Eventually epidemiologists established that HIV was not airborne, spreading instead through unprotected sex, blood transfusions, and shared needles. Yet many people became convinced that a touch was enough to pass the virus along. So was sharing a sandwich or—most magical of all—breathing the same air.

At the start of the 1985 school year in New York City, a second grader diagnosed with HIV led parents to keep eleven thousand students home. A local politician, State Assemblyman Frederick Schmidt, came to a demonstration to egg on the ostracism. He suggested that HIV could travel through the air. "There is no medical authority who can say with authority that AIDS cannot be transmitted in school," Schmidt declared. "What about somebody sneezing in the classroom?"

As fears about HIV grew, the New School for Social Research in New York planned a conference about the epidemic. The organizers hoped the meeting would "lead the way to a calmer and more effective public response to the problem." In January 1988, they brought disease experts to discuss HIV, along with sociologists to explain how social forces worsened epidemics.

One of the speakers was Lederberg. He started off his lecture by explaining that HIV was not a freakish alien, but very much part of the natural world of pathogens. He explained to the audience the huge range of pathogens' diversity and their impressive ability to evolve. New diseases could readily emerge as pathogens shifted from animals to people, and we should not assume we would be able to vanquish them. "Our principal competitors for dominion, outside our own species, are the microbes: the viruses, bacteria, and parasites," Lederberg said. "They remain an interminable threat to our survival."

For Lederberg, HIV was the epitome of viral threats. "There is nothing in the natural history of AIDS to point either to a cure or to a vaccine," he warned bleakly. And as bad as it might be as a blood-borne virus, Lederberg could imagine it getting worse by becoming airborne. If the conference organizers expected the speakers would offer restrained reflections based solely on the facts, they must have been sorely disappointed by Lederberg's speech. He gave a scientific sheen to the fears spread by the likes of Assemblyman Schmidt.

"What are the odds of its learning the tricks of airborne transmission?" Lederberg asked. "The short answer is, 'No one can be sure.'"

○ ○ ○ ○

It's not clear where Lederberg got the idea of airborne AIDS, or why he liked to talk about it. What is clear is that he had learned the value of a good story.

People had been telling stories about apocalyptic germs for nearly a century. In the 1890s, H. G. Wells concocted a tale of terrorists bent on releasing plagues. Six decades later, Alexander Langmuir went on television to walk the American public through Cold War biological warfare scenarios. And in the 1960s, the novelist Michael Crichton echoed Lederberg's own warnings about extraterrestrial diseases in a bestselling novel.

In *The Andromeda Strain*, a satellite is sent into the upper atmosphere to collect microbes that the US military can use as biological weapons. It collides with a meteor carrying germs, and when it subsequently crashes in Arizona, the microbes escape and kill off an entire town. What makes the deadly new disease so frightening is how it spreads. A scientist who examines the microbe finds that it can kill a rat without any direct contact. "Interesting, he thought. Airborne transmission," Crichton wrote.

Crichton never publicly admitted to borrowing the idea for *The Andromeda Strain* from Lederberg. But Lederberg could see some clues embedded in the book itself. One of the characters is a Nobel Prize–winning scientist from Stanford—where Lederberg had been teaching since 1958. When Universal Studios started work on a movie version, Lederberg wrote to the producers to ask that they make sure no one confused him with Crichton's character.

Lederberg didn't want Crichton's science fiction to stain his authority as a scientist. But he also recognized, as he once wrote to a colleague, that "you might find ways to use its dramatic impact." An imaginative story—like how airborne AIDS might destroy the world—could motivate the public in a way no monograph ever could.

In May 1989, Lederberg took his tale of airborne AIDS from New York to Washington, where he attended a meeting of two hundred leading scientists. They had gathered to discuss the threat of emerging viruses. In his lecture, Lederberg warned his audience that the survival of the human species was not guaranteed. And he once more brought up the possibility that HIV could mutate into an airborne form. If it did, he predicted, it might rival the Black Death in its destruction.

When Lederberg finished, a fifty-four-year-old biologist with a gray thatch of hair and thick-framed glasses stood up in the audience. Howard Temin had won a Nobel Prize of his own—for discovering one of the proteins that HIV uses to hijack our cells—and he was having none of it. "I think that we can very confidently say that this can't happen," Temin told Lederberg.

Temin argued that any mutations that would make HIV airborne would also tame it. To gain a new transmission route, the virus would be forced to lose its ability to crush the immune system. "So it would be worse in the sense that it might be more contagious, but it might just be another cold virus," Temin said.

"I don't share your confidence about what can and cannot happen," Lederberg shot back. After all, the lung was loaded with immune cells similar to the kind that HIV infected in the blood. A few mutations could let the virus shift to new cellular hosts.

The argument, one observer later said, "certainly induced a lot of adrenaline." In the end, Temin granted that he couldn't categorically rule out airborne AIDS. But the chances were virtually nil. "You don't have to stay up nights worrying about it," Temin told Lederberg.

"I'm glad I worry enough for both of us, Howard," Lederberg replied.

○ ○ ○ ○

The scientists gathered at the Washington meeting could take some comfort in Temin's argument that airborne AIDS might never evolve. But at the conference, they heard plenty of stories about real pathogens that were frightening in their own right. Influenza experts explained how pandemics had their evolutionary origins in ducks and other aquatic birds, which harbored a vast diversity of flu viruses. They replicated in the guts of the birds, which usually suffered only mild symptoms. But those mild strains could evolve into deadly ones, which could then evolve into viruses that could infect human airways. It was only a matter of time before a new bird flu spilled over, and then there was no telling how bad the next influenza pandemic would be.

Other scientists spoke of less familiar diseases that might someday produce huge epidemics. In 1976, for example, the residents of a remote village in Zaire had succumbed to a devastating fever. Some of its victims even bled from their eyes. Most of the infections quickly led to organ failure; out of 318 reported cases of the mysterious disease, 280 people died—an astonishingly high mortality rate of 88 percent. Blood drawn from the sick villagers was shipped to a lab in Belgium. Researchers isolated an unfamiliar virus with a sinister serpentine shape. It was named for the river that flowed near the village where it first struck: Ebola.

In the years that followed, the Ebola virus periodically flared across the central belt of Africa, from Gabon to Sudan. It appeared to spread through the fluids that victims unleashed—their vomit, feces, and blood. In a particularly cruel turn, family members who washed a victim's cadaver for burial put themselves at high risk of infection. Most victims, it turned out, did not suffer uncontrollable bleeding. And proper medical care dramatically lowered the mortality rate. But it was difficult to study

Ebola. Outbreaks would flare up in villages, claim a few dozen lives, and burn themselves out. The virus did not vanish, however; it hid in some animal—perhaps a bat—until a new series of events triggered a fresh outbreak.

"I can tell you we were holding our breaths there for a while," one expert later told Rick Weiss, a reporter at *Science News*. "We thought we were looking at the Andromeda Strain."

Ebola even drew the attention of the American military. The thousands of scientists still working at Fort Detrick on biodefense studied a catalog of pathogens that other countries might use as weapons against the United States. In 1979, the United States Army Medical Research Institute of Infectious Diseases (USAMRIID) added Ebola to the catalog.

To better understand the threat, researchers at Fort Detrick grew stocks of Ebola virus and injected them into guinea pigs. They began developing vaccines and tested antivirals. And they also created sprays of droplet nuclei laced with viruses. In this new airborne state, Ebola could infect monkeys that inhaled it.

Lederberg had something new to worry about. He did not trust that the Soviet Union had given up its germ warfare ambitions after signing the Biological Weapons Convention in 1975, and he worried that it would look at new pathogens as potential weapons. Lederberg's hunch about Ebola would later turn out to be right. In 1986, the Soviet leader Mikhail Gorbachev secretly signed a five-year plan to develop new airborne weapons from a list of the world's most dangerous pathogens. The list included Ebola.

Even if the Russians never dropped a single Ebola virus on the United States, Lederberg believed studying the disease was still an urgent priority. Under the right conditions, it might spread to cities and rage uncontrollably. Perhaps it might evolve to transmit more efficiently from one human to another. If HIV could become airborne, as Lederberg feared, perhaps Ebola could as well.

Lederberg went around Washington trying to drum up support for more biodefense research on pathogens like Ebola. He found it hard to get the Bush administration to take the threat seriously. He needed a way to get them to stay up at night with him. When a *New Yorker* writer named Richard Preston met with him to research a story on emerging viruses, Lederberg saw an opportunity. He told Preston about how Ebola had arrived in the United States.

In October 1989, a shipment of monkeys was delivered from the Philippines to the United States for lab research. The long-tailed macaques first went into quarantine at a facility in Reston, a suburban Virginia town. Two of the monkeys were already dead by the time they arrived. That wasn't very unusual, but what happened next was: over the next few weeks, more of the monkeys died. USAMRIID researchers inspected the animals and discovered they carried a previously unknown strain of Ebola virus. The scientists killed the remaining monkeys and decontaminated the facility.

None of the staff at Reston got sick, but the outbreak was deeply unsettling. The virus appeared to have spread between monkeys in separate cages. They were so far from each other that they could not have passed the virus along by direct contact, and it was hard to imagine short-range droplets from coughs spreading it. An awful possibility emerged: Ebola had taken to the air.

○ ○ ○ ○

As Preston set off to investigate the Reston outbreak, Lederberg went on lobbying for biodefense research. He warned that the Soviet Union was probably not alone in developing biological weapons. Intelligence analysts were starting to suspect other countries were building up their own arsenals. At the top of their list was Iraq.

Not long after Saddam Hussein seized power in 1979, he ordered the creation of programs to develop nuclear, chemical, and biological weapons. The biological weapons effort got underway around 1985: scientists secretly produced thousands of liters of anthrax spores, grew pathogens to kill wheat, and investigated other potential weapons, including smallpox. Saddam deployed chemical weapons during the Iran-Iraq War, and after the war he continued to develop his arsenal to use against other enemies.

American CIA agents managed to gather only a few hints of what Saddam was up to, leaving experts in the late 1980s to debate just how much of a threat Iraq posed. The United States had a complicated relationship with Iraq, supporting it against Iran while also warning it not to develop chemical, biological, and nuclear capabilities to use against Israel. Tensions rose drastically in early 1990, when Saddam began preparing an invasion of Kuwait, the tiny oil-rich American ally. Saddam personally authorized using biological weapons against the United States, as well as Saudi Arabia and Israel. In its massive buildup of forces along the Kuwait border, Iraq secretly deployed 166 bombs and twenty-five missiles with anthrax and nerve gas, ready to be unleashed on Saddam's order.

Lederberg did not know these details, but what little he gleaned from reports worried him enough. He lobbied officials to prepare for a biological weapons attack—not just in Kuwait, but on American soil too. The Bush administration responded only tepidly. It set up a few stockpiles of antibiotics around Washington and vaccinated 150,000 soldiers against anthrax.

When Saddam's forces invaded Kuwait in August 1990, those precautions turned out to be unnecessary. There was no attack on Washington as Operation Desert Storm swiftly pushed Saddam's army into retreat. As his troops fled Kuwait, they set fire to oil wells, which lofted toxic smoke into the air. But they put no anthrax in the air.

Lederberg did not sleep any more easily, however. Within months of the Kuwait war, the Soviet Union dissolved, and no one in the West could tell what had happened to its biological warfare program. It was conceivable that Soviet scientists were fanning out across the world, offering their services to the highest bidders.

When Bill Clinton was elected president in 1992, he proved more receptive to Lederberg's worries. Clinton appointed him to a White House panel that examined the threat of weapons of mass destruction. In the panel's assessment, germ warfare stood out. "In principle, biological weapons efficiently delivered under the right conditions against unprotected populations would, pound for pound of weapon, exceed the killing power of nuclear weapons," Lederberg and his colleagues warned in 1993.

The advisory panel picked out anthrax to drive their point home. The United States had plenty of experience in turning woolsorter's disease into a deadly weapon, and American agents suspected that the Russians had released anthrax at Sverdlovsk. On a clear, cool night, the panel concluded, a hundred kilograms of anthrax spores released into the Washington sky would form a cloud that might kill more than a million people. By laying out an aerobiological nightmare, they repeated the ritual that Alexander Langmuir had performed on television a generation earlier.

Over the course of the 1990s, the public was provided with a lot of fodder for such nightmares. Lederberg's tip to Preston about the Ebola scare in Reston had pupated into a sensational *New Yorker* story entitled "Crisis in the Hot Zone." Appearing in October 1992, it introduced many readers to the threat of emerging viruses and to the scientists who studied them in pressurized body suits. Preston made much of the possibility of Ebola going airborne. He quoted Stephen Morse, a virologist at Rockefeller University, musing about how HIV might fly. He interviewed Lieutenant Colonel Nancy Jaax, who had infected monkeys with Ebola in 1980 to test antiviral drugs. The monkeys she infected all died, but so did a pair of healthy monkeys she kept on the other side of the lab as controls.

"That was when I knew that Ebola could spread through the air," Jaax told Preston.

Preston then expanded his feature into a best-selling book. In *The Hot Zone*, published in 1994, he added fresh speculations about airborne viruses, raising the prospect of "a kind of airborne Ebola flu." He called it a thing so horrible that "you could no more imagine it than you could imagine a nuclear war."

The following year, a fictionalized form of Ebola became the star of the hit movie *Outbreak*. Loosely based on Preston's work, the movie opens by quoting Joshua Lederberg: "The single biggest threat to man's continued dominance on the planet is the virus." The movie follows a deadly virus as it spreads from Africa to California, mutating along its journey until it starts soaring from host to host—in one scene flying through the ventilation system of a movie theater and plunging into a victim's laughing mouth. "It's gone airborne!" Dustin Hoffman declares.

Lederberg could not have asked for a greater return on the investment of his tip. Washingtonians who read *The Hot Zone* or watched *Outbreak* could picture a horrific airborne virus wafting into their own offices. But the true nature of diseases like Ebola got lost in his lobbying.

There was no firm evidence that Ebola could spread among humans in droplet nuclei. The Reston strain, on which Preston had based his book, did not actually look like the kind of virus that could end civilization. When three lab workers later got infected with it, they experienced no symptoms at all. And in the full scope of global health, Ebola was not a major threat. By the 1990s, it had managed to kill only a few hundred people in total since its discovery. Meanwhile, tuberculosis—a disease that Wells had clearly shown was airborne—was killing more than 2 million people every year but attracting woefully little attention. Poor people dying slowly of a preventable disease didn't make for good ticket sales.

○ ○ ○ ○

The mid-1990s also saw a fresh surge of fear about biological weapons. In 1995, members of a death-obsessed cult in Japan called Aum Shinrikyo were arrested for pumping nerve gas into Japanese subways, killing thirteen people and injuring hundreds more. Police discovered that they had previously tried to rain death on the streets of Tokyo by releasing spores of anthrax from a rooftop. Fortunately, that effort had failed. To President Clinton, the Aum Shinrikyo attacks marked a new chapter in warfare. They demonstrated, he said, "the dangerous link between terrorism and weapons of mass destruction."

New details were also emerging at the time about how far the Soviet Union had gone in its biological weapons program before its fall. Matthew Meselson's team confirmed that an anthrax factory at Sverdlovsk had unleashed death in 1980. Ken Alibek, who came to the United States in 1992, began sharing even more stunning tales of the Soviet biowarfare program he had left behind. Reports also surfaced that Aum Shinrikyo had sent members to Soviet labs to learn about biological warfare. Clinton worried that other terrorists might get hold of pathogens.

Along with terrorists, Iraq also aroused Clinton's suspicions. He pushed for sanctions to force Saddam to come clean about his arsenal. Iraqi officials eventually admitted to deploying bombs and missiles in advance of the Kuwait war, but they insisted the weapons were never used. After the war, they said, Saddam ordered the arsenal destroyed so that no one would know it ever existed. Iraq's anthrax stocks were poured into the ground, and its main germ warfare lab was buried under a heap of earth.

Clinton was still not satisfied, because Saddam continued to resist inspections from the United Nations. He behaved as if he was still up to something. That uncertainty was intolerable, Clinton once told Yeltsin, because even a tiny supply of biological weapons "can do a lot of damage, as we saw in the Tokyo subway attack a couple of years ago."

The United States stepped up its pressure on Saddam to let United Nation inspectors into Iraq. To build public support for the effort, Clinton dispatched his secretary of defense, William Cohen, to appear on television. On the ABC News program *This Week* on November 16, 1997, Cohen held up a five-pound bag of Domino sugar. He warned that if the bag was full of anthrax spores, it would be enough to kill half of Washington. That was a hugely exaggerated number, but at a time when cults were spraying anthrax off rooftops, it seemed prudent to err on the side of hyperbole. The alarm at the White House continued to rise throughout 1998, thanks to a secret American intelligence report claiming that Iraq was probably concealing smallpox viruses for military use. Colonel David Franz, a Fort Detrick scientist, appeared on an hour-long prime-time special to warn that bioterrorism was no longer hypothetical. It was now practically inevitable. "The likelihood of there being an attack someplace in the United States is—is fairly high, probably in the next five years," Franz warned.

The year 1998 also saw the publication of Richard Preston's next book—this time, a novel. In *The Cobra Event*, Preston dramatized the link that government officials were making between emerging diseases, biological warfare, and terrorism. He imagined a villainous scientist inventing an airborne form of smallpox that attacks the brain. Clinton devoured the book and recommended it to his cabinet as well as to Newt Gingrich, the Republican Speaker of the House.

A few skeptics tried to tamp down the panic. The terrorism expert Bruce Hoffman pointed out that no one since World War II had managed to use biological weapons to kill on a massive scale. Aum Shinrikyo might have gained headlines for trying, but they had failed miserably in causing mass destruction.

"I don't know how many times I would go to talks and there would be a high-level government official who would say that the Aum attack demonstrated that 'we've crossed the threshold,' and also that the Aum attack demonstrated the ease of use," Hoffman later recalled. "But when you parsed such claims, this showed how difficult it really was. We were drawing all the wrong conclusions."

The skeptics failed to slow the spiral. As Clinton went on talking with experts like Lederberg, he grew more concerned. "Everything I heard confirmed that we were not prepared for bio-attacks," the president later wrote in his memoir. His administration launched a secret research program that pushed dangerously close to violating the 1972 Biological Weapons Convention. US scientists reconstructed a Soviet microbe bomb, created a vaccine-resistant anthrax strain based on Alibek's accounts, and even built a factory to make new spores.

Publicly, Clinton announced that the threat of bioterrorism required that public health and national defense blend together even more than they already were. In a May 1998 speech at the US Naval Academy, he announced a major new policy "to aid our preparedness against terrorism, and to help us cope with infectious diseases that arise in nature." Clinton requested $300 million in the 1999 budget for defense against biological and chemical terrorism.

Clinton followed up a few months later on January 22, 1999, with a speech at the National Academy of Sciences. There he announced even more funding for research on pathogens, vaccines, drugs, and diagnostics, along with "the first ever civilian stockpile of medicines to treat people exposed to biological and chemical hazards." At his side was Joshua Lederberg. In a decade, Lederberg had gone from theoretical debates about breathable AIDS to standing next to a president as he made a vast investment against aerobiological threats. In his speech, Clinton thanked Lederberg for bringing scientific respectability to the fear of biological terrorism "because I then had experts to cite for my concern and nobody thought I was just reading too many novels late at night."

The biodefense experts who received that money made sure that Clinton's team didn't lose sight of the danger of germ warfare. Instead of movies or novels, they created a game. The military had a long tradition of playing war games to prepare for future conflicts, and now the biodefense experts invited senators and cabinet secretaries to fight pathogens. In the first edition, played in 1999, the disease was anthrax. In the following year, the players faced an outbreak of pneumonic plague, the airborne form of *Yersinia pestis* that devastated Manchuria in 1910. The games always ended badly, as the players discovered how unprepared the United States was for an attack from the air.

○ ○ ○ ○

When George W. Bush was sworn in as president in January 2001, biodefense experts discovered that the new administration did not share Clinton's worries—at least not at first. Kenneth Bernard, Clinton's top biodefense advisor, prepared a transition memo only to discover that his job had been eliminated. "They just threw the memo away, and said, 'This is not a national security issue. This is one of those Clinton things,'" Bernard told NBC News.

To win over the Bush administration, the biodefense experts tried staging a new game in June 2001 with smallpox. In Dark Winter, as the exercise was called, the players were informed that Iraq was believed to have reconstituted its biological warfare program. A suspected al Qaeda lieutenant was caught in Russia shopping for weaponized pathogens along with plutonium.

The players then learned that a biological attack was underway in malls in Oklahoma City, Philadelphia, and Atlanta. The early reports suggested that someone had sprayed smallpox in the air. Three thousand people were infected, and in the days that followed, the first victims spread the disease to others. The players scrambled to administer vaccines to a country that had not seen a smallpox case for decades. They were too slow to act, and the attack killed a million Americans. Despite the hypothetical carnage, however, Dark Winter failed to impress the Bush administration. At the end of the summer of 2001, its senior members still showed little concern about airborne slaughter.

A few weeks later, the United States suffered its first large-scale surprise attack since Pearl Harbor. It did not arrive as a shower of smallpox viruses or a cloud of anthrax spores. It came in the form of hijacked planes.

○ ○ ○ ○

On September 11, 2001, two planes crashed into the World Trade Center. A third hit the Pentagon, while a fourth destined for the White House crashed in a Pennsylvania farm field. Nearly three thousand people were killed. Some government officials feared that the planes had been loaded with biological weapons, and so the air at the crash sites was sampled and sent to Fort Detrick to check for anthrax. The tests all came back negative.

Yet the fear of germ warfare endured. Perhaps 9/11 would be followed by an attack with biological weapons. The FBI discovered that the al Qaeda hijackers had taken flying lessons in Florida. They asked their teachers a lot of questions about crop dusters, raising fears they might spray pathogens over cities. Michael Osterholm, a University of Minnesota epidemiologist who had lobbied Congress to support Clinton's bioterror plan, told a reporter that hundreds of thousands of people might die in a coming attack. "Smallpox could bring us to the brink of a collapse of society as we know it," he warned.

Now the Bush administration took a sudden interest in the Dark Winter game. When an aide to Vice President Richard Cheney showed him a video of the exercise, he was struck by how smallpox viruses had the capacity to cause a nuclear-level disaster. Cheney brought the Dark Winter video to President Bush to watch. Bush was briefed on how a smallpox attack on New York could infect up to 3 million people. He and Cheney reportedly discussed carrying out a massive vaccination campaign to prepare for such an assault. And within a few weeks, it looked like the follow-up attack to 9/11 had begun.

It began on September 27 when a photo editor named Robert Stevens climbed into the back of his car. He and his wife were driving from their home in Florida to visit their daughter in North Carolina when Stevens was suddenly slammed by symptoms that felt like the flu. His wife begged to take him to a hospital, but he told her to just drive to their daughter's house. There he slept well through the night. The next day, he felt well enough to drive back home with his wife. But once they got back to Florida, he started vomiting. A spinal tap revealed Stevens was infected with *Bacillus anthracis*. He had woolsorter's disease, caused by inhaling airborne spores. Robert Stevens died on October 5.

Investigators discovered that he had breathed in spores released from a letter that came to his office. Tracing its journey to Stevens's desk, investigators discovered that spores had leaked out of the envelope along the way. They sickened some of his office mates, as well as postal workers. Soon more anthrax letters were arriving at other offices, including at NBC News in New York and at government buildings in Washington, DC. The United States Senate shut down for decontamination. "You can not stop us. We have this anthrax," one of the letters read. "Allah is great."

"I was struck by a sickening thought," Bush later recalled in his memoir. "Was this the second wave, a biological attack?"

On October 25, a Senate subcommittee met to consider the lessons Dark Winter had to offer as the country reeled from the anthrax letters. "It appears to have foreshadowed this event," Senator Mary Landrieu of Louisiana declared. "At that time, nobody imagined that we would be facing a biological weapon terrorist attack right here in the United States Senate and elsewhere in the Nation, using the US Postal Service as the means of delivery. Now all Americans understand how important it is to think through and plan ahead for such once unthinkable contingencies. We are living through one today."

Many officials suspected al Qaeda was to blame. Cheney aggressively pushed the link, tying the language in the letters to fragmentary intelligence reports about their training camps in Afghanistan. "We have copies of the manuals that they've actually used to train people with respect to how to deploy and use these kinds of substances," Cheney said in an October 12 television interview.

Meanwhile, the Bush administration gathered any story from intelligence sources—no matter how sketchy—that fit their belief that Saddam was preparing for germ warfare. Two Iraqi defectors claimed that Iraq had a fleet of mobile labs in which it could brew anthrax and other biological weapons of mass destruction. The Bush administration tied the stories about Iraq and the ones about al Qaeda into a feverish aerobiological plot. Saddam might have been providing anthrax to the terrorists, perhaps even the anthrax that was spreading through the United States mail. All the while, more spore-packed envelopes were arriving in American offices and homes. The string of attacks, which came to be known as Amerithrax, exposed twenty two people by the end of 2001. Five of them died.

The hunt for the Amerithrax killer dragged on for weeks, then months. All the while, the Bush administration began preparing for Dark Winter–style attacks that could endanger the entire nation. Kenneth Bernard found himself back in the White House as a bioterror advisor. Half a million troops were ordered to get a smallpox vaccine. Bush got one in December 2002 to promote the effort. Anthony Fauci, the HIV expert who had gone on to lead the National Institute of Allergy and Infectious Diseases, helped the White House put together a six-billion-dollar effort to expand Clinton's bioterror stockpile with vaccines and treatments for airborne terror threats such as anthrax, Ebola, plague, and smallpox. Fauci thought up a name for the program, inspired by Operation Desert Shield, the campaign that Bush's father had launched to defend Kuwait in 1990. Fauci called the new operation Project BioShield.

On February 3, 2003, Bush traveled to Bethesda, Maryland, to unveil Project BioShield at the National Institutes of Health. "America's war on terror has tested this nation, has tested our resolve, our will, our determination, and I'm confident that we can call upon our resources and strengths to prevail," Bush said.

Two days later, Secretary of State Colin Powell paid a visit to the United Nations to give a speech of his own. A well-respected general with a reputation as a reluctant warrior, Powell had a mission to bring the world together in opposition to Saddam Hussein. In his televised speech, Powell accused Iraq of hiding chemical and biological weapons from inspectors, of failing to account for what had happened to the germs they already admitted stockpiling in the 1990s. He showed off artists' renderings of mobile biological labs. And like William Cohen and Alexander Langmuir before him, Powell stoked fear of biological warfare with a televised prop. He raised a vial of white powder.

"Less than a teaspoon full of dry anthrax in an envelope shut down the United States Senate in the fall of 2001," Powell said. "This forced several hundred people to undergo emergency medical treatment and killed two postal workers just from an amount, just about this quantity that was inside of an envelope."

Saddam had never accounted for twenty-five thousand liters of anthrax that UN inspectors thought he had retained. If he still had them, Powell warned, he could make "tens upon tens of thousands of teaspoons." He could use his fleet of mobile biological labs to make many thousands more. "Leaving Saddam Hussein in possession of weapons of mass destruction for a few more months or years is not an option, not in a post–September 11th world," Powell declared.

The following month, the United States invaded Iraq. Once its army had routed Saddam, investigators spread across Iraq to search for the stockpiles. They found none. The mobile labs were a fiction. Saddam had actually destroyed his arsenal of biological weapons over a decade earlier. But he made a disastrous choice to leave the United States imagining that he still had it.

More than a half million people are estimated to have died over the course of the Iraq War. Some were killed by bullets and bombs, others by the collapse of Iraq's infrastructure. Not one died from an airborne biological weapon. A year after Powell gave his speech at the United Nations, he told a Senate hearing that the Bush administration was right to start a war based on their belief that Saddam had biological weapons. "We believed it because the intelligence information available to us said the stockpiles were there," he said.

The search for the Amerithrax killer proved almost as fruitless. When FBI researchers examined the spores sent through the mail, they saw signs that they had been carefully prepared in a lab so that they would readily fly into the air, allowing them to be inhaled. The DNA of the bacteria turned out to be similar from envelope to envelope, suggesting that they all came from a common source. But they did not descend from bacteria brewed in an al Qaeda camp in Afghanistan or a mobile lab rolling down an Iraqi highway. They bore the genetic signature of spores grown in one particular flask at Fort Detrick.

The FBI narrowed their investigation to an anthrax expert at Fort Detrick named Bruce Ivins. After the FBI informed him they would be pressing charges, Ivins died by suicide in July 2008. In later years, some outside scientists questioned the strength of the evidence against him.

But the US government decided against any further investigation of the Amerithrax murders.

That turn of events did nothing to slow down the biodefense spree, which continued to be justified by the imminent threat of airborne attacks. The Department of Homeland Security rushed to set up the country's first early warning system for biological attacks that expanded to six hundred sensors in thirty cities. The network was supposed to quickly detect clouds of anthrax or other pathogens floating on the wind. But this was not the realization of Alexander Langmuir's dream to protect Americans. It was a poorly designed project that lacked evidence that it could work reliably. Between 2003 and 2014, BioWatch detected a hundred forty-nine "actionable results," but not one of them turned out to be a biological attack, or even a natural danger. The Department of Homeland Security tried to improve BioWatch, but mainly just swelled its budget by billions of dollars. In 2024, more than twenty years after the project's launch, a bipartisan commission gave the program a blunt assessment: "Put simply, BioWatch is a waste of money."

The Georgetown University terrorism expert David Koplow bemoaned how the United States had become trapped in an expensive, endless nightmare. "The American people are no safer, and feel less secure, than we were before all the hysteria began," he declared. "Far too little has been done to address the genuine biological threats to Americans and to suffering people around the world—the quotidian scourges of AIDS, tuberculosis, malaria, measles, and cholera—that not just 'threaten' us in the abstract, but that actually kill and incapacitate millions of people annually."

By the start of the twenty-first century, aerobiology had drifted far from the hopes that its pioneers had had for it in the 1930s. Public health experts largely dismissed it, while others used it to stoke geopolitical nightmares. And yet the new century would bring glimmers of a revival to aerobiology. Some researchers rediscovered the Wellses as new diseases emerged. And more scientists began to follow Fred Meier's example. Instead of treating aerobiology as a way to rain down death, they looked up: up at the sky and the life it held, up to the clouds and beyond.

PART 4

Resurrection

thirteen

SEA, LAND, FIRE, CLOUDS

In the final days of the twentieth century, a pair of aerobiologists gave their field a harsh report card. Writing in the December 1999 issue of the journal *Aerobiologia*, Paul Comtois and Scott Isard bemoaned "the low visibility of aerobiology as a complete science by itself."

Aerobiologists had only themselves to blame. Their field was now over sixty years old, and yet it remained scattered and disorganized. Aerobiologists had gotten very good at collecting this kind of pollen or that kind of fungal spore. But in 1999, Comtois and Isard complained, they were still not grappling seriously with deep questions about the planetwide movement of life through the atmosphere. They were stuck, the scientists said, on a "stagnant plateau."

But Comtois and Isard believed it was possible to kick-start aerobiology again. "We have the tools to forecast the flows of biota over the entire Earth," they wrote. It was just a matter of using those tools. If aerobiologists learned how to harness the new satellites entering orbit or the computer networks getting wired across the planet, "aerobiology will finally be coming of age in this next century."

Comtois and Isard declared it was high time for the new age to begin: "We believe that it is important to refocus aerobiology research NOW."

Aerobiology did indeed get refocused at the start of the twenty-first century, and it started growing up at last. For one thing, aerobiologists got braver. They launched more ambitious expeditions, and they learned how to collaborate in bigger teams to merge their data from across wider stretches of the sky. And they seized new kinds of technology and adapted them for the air.

When Fred Meier gathered organisms in the stratosphere in the 1930s, he didn't have many ways to figure out what he had caught. He could try to identify them by shape under a microscope. If he was lucky, he could figure out the nutrients they needed to grow in a dish. Generations of aerobiologists worked under the same constraints. But in the 1990s, geneticists pioneered a new way to identify living things. They created tools to tear open cells, pull out their genes, and read their sequences.

Now it became possible to recognize a baobab or a green mamba by its unique genetic signature. It also became possible to scoop up a spoon of soil and take a complete census of all the life-forms it contained. Scientists might find the DNA of animals like ants or nematodes or of plants like dandelions or sphagnum moss. They also identified spores of fungi, amoebae, bacteria, and viruses. DNA revealed that the diversity of microbes was overwhelming: that spoon of soil might contain thousands of species, most new to science. Looking for microbes with nothing but microscopes and dishes full of food had left the vast majority of species unseen. Now their genes could testify to their existence.

Before long, some scientists started using DNA as a way to track airborne life too. In the 1990s, a team of scientists from the Australian National University climbed Mount Melbourne, a nine-thousand-foot volcano in Antarctica. As they reached the peak, the snow and ice there was melted away thanks to the subterranean heat. The warm, wet ground served as a refuge where patches of moss could grow. The researchers clipped a few shoots and brought them back to the university. On a separate expedition, they climbed Mount Erebus, more than two hundred miles to the south, and found more patches of moss. The scientists then examined bits of DNA from the moss from the two mountains. In 2001, they reported that the plants were close cousins. It's likely that a single moss plant, perhaps from as far away as New Zealand, released a cloud of spores. Most died on the unforgiving Antarctic landscape. But the ones that landed on the two volcanoes found lucky warm spots where they could grow.

Soon scientists figured out how to read the DNA from microbes snatched from the air itself. In 2000, a team of scientists in Salt Lake

City, Utah, pumped 370,000 gallons of air through a modern version of an aeroscope. When they pulled the DNA out of the cells trapped in the device, they identified more than three hundred distinct genetic sequences, each possibly coming from a different species of bacteria. The air, it turned out, was a zoo.

○ ○ ○ ○

With tools like DNA sequencing, aerobiologists began getting a much better idea of what kinds of life float in the air and where they come from. A substantial fraction of organisms rise from the oceans that cover 70 percent of the planet. In fact, it's from the ocean that life probably first rose to the air billions of years ago.

To learn how the sea sends organisms into the sky, I paid a visit to the Pacific Coast north of San Diego. There I met two scientists, Kim Prather and Grant Deane, on a hill that sloped steeply down toward the beach. We walked onto the Ellen Browning Scripps Memorial Pier, a concrete span that extends a thousand feet over the steel-blue ocean. It looked like a transpacific bridge from California to Taiwan abandoned just after construction had started.

As we walked down the pier, huge waves crashed below us, some carrying surfers toward land. We made our way past parked forklifts and powerboats on trailers. Gray metal boxes the size of elephants were loaded with instruments. At the end of the pier, anemometers twirled in the winds coming off the Laguna Mountains. A hose sucked in air and pumped it through a device to measure the level of carbon dioxide. Each day the device added a new data point to a record that had started in 1957. Nowhere on Earth has a better record of how we have brought global warming upon ourselves. Pipes ran the length of the pier, delivering endless rivers of seawater inland to the Scripps Institution of Oceanography. Prather and Deane got some of that flow; their portion traveled to the Hydraulics Laboratory, a barn of a building in which they had built a giant wave machine. The water filled a hundred-foot-long channel with clear plastic walls. A giant paddle kicked up waves a few feet high. They rippled and crested while fans blasted winds across their surface.

On my visit, we decided to first look at real waves instead. Prather and Deane had spent a lot of their lives out on the open ocean, which you could see on their sea-blown faces. On the day of my visit, the sun shone down through high clouds. Prather and Deane were happy to spend a morning looking at the ocean booming below us. Long strips of heaving water rolled toward the shore. As they crested, Prather and Deane pointed them out to each other the way ornithologists point out vireos.

"Look at this foam!" Prather shouted to Deane. "The spray is just huge today. It's really putting on a show."

For a while we watched one big wave take shape. The winds coming over the cliffs blew against its crest and peeled off a shimmering spray. As the wave started to curl, Deane had me look not at the water but at the space just below. "It encloses a tube of air," he said.

As the wave crashed, it pushed that tube of air into the ocean below, turning it into a fizz of bubbles. "You know when you look at a stream of water from a tap, how it breaks into droplets?" Deane asked. "There's a similar kind of process that occurs in that tube, which breaks into bubbles."

The bubbles plunged down as much as several feet underwater. As the wave traveled on, the bubbles started to rise back up to the surface. As they reached the air, they gave the water the look of marine champagne. The bubbles wandered around the top of the ocean and found one another, joining together into filigrees of foam.

"That foam is a sign of life," Prather said.

The foam was laced with biological molecules—oily lipids, proteins, fragments of DNA—and it also harbored living things such as algae, fungi, bacteria, and viruses. As bubbles in the foam popped, they lofted some of the life into the air. Some was rising over the pier, headed for the clouds. Life had been forming that ocean foam and taking flight ever since microbes established themselves in the ocean more than 3.5 billion years ago.

"They've always been traveling around," Prather said.

To study those travels, Prather has flown in planes to capture droplets of seawater and the cells they contain. Deane goes on sea voyages to bottle the spray from waves crashing hundreds of miles from shore. "I've been out there in storms, strapped to the side of a ship, chasing bubbles," he told me. And the two scientists have also studied the process in their barn, where they face fewer risks to their lives and enjoy more control over their waves. They've used their wave machine to reconstruct the moment that marine life takes flight. Cameras and an assortment of devices precisely recorded the spray that came off the breaking surf. Through ports studding the length of the channel, the scientists drew off air and isolated floating droplets for closer analysis.

In one experiment, they poured glowing green bacteria into the water. They stirred the water into crashing waves, which churned up bubbles that then broke apart into droplets. Afterward, the scientists sucked some of the air out of the wave machine. When they inspected the droplets one by one, they could see that some of them glowed green: a sign that the marine champagne can loft living things into the atmosphere.

The steps by which the ocean gives life to air have become clearer thanks to studies like this one. As waves crash, they drive bubbles down through the ocean's richest zone of life. The top few feet of the sea are bathed in light from the sun, so they are home to vast numbers of photosynthetic bacteria and algae that serve in turn as food for grazing organisms that can also consume the oxygen in the surface water. Those grazers are in turn prey for predators, and predator and prey alike are hosts for deadly viruses. When infected cells die, they rip open and dump their organic matter for scavengers to slurp.

As an air bubble plunges into that zone, it exerts a magnetlike pull on the biological material floating around it. By the time a bubble rises back to the surface of the ocean, its walls are laden with organic molecules. Some bubbles carry entire cells in their walls.

Once bubbles rise from the depths, they sit like domes on the ocean's surface, sometimes for hours. As they drift together, they may form a slab of foam that rides a wave onto the beach. Most bubbles don't make it to shore, though. The walls of their domes erode until holes pierce through, breaking them apart into tiny droplets that blow away in the wind. All that remains of the broken bubble is its base, which looks like a miniature crater on the ocean. The crater's rim flows down, creating a miniature circular wave. When the wave converges on the lowest point in the crater, it crashes in on itself, producing a jet of larger droplets that leap into the air.

The droplets released by bursting bubbles contain not just water but minerals, gases, and a lot of biological matter. Mucus-like molecules organize themselves into a shell that encases the airborne droplets, squeezing them into odd beanlike shapes. Smaller droplets often carry only fragments of cells, but the bigger ones can accommodate entire organisms. They are the cargo planes of the aerobiome.

As Prather and Deane have worked out the steps by which ocean life gets into the air, other researchers have made estimates of how many cells actually make the journey. In 2010, the research ship R/V *Hespérides* took a voyage around the world packed with equipment for cataloging the oceans' diversity, from their dark depths to the atmosphere overhead. At 118 sites across the planet, samplers on the ship's top deck sucked in air. Each cubic meter contained thousands of floating cells. Based on the cells they collected, the *Hespérides* scientists estimated that more than a billion trillion organisms rise from the world's oceans each year.

Some of those organisms fly only for seconds, others for minutes. But about 10 percent of marine microbes are still airborne after four days. That's enough time to cross the entire Atlantic Ocean by air—a journey that can take years for microbes to make on ocean currents. Many of the organisms that make long-distance journeys through the air die along the way. But with such a vast flock taking flight, the aerobiome is able to seed distant waters, scrambling the diversity of marine life from ocean to ocean.

○ ○ ○ ○

The oceans may hurl huge numbers of microbes into the air, but it's the land from which most of the aerobiome rises. Even the harshest regions on Earth, like the Dry Valleys of Antarctica or the Atacama Desert of Chile, are coated with living films of microbes. These so-called biological soil crusts cover about 12 percent of dry land—an area roughly equivalent to South America. They endure thanks in part to pigments that the microbes evolved to shield themselves from the sun's ultraviolet rays.

The microbes squirt out a rubbery goo that binds them together in a sponge-like carpet that absorbs what little water falls on them. When winds blow over a biological soil crust, they can peel off bits and hurl them for miles. As the air heats up over land each day, it can produce violent updrafts that can rip up even bigger chunks, like the paper meteorites that Christian Ehrenberg first recognized as alive in the 1840s. Immense disturbances in the atmosphere produce dust storms that lift billions of tons of dust into the air each year. Each grain of dust may ferry a huge number of microbes into the sky. In 2018, German scientists inspected individual grains of sand they had picked up from a beach. A single grain could harbor 100,000 cells belonging to 6,000 different species.

Biological soil crusts might have first spread onto land 3.2 billion years ago, judging from the oldest terrestrial fossils. The first microbial mats likely grew along a braidplain of streams flowing across an arid landscape. Those streams would have periodically dried out, so the microbes would have needed to evolve adaptations to survive the direct exposure to the air. Their adaptations probably also helped the microbes survive when winds scoured them from the land and carried them miles into the sky. The winds might have helped spread biological soil crusts across entire continents, dropping them in new habitats. As life flew through the air, in other words, the land came to life.

Most of the biological soil crusts gave way to plants, which evolved from algae and spread onto land about half a billion years ago. Mosslike forms evolved first, followed by plants with roots, stems, leaves, and seeds. By about 380 million years ago, they grew into the first forests. Fungi moved onshore with plants. Some fed on dead plants in the soil, and others formed partnerships with living ones. The fungi broke down minerals that they delivered to plant roots; in exchange, the plants provided the fungi with carbohydrates. Both plants and fungi became staggeringly successful, and a good part of that success was due to their ability to use the air to their advantage.

Mosses, for example, grow a stalk with a pouch of spores at the tip, which they release like puffs of smoke into the air. As many as six million moss spores may fall on a single square meter of bog over the course of one summer. Some spores that get swept up by strong winds may travel huge distances—far enough to grow on top of volcanoes in Antarctica.

Many trees and flowers also use the air for sex, releasing pollen grains to fertilize the seeds of other plants. To increase their odds of finding another member of their species, plants can release vast amounts of pollen into the wind. A single white mulberry tree can release 486 billion pollen grains each spring.

But plants don't rely on numbers alone for their reproductive success in the air. They are sensitive to the weather, typically releasing pollen grains when the air is most likely to be moving fast enough to carry the pollen away. Some plants even speed up their pollen grains as they release them. Lamb's-tongue, a common weed, grows tiny flowers that resonate in even the faintest breeze. As they flail back and forth, the flowers shake pollen into the air.

As plants came to dominate the land, the fungi that depended on them thrived as well. Underground, fungal threads grew into networks that stretched for miles below forests. They supplied trees with nutrients and even began to shuttle carbon from one tree to another. Other fungi found success in attacking plants. Rust drifted through ancient forests, landing on trees and killing off their leaves and branches.

Fungi became especially adept at using the air to spread. Some fungi pack spores into little tufts called conidia, which blow away to produce clones of themselves. Other fungi, such as honey mushrooms, wait to have sex before making their spores. Fungal sex occurs when the threads from two organisms cross paths. If they belong to different mating types, they may fuse their cells to create a joint offspring. Those cells may then multiply into a structure that rises toward the surface, pushes through the dirt and the leaf litter, and unfurls itself into a mushroom. The gills of the mushrooms are lined with spores that can get carried away by currents of air.

Many species of fungi have evolved adaptations to improve the odds that the spores they release will blow far away. When rust grows on barberry plants, its spores develop inside rigid tubes. Water seeping into the tubes causes the spores to swell until the pressure gets so great that they pop out of the tubes like cannonballs. Other fungi shoot up a wave of spores with so much force that they pull up a wake of air. They then release a second wave of spores that drafts behind the first, rising even higher.

○ ○ ○ ○

The aerobiome is largely made up of species that always remain tiny, like bacteria and algae. It also includes species that shrink from big to tiny over the course of their life cycles. A loblolly pine can grow to a height of one hundred fifty feet and weigh more than eight tons. In that form, it cannot fly. But the tree can squeeze its genetic legacy into a pollen grain light enough to float for twenty-five miles before delivering its genes to another tree.

Animals are another matter altogether. They don't float. There are scattered reports of animals getting swept up by storms, but they never end well. Around the year 200 AD, for instance, the Greek writer Athenaeus recorded a rain of frogs that came down in what is now the Balkans. "There were so many of them that they filled the houses and the streets," he wrote. "For the first few days the people coped by killing them and keeping the doors of their houses shut. But this got them nowhere: their pots and jars were full of frogs; they found frogs being stewed and roasted along with their food; on top of this, it became impossible to drink the water, or even to put their feet on the ground, because of the heaps of frogs; and since the stench of the dead ones was making them miserable, they abandoned the place."

Some animals, however, can use their muscles to fight gravity and stay aloft. The ancestors of insects came on land about 425 million years ago, and about 100 million years after that, some of them took flight. Some species zipped from tree to tree, driving their mouthparts into the soft tissues of the plants to drink their fluid. Dragonflies darted over open spaces, looking for other insects to kill. Only much later did the flying insects we are most familiar with—such as bees, mosquitoes, and butterflies—evolve.

It wasn't until the late 1900s that scientists began to appreciate that flying insects can migrate extraordinary distances. At first, scientists could study the journeys of insects only by noting how the same species showed up in different places. A few followed Fred Meier's example and caught insects from the open cockpits of planes. Then, in World War II, the invention of radar made it possible to bounce radio waves off flying swarms. One night in June 2019, meteorologists spotted a vast cloud moving across California but saw no sign of rain. They called an amateur weather spotter directly under the cloud who told them he was surrounded by falling ladybugs. The swarm spread over eighty miles, with a dense core ten miles wide floating five thousand feet to nine thousand feet in the air. Despite the swarm's vast size, the meteorologists lost track of it, and the ladybugs disappeared into the night.

Some scientists have started making systematic surveys of insects across large areas. In the 1980s, researchers established radar stations along a four-hundred-mile stretch of farmland in eastern China to track insects, and in more recent decades, they upgraded the equipment to run automatically. In 2024, the researchers reported that 9.3 trillion insects passed over the region every year—fifteen thousand tons of airborne biomass.

Vertebrates took longer to take to the air. The first to fly were bristle-covered reptiles called pterosaurs. They became extinct 66 million years ago, but by then birds were flying as well. Today ten thousand species spend much of their life in the air on feathered wings. Bats evolved about 50 million years ago, using their leathery membranes to stay aloft. Today, with fourteen hundred species, bats are the most diverse kind of mammal on Earth. Both bats and birds also help other species fly. After feeding on fruit, they may travel for miles before releasing the seeds in their droppings. Birds flying thousands of miles to isolated islands may deliver spores stuck to their feet. And migratory waterfowl have spread influenza viruses across the planet.

○ ○ ○ ○

Along with bogs and bumblebees, along with grasslands and flocks of geese, life on land gave rise to fire. Plants released oxygen into the atmosphere, altering its chemistry so that a lightning bolt could turn them to flame. Ancient charcoal suggests that fires started burning by 430 million years ago. As plants became more successful, they created more fuel.

Today, roughly 1.5 million square miles of land burn each year—an area bigger than India.

In 2015, Leda Kobziar was watching the smoke rise from a burning pine forest in Florida when a question came to her mind: was anything alive in the plume? At the time, she was a professor at the University of Florida, studying how forests are rejuvenated by fires. She knew that the soil of forests is dense with microbes. The leaves and bark of trees are also coated in an invisible skin of bacteria and fungi that promotes the health of the plants. Kobziar wondered if fires could loft those organisms into the air. She decided to play the part of Fred Meier but sample smoke instead of clouds.

The University of Florida regularly set small fires on forest land it owned to manage the trees and to allow scientists like Kobziar to study them. In advance of a burn, she and a postdoctoral researcher stuck Petri dishes to the ends of painter's poles. When the fire started and flames rose six feet from the ground, the scientists held the dishes as close as they could to the blaze. They brought the dishes back to Kobziar's lab, where a remarkable burst of microbes grew on them. Kobziar found that the air near the fire had more microbes, and more species of them, than ordinary air.

Kobziar was so puzzled by that fiery life that she turned all her attention to it when she moved to the University of Idaho. Out west, managed burns were far bigger than those in Florida. In June 2019, for example, teams of Forest Service workers walked with fire throwers through stands of Engelmann spruce and subalpine fir in Fishlake National Forest in Utah, setting the trees ablaze. Helicopters thumped overhead, shooting flames of their own at the tops of the trees. As the Fishlake fire grew, its intense heat generated winds as fast as thirty miles an hour that lofted smoke high into the sky.

To study such a vast burn, Kobziar couldn't hold out a painter's pole. She deployed a squad of fireproofed drones instead. Guided by a team of pilots standing at the edge of the fire, the drones flew over the inferno, gulping smoke along the way. After a few flights through the blaze, the drones landed so that Kobziar and her team could pull filters out of them. They put the microbes in a sterile Tupperware box with gloves attached to holes on the side. Putting their hands into the gloves, the scientists shucked cells and DNA off the filters and into vials.

Despite the fire's scorching heat, Kobziar and her colleagues found 133 living species of fungi floating in the Fishlake smoke. The bacteria totaled more than a thousand species. Their genetic signatures told Kobziar where some of them had come from. The violent updrafts stripped the trees of their microbial coats and pulled other organisms out of the ground. Some of the cells probably died as they drifted into the tongues of fire. But many of them escaped destruction, catapulting quickly above the flames and out of danger.

Kobziar found many more species above the Fishlake fire than occur in ordinary forest air. And when she counted up the cells, she found four times more of them in the plume. By her calculations, a square meter of burning forest liberates 170,000 cells every second. If Kobziar's numbers are correct for other fires around the world, then flames may be responsible for a huge fraction of the entire aerobiome.

○ ○ ○ ○

All told, from fires and dust storms, from crashing ocean waves and hurricanes, a trillion trillion bacteria cells are emitted into the air each year—a mass of more than 100,000 tons. Close to the same number of fungal spores are released as well, and with their bigger size, they weigh in at about 50 million tons. What comes up sooner or later comes down. The aerobiome is a peculiar realm: an ecosystem of visitors. A flea may hop into the air for a second, a diatom may be carried by the wind for days before falling back into an ocean, and a common swift can fly for ten months before landing to build a nest. But sooner or later, they all return to Earth. By one estimate, a single square meter of ground may be pelted with 100 million bacteria during every hour of a rainstorm.

While the aerobiome is transient, scientists now recognize it as a distinct zone of life, one that follows its own ecological rules and that exerts a powerful influence on the planet below. One of the best places to appreciate its power is a mountain in France called Puy de Dôme. Since 2003, a scientist named Pierre Amato and his colleagues have regularly climbed to its peak to study the life that floats through the air.

Puy de Dôme formed about eleven thousand years ago when a fist of magma punched up into the rolling hills of central France. It created a volcano that spilled out lava before going dormant. As a graduate student at the nearby University of Clermont Auvergne, Amato would visit the mountain when clouds hid its peak. He would reach the summit and then climb onto the roof of a weather station. Surrounded by a blank white glare, Amato would open sterile metal tubes to capture the mist. He would then ferry the cloud water back down the mountain to the university.

When Amato analyzed the droplets, he discovered that they contained microbes. After finishing his degree, Amato became a staff scientist at the university and began bringing students of his own to the mountaintop. Over the years, Amato got better at counting microbes. He learned how to spot the proteins each cell was making in the air. He began to sequence the DNA of the organisms to determine which species they belonged to.

Clouds, Amato demonstrated, are alive. Every teaspoon of mist floating over Puy de Dôme contains several thousand microbes. While many are dead airborne husks, some are still alive. They make new proteins and destroy old ones. They grow in the clouds and even divide in two. Their DNA has revealed that some belong to familiar species, but many are new to science. Scientists who use DNA to identify species have to stay perpetually anxious about contamination, and Amato is no exception. A hawk soaring over Puy de Dôme might fly over Amato's tubes and shake microbes off its feathers. In Amato's lab, a graduate student may exhale germs into a test tube. Over the years, Amato has rejected thousands of potential species, suspicious that he or his students smeared skin microbes onto the equipment. Yet a single cloud, by his estimation, can contain thousands of species.

Amato and other scientists who study clouds suspect that they may be particularly good places for bacteria to survive—at least for some species. "Clouds are environments open to all, but where only some can thrive," Amato and a team of colleagues wrote in 2017. The successful species are the ones that can thrive on the food in the atmosphere. Every organism on Earth has to make the same kind of fuel: a molecule called adenosine triphosphate, or ATP for short. As organisms ascend thousands of feet, they use up the fuel they produced before the flight when they were sitting happily on a pine needle or immersed in a bog. But Amato has found that bacteria in clouds have a healthy supply of ATP—one that they must be making as they float. Cloud droplets contain many different molecules, and some microbes are able to use them to grow. In other words, bacteria eat clouds. Worldwide, by one estimate, cloud microbes break down a million tons of organic carbon every year.

Findings such as these suggest that the aerobiome is a force to be reckoned with—one that exerts a powerful influence on the chemistry of the atmosphere. The aerobiome even alters the weather.

As a cloud forms, it creates updrafts that lift water-laden air to high altitudes that are cold enough to turn the water to ice. The ice then falls back down. If the air near the ground is cold, it may land as snow. If it is warm, it turns to rain. But it can be surprisingly hard for ice to form in a frigid cloud. Even at temperatures far below the freezing point, water molecules can remain liquid. A seed of impurity is required. As water molecules stick to its surface, they bond to one another. Other water molecules then lock onto them and assemble into a crystal structure. Scientists have found that fungi, algae, pollen, lichens, insects, bacteria, and viruses are especially good at encouraging water to freeze. The life that floats in clouds seeds much of the rain and snow that falls back to Earth.

It's possible that clouds and life are linked in an intimate cycle. It turns out that one of the best rainmakers is a type of bacteria called *Pseudomonas*. Scientists are not sure why those bacteria in particular are so good at forming ice in clouds, but it could have to do with the way they grow on leaves. When cold rain falls on a leaf, *Pseudomonas* may help the liquid water turn to ice. As the ice cracks open the leaves, the bacteria can feast on the nutrients inside.

Some scientists have even speculated that plants welcome bacteria like *Pseudomonas*, despite the damage they cause. As the wind blows the bacteria off the plants and lofts them into the air, they rise into clouds overhead. Clouds seeded with *Pseudomonas* pour down more rain on the plants below. The plants use the water to grow more leaves, and the leaves support more bacteria, which rise into the sky and spur clouds to rain down even more water to nurture life below.

Research on the life in clouds also raises the possibility that airborne organisms might exist on other planets—even ones that might seem the worst places for life to survive. Venus, for example, has a surface temperature hot enough to melt lead. But the clouds that blanket Venus are much cooler, and at an altitude of thirty miles, they have the same temperature and pressure as clouds on Earth. Sara Seager, an astrobiologist at MIT, has speculated that life might have arisen on the surface of Venus early in its history, when it was cooler and wetter. As the planet heated up, some microbes found refuge in the clouds. Instead of sinking back to the surface, they might have bobbed up and down in the atmosphere, riding currents for millions of years.

Above the Earth's clouds, the aerobiome ebbs into the unknown. In the decades since Fred Meier captured stratospheric life, a few teams of scientists have searched at even higher altitudes. In 1974, Soviet scientists fired rockets from the steppes of Kazakhstan into the upper atmosphere. The rockets broke apart to release sterile microbe traps. The traps parachuted down to Earth, and inside them the scientists found bacteria and fungi. One of their four rockets found life forty-eight miles above the planet, more than three times higher than *Explorer II* reached.

Aerobiologists today are leery about embracing that record. The Soviet rockets actually went beyond the stratosphere to another layer, called the mesosphere, where meteors falling toward Earth burn into shooting stars. It hardly seems like a place where life could endure. But more recent studies have confirmed Meier's finding that the stratosphere is alive. Over the years, NASA has launched a number of balloons that have found life as high as twenty-five miles.

Even if microbes don't ascend to the mesosphere, getting twenty-five miles into the air is an impressive feat. For one thing, the physics of the atmosphere makes it hard for a microbe to reach that height. The wind that stirs microbes up from land and sea is almost entirely confined to the troposphere, the lowest layer of the atmosphere.

It's possible that massive thunderheads can ram their way above the troposphere and spray microbe-loaded droplets into the stratosphere above. And some microbes may get catapulted there by volcanoes. A NASA research jet once captured microbes as it screamed across the lower stratosphere. Among the organisms the jet caught, microbiologists identified a species called *Bacillus luciferensis*, which gets its diabolical name not from the devil but from Lucifer Hill, a very active South Atlantic volcano on Candlemas Island. The bacteria that NASA trapped in the stratosphere were 99 percent genetically identical to ones growing on top of Lucifer Hill, which were first found by a team of researchers with the British Antarctic Survey in 1996.

However microbes get to the stratosphere, they end up in what may be the most ruthless environment on Earth. Gases become wispy, water practically nonexistent. In the stratosphere, microbes can be ravaged by ultraviolet light, fast-moving subatomic particles blasted out from the sun, and cosmic rays streaming in from other parts of the galaxy. The collisions can destroy genes and proteins alike. It's possible that the microbes that manage to reach the stratosphere are equipped with proteins that repair radiation damage. They may also survive by hiding on the shady side of dust grains. And they may then return to Earth, to the land or the sea, to continue to multiply and create more life that may have a chance to rise back up into the air. Whatever their secret, those stratospheric voyages mark the outer limits of the aerobiome—and thus of life as we know it.

fourteen

WE'RE ALL GOING TO GET IT

Aerobiology's revival at the start of the twenty-first century came about not only because scientists could start grabbing genes from the air. A frightening new disease also emerged, the kind of threat that Joshua Lederberg and Stephen Morse had been warning about for years. And when researchers tried to understand how it spread so quickly, some of them rediscovered the lost work of William and Mildred Wells.

The new pandemic began with a cough. In November 2002, in the province of Guangdong in southeastern China, a middle-aged businessman began to feel ill. He coughed, struggled to breathe, and spiked a fever. When he went to a city hospital, doctors there could find no sign of a familiar respiratory disease. Epidemiologists would later describe the man's condition as "atypical pneumonia." He passed on whatever germ he harbored to five members of his family. They suffered similar symptoms, but eventually all six recovered.

The name of the Guangdong businessman has slipped out of recorded history. In late 2002, no one considered his illness worthy of note. But whatever made him sick did not vanish: a few weeks later in another part of Guangdong, a thirty-six-year-old cook named Huang Xingchu became ill with the same symptoms. He stayed with his parents in the city of Heyuan to recover, but his lungs became only more clogged. When his parents took him to a city hospital, the doctors there wondered how a young, healthy patient could have been so ravaged by pneumonia. The disease seemed like it might have been caused by some kind of bacteria, but antibiotics failed to clear it up. Huang's pneumonia was certainly atypical.

Huang recovered, but only after spending three weeks on a ventilator and infecting seven hospital workers. Word of the outbreak spread across Heyuan, causing parents to pull their children from schools and customers to clear store shelves of antibiotics. To calm the public, Heyuan authorities stressed that no one had died from the atypical pneumonia and claimed that it was not spreading. In early January 2003, a team of provincial epidemiologists inspected Heyuan and pronounced the city safe.

Despite the soothing words, the atypical pneumonia was indeed spreading. Soon it reached Zhongshan, another city in Guangdong. Twenty-eight people there fell ill, including thirteen medical workers. The Zhongshan outbreak finally prompted the provincial authorities to contact Beijing. A team of experts carried out an in-depth study of the Zhongshan outbreak and released a report on January 23, 2003. They concluded that the atypical pneumonia had likely been caused by a new virus. They suggested giving corticosteroids to patients to ease the inflammation in their lungs. To block the virus's spread, they recommended disinfection, handwashing, ventilation, and masks.

The report was sent out to Guangdong's hospitals, but it went largely ignored in the frantic preparations for the Lunar New Year. To celebrate the holiday, people traveled long distances to family gatherings. They brought the virus with them, spreading it to new places, including Guangzhou, the biggest city in the province. And yet, despite the growing threat of the new pneumonia, the province's schools and businesses stayed open. Over the previous decade, Guangdong had become an industrial powerhouse, packed with factories that supplied the West with cheap goods. Closing down the province to stop the virus would have meant missing production targets.

On February 5, 2003, when Colin Powell was holding up a vial of powder at the United Nations, Guangdong's hospitals were filling with patients. Masked police mysteriously appeared at the Guangzhou train station, breaking up crowds without explanation. The government throttled news of the outbreak, but they couldn't stop rumors from circulating by text. When the World Health Organization learned about the digital chatter, Chinese officials assured them that Guangdong had experienced a local outbreak that had burned itself out. But in fact, the province was seeing dozens of new cases every day.

Guangdong's doctors and public health authorities had to contend with the outbreak largely on their own. By the end of February, they were starting to rein it in. But the government's continued secrecy meant that their hard-earned knowledge of the disease did not reach other parts of China, even as the virus did. The growing confusion was compounded on February 20, when the Chinese Ministry of Health mistakenly announced that the Guangdong outbreak had been caused not by a virus, but by the bacteria *Chlamydia*.

Meanwhile the atypical pneumonia moved beyond China. A single person may have ferried it across the border and seeded a global outbreak.

Liu Jianlun worked as a nephrologist at a Guangzhou hospital that had experienced an outbreak. When he came down with flu-like symptoms, he took antibiotics and soon felt well enough to attend his nephew's wedding in Hong Kong. He boarded a bus with his wife and traveled to the Hotel Metropole. That night Liu got ill, and the next day he was admitted to Kwong Wah Hospital. He died there on March 4.

During his short time at the Metropole, Liu infected seventeen other people on the ninth floor. Later, the World Health Organization would speculate that he might have vomited or coughed by the elevator, passing on the infection to the others. Those newly infected people left Hong Kong for their homes in Singapore, Vietnam, and Canada. Ten days after Liu died, the World Health Organization issued a global alert that hundreds of people were infected, and six were dead. They also gave the disease a name: severe acute respiratory syndrome, or SARS for short.

By then, scientists at the University of Hong Kong were closing in on the cause of SARS. They ruled out *Chlamydia*, favoring some type of virus instead. One prime suspect was influenza. In the early 2000s, virologists were worried about a lethal new strain of influenza known as H5N1. It spread quickly among birds and sometimes jumped into people. If H5N1 had evolved the ability to spread from person to person, it might cause the next pandemic.

But before long, the Hong Kong researchers eliminated H5N1 from suspicion as well. Through a microscope, they finally laid eyes on the virus that caused SARS. On its surface, it bore a distinctive crown of proteins that betrayed its identity. SARS, they realized, was caused by a coronavirus.

○ ○ ○ ○

Virologists first discovered coronaviruses in the 1960s. They found them infecting many animals, including chickens, pigs, cats, and dogs. A few types of coronaviruses also infected people, but they seemed mainly to cause mild colds. Coronaviruses did not keep Joshua Lederberg up at night. The biodefense experts at Fort Detrick did not look at coronaviruses as biological weapons in waiting. Veterinary scientists created coronavirus vaccines for livestock and pets, but none were made for people.

That complacency turned out to be a mistake. A dangerous new human coronavirus was on the loose. It inflamed lungs and filled them with mucus. In many coronavirus victims, X-rays revealed a haze of lung damage known as ground-glass opacities. Epidemiologists estimated SARS had a mortality rate in the neighborhood of 10 percent.

Many SARS patients failed to pass on the virus to anyone else, but some of them could infect dozens of others. In emergency rooms, they felled patients and hospital workers alike. Epidemiologists puzzled over the so-called super-spreaders. One possibility was that certain medical procedures triggered viral explosions. Hospital workers sometimes had patients inhale medicine to break up phlegm in their lungs. As the patients coughed up sputum, they might exhale a viral mist that put people at close range at risk.

In Hong Kong hospitals, medical workers began walking around with N95s on their faces. Before SARS, the staff had worn the masks only when they had to treat a tuberculosis patient. Now they hoped N95s would protect them from the coronavirus.

○ ○ ○ ○

People outside hospitals tried to protect themselves as well. Those who could not get hold of N95s resorted to surgical masks. People wore them on the streets of Hong Kong, Taipei, and Tokyo, and Westerners recoiled at the eerie, apocalyptic sight. "It is beginning to feel a lot like Halloween here, with more and more people wearing masks and the health news getting scarier by the day," a *New York Times* correspondent reported from Hong Kong.

For Westerners, cities full of masks belonged to ancient history. After the 1918 pandemic, the only people who wore masks did so in hidden places: operating rooms, tuberculosis wards, coal mines. In 1928, when influenza cases spiked in the United States, the surgeon general discouraged the public from donning masks again, saying he doubted their value.

But in many Asian countries, masks became a familiar feature of everyday life in the twentieth century. In Japan people tied them on for every new flu outbreak. They did not need a government edict to take protection; it had become a medical courtesy: sick people used masks to avoid coughing their germs on others. In the 1950s, the Japanese government created a new reason to wear masks when it cut down forests of native oak and maple, replacing them with plantations of sugi trees. When the trees matured in the 1970s, they flooded the country's air with pollen, and more than a quarter of the people in Japan began suffering from hay fever. Each spring, victims of sugi allergies put on surgical masks to block the pollen grains.

Masks had debuted early in China, during the Manchurian plague outbreak of 1910. Wu Lien-teh launched the custom in draconian style, dispatching police to make sure people were wearing masks and to punish the offenders. In later years, China's ruling Nationalist Party relaxed its rules about masks to tamp down anti-government uprisings. Even when new outbreaks hit China—plague in 1920 and meningitis in 1929—the government didn't bring back Wu's harsh mandates. Forcing people to wear masks might have led to rebellion.

Instead of mandates, the Chinese government found better luck in encouragement. Its softer touch made masks commonplace in China, and by the 1930s covering one's face had become downright fashionable. One pharmacy advertised the high style of its anti-plague handkerchiefs. "Using it to cover your face will be much more elegant than wearing an ugly mask," it promised. Masks fell out of favor and fell back in over the next few decades. When China's booming economy at the end of the 1990s led to a boom of coal-fired plants, people had a new reason to wear masks: to protect their lungs from smog.

SARS sent a new wave of people in China scrambling to find masks. Wearing them became a way to cope with the uncertainty of a strange new threat. Chinese authorities discouraged the practice at first, unhappy that people were making it plain for all to see that something was not right. But as the epidemic worsened, the officials changed course, requiring people to mask up in public places. In China and other parts of East Asia, the SARS outbreak strengthened the cultural practice of masking. It became a signal between people that they shared the same uncertain, frightening fate. They shielded themselves from infection but also protected one another against a threat that came from within their own bodies.

° ° ° °

In late March 2003, SARS demonstrated a dreadful new capacity for spreading. Hong Kong's United Christian Hospital admitted fifteen patients with SARS who all lived in a single building. Until then, SARS had been known to explode only in hospitals. Now it seemed that the disease could also surge through homes with the same ferocity.

The building belonged to a cluster of nineteen orange-and-beige high-rises collectively known as Amoy Gardens, where nineteen thousand residents lived in tiny apartments. Inspectors swarmed the building where most of the patients lived, known as Block E, and discovered that the outbreak was still growing. Dozens of residents in Block E were falling ill every day, and new cases were turning up in other buildings.

In the middle of the night on March 31, police in surgical masks surrounded Block E and closed it off with metal barricades and tape. Health workers in full protective gear locked the doors and prevented anyone from entering or leaving the building. Officials piled food and supplies near the entrance for the people trapped inside. But the outbreak kept growing, and so the authorities took an even more drastic move: they evacuated Block E. On April 1, dozens of minibuses filled up with residents and took them to holiday camps on the outskirts of the city.

By the time the Amoy Gardens outbreak finally ended, 329 residents had come down with SARS, and forty-two had died. Afterward, experts tried to reconstruct what had happened by visiting the buildings, talking to residents, and judging how they had come into contact with one another. They inspected water tanks, hunted for rats, caught cockroaches. Among their ranks was a mechanical engineer named Yuguo Li. Li had come to study the air.

Before the SARS outbreak, Li had been a professor at the University of Hong Kong, where he made a name for himself studying how currents flow inside buildings as cool air sinks and hot air rises. He also studied the air in hospital isolation rooms, ensuring that ventilation didn't accidentally spread pathogens it was supposed to shunt away. When the news of Amoy Gardens broke, Li was startled by how similar Block E was to his own apartment building. They both featured a large light shaft that ran up one wall, providing a little sunlight to interior rooms. Li, as a ventilation expert, didn't just see light in the shaft. He saw a column of air warmed by the daytime heat and shooting up toward the top of the building. "To me it was a chimney," Li told me.

Li informed his dean that he suspected the chimneylike air shaft might have helped spread SARS through Block E. He soon found himself on the team of researchers dispatched to Amoy Gardens to figure out how the outbreak had occurred.

Epidemiological detective work revealed that a single person had unleashed the virus in Block E. Kaixi Wang lived twenty miles away, just over the Chinese border in Shenzhen. Twice a week he traveled to Hong Kong for dialysis at the Prince of Wales Hospital. On his visits, he stayed with his brother on the sixteenth floor of Block E. In March 2003, Wang received dialysis in a section of the hospital called Ward 8A, where another patient was suffering from pneumonia—which would later turn out to be SARS. The outbreak at the hospital sickened 138 people there, including doctors, nurses, medical students, visitors, and patients. One of those patients was Wang.

After he finished his dialysis, Wang went back to his brother's apartment, and it was there he started to feel ill and developed diarrhea. On a return trip to the hospital for dialysis, his symptoms got even worse, and his doctors diagnosed him with SARS. He had to stay there for months to recuperate.

The Amoy Gardens investigators concluded that Wang had started the outbreak in his brother's bathroom. Viruses present in his diarrhea entered one of the building's vertical pipes. They spread through the pipe and reached broken floor drains in other apartments. When people turned on their bathroom exhaust fans, they pulled virus-laden droplets from the drains into the air, which then landed on surfaces in the bathrooms. Block E residents picked up the virus and then spread it farther through the building by contaminating more surfaces—by turning doorknobs, pushing elevator buttons, and so on.

Yeoh Eng-kiong, the health secretary, held a press conference to unveil the results of the investigation. The people of Hong Kong should take some comfort in the results, he argued, because they indicated that SARS was not spreading long distances through the air like smoke. If it had, it would have wafted into many more apartments and turned Amoy Gardens into a high-rise graveyard. "We're not talking about 300 cases," Yeoh said. "It would be thousands, tens of thousands."

Yuguo Li found that reasoning flawed. He understood why Yeoh and other public health officials would want to assure people that SARS was not airborne—"to avoid a panic situation," he told me. But that did not make it true. Li continued to investigate the Amoy Gardens outbreak. And before long, he also started looking into the outbreak that had seeded Amoy Gardens with SARS: the earlier outbreak at the Prince of Wales Hospital.

Ignatius Yu, a hospital epidemiologist, invited Li to help him study what had happened in Ward 8A. Li reluctantly agreed to put on a mask and go to the hospital to look at the rooms in person. The ward contained four cubicles, each holding ten beds. When Li later reviewed the cases at the hospital, he began to doubt the virus had moved only through close contact. To solve the mystery of Ward 8A, he decided he would have to reconstruct the flow of air through the rooms.

Li had never solved a problem quite like this one before, so he looked through the scientific literature for guidance. "Anybody doing research, you go back to history to see what happened," Li said. "Otherwise, you reinvent the wheel." His search led him back seventy years, to an American engineer and his doctor wife, whom Li had never heard of.

Li began reading papers by William and Mildred Wells. He came across references to *Airborne Contagion*, but the University of Hong Kong did not have a copy. The book had become so rare that Li had to arrange for an American library to photocopy its edition and mail the pages to him. For the first time, Li encountered the tortured prose that had vexed Wells's editors and disappointed his reviewers. "I found the book was not easy to read," Li told me.

But he soldiered through, and he gradually came to understand Wells's theory of airborne infection. "And then you realize this is all about droplets," Li said.

○ ○ ○ ○

One of the people who got infected at Hong Kong's Hotel Metropole was a seventy-eight-year-old grandmother who had come from Canada for a vacation. On February 23, she and her husband flew back home to Toronto and returned to the apartment they shared with their two grown sons, a daughter-in-law, and their five-month-old grandson. After two days, the grandmother spiked a fever.

Her doctors did not make much of her illness. They started her on a course of antibiotics, but the drugs did her no good. She grew weaker over the next week and died at home on March 5. Her death certificate listed a heart attack as the cause of her demise.

Before she died, the grandmother infected other members of her family. Her son grew so ill that he headed to Scarborough Grace Hospital, where he sat in the crowded emergency room for over sixteen hours, breathing laboriously without a mask, before he was admitted to a room. His doctors, not yet aware of the crisis in Hong Kong, thought he might have tuberculosis. They moved him into an isolation room so that the bacteria wouldn't spread by air to infect anyone else. He died on March 13.

Despite those precautions, the son had infected patients and staff in the emergency room. They passed on the coronavirus to others, and the first outbreak of SARS outside of Asia was now underway. SARS spread out of Scarborough Grace Hospital, and before long the World Health Organization added Toronto to a list of cities to avoid. Toronto hospitals struggled to care for the sick as their staff got infected. Allison McGeer, the director of infection control for Mount Sinai Hospital, came down with chills and aches on March 30 and had to go home to recover in isolation. So did six other health officials who had spent time with her at the hospital. A reporter reached McGeer at home and asked for her prediction about the future of SARS. "If we don't have a vaccine, yes, we're all going to get it," McGeer said.

Research on a SARS vaccine was already underway when McGeer made that prediction. But there was no telling when the vaccine experts at the US National Institutes of Health would be done with the job. The record at the time for the fastest development of a vaccine—for mumps—had been four years. And there was no guarantee the vaccine would even work, since no one had ever made a coronavirus vaccine for people before. In the meantime, the only thing to do to fight SARS was to shield as many people as possible from infection.

But Canadian doctors were divided over how best to protect the public, because the coronavirus spread in baffling ways. In some cases, it seemed to travel long distances from room to room. In other cases, a single patient infected dozens of people. Some infectious disease experts raised the prospect that SARS was airborne. Others argued that it was far more likely that the coronavirus did not have to travel so far. "There is compelling evidence that the SARS coronavirus is spread through droplet and contact transmission," the Canadian infectious disease expert John Conly later declared—referring to the short-range spread of large droplets in coughs and sneezes. He did not see strong evidence that SARS spread over long distances in droplet nuclei.

"Part of the heated debate during the SARS outbreak was over whether N95 respirators were really necessary," the authors of a Canadian government report later wrote. If droplet nuclei wafting through the air were causing a lot of infections, N95s might stop them. But putting the respirators into widespread use would have been a burden to hospitals. Not only would they have had to pay for the N95s, but they would have had to instruct their staff about how to use them properly. Some hospitals decided surgical masks were good enough. Other hospitals chose to be more cautious and treated SARS as airborne. "It was eerie—like you were on Mars or on a new planet," one doctor later recalled. "You sit in meetings, everyone around the table is wearing an N95 mask."

Americans to the south were slow to turn their attention to SARS, distracted by their army sweeping through Iraq. Only after US forces reached Baghdad and started their fruitless search for weapons of mass destruction did Americans grow more aware of the coronavirus. With hundreds of people falling ill just over the border in Canada, it seemed like only a matter of time before SARS moved south.

On April 24, an editorial cartoonist named Bill Schorr captured the moment with a drawing of two men on a park bench. A paunchy man in a plaid jacket talks to a skinny man next to him wearing a gas mask to protect him from the air. The skinny man's head looks like that of a black double-beaked bird. He holds a newspaper. Its headline reads, "SARS SPREADS."

"Still worried about terrorists?" the paunchy man asks.

"No . . . Canadians."

During the Canadian outbreak, a total of three hundred seven-five people got infected around Toronto, and forty-four died. The city managed to stop the coronavirus from spreading farther by quickly identifying people with SARS and isolating them. Thankfully, it was only after people showed symptoms that they started spreading the coronavirus. In places like Hong Kong, where the coronavirus had spread farther and caused more deaths, people retreated into their homes. By avoiding contact with others, they deprived the virus of the chance to spread.

"As April turned to May, and warmer days commenced, we looked around and realized that we were still alive," the Hong Kong journalist Karl Taro Greenfeld later wrote. Greenfeld spent weeks during the SARS epidemic staring at people in masks. Now he began to see full faces again. "One day, I found myself sitting in a steamy chicken-and-rice place full of other customers," Greenfeld recalled. "Oh, I thought. This is what life is."

The United States never did get SARS from Canada, or anywhere else for that matter. The disease that seemed like it would sicken every person on Earth caused 8,096 cases worldwide and 774 deaths. In the realm of infectious diseases, this was a minuscule toll. HIV was killing hundreds of people every hour throughout the SARS pandemic.

In 2006, the World Health Organization published a report on SARS. They summed up the experience in five words: "We were lucky this time."

○ ○ ○ ○

The world eventually forgot about SARS, but the disease left some marks behind. Masks became a more common sight in East Asian cities like Hong Kong—not only during flu seasons but as a precaution people took before they shared indoor air on subways or planes. In the West, which was largely spared the terror of SARS, masks remained rare.

A scene in the movie *The Big Short* captured the cultural split that emerged. Brad Pitt plays an eccentric financial trader who retires so that he can prepare for the collapse of civilization. Two young friends ask him to come to New York to help them figure out how to make a fortune on the housing crisis that led to the 2008 financial crash. Pitt is filmed riding down an airport escalator. He locks eyes with an Asian man riding up past him, wearing a surgical mask. Pitt offers a salute. The camera pulls back, revealing that he is wearing an N95. Pitt stands out in a sea of maskless, clueless Westerners.

Virologists also did their best to keep the memory of SARS alive. They continued to grow stocks of the virus that caused it, known today as SARS-CoV. They studied SARS-CoV for lessons about how new pathogens spill over into our species. More clues to the origin of SARS came from live-animal markets in China, where researchers found SARS-CoV infecting palm civets and other wild mammals sold for food. Years later, researchers exploring caves in China found closely related coronaviruses harbored by bats. Bats, the research showed, were the original source of SARS-CoV. They had passed the ancestral virus to one another as a harmless gut infection for years. It likely moved from a bat to a mammal that was brought to a Chinese market to be sold for food. And after incubating in those animals, it jumped to people.

Epidemiologists also studied SARS after its disappearance, trying to reconstruct the devastation it caused. In its 2006 report, the WHO concluded that it was a short-range disease, stating, "The main mode of transmission is through respiratory droplets that require close contact or transfer through fomites." But the more Yuguo Li looked into what had happened in Ward 8A and Amoy Gardens, the less he agreed with that conclusion.

On the day of the outbreak at the Prince of Wales Hospital, sixty-six medical students had paid a visit to Ward 8A. Li tracked their movements and found that many of the students who developed SARS never got close to the infected patient. He was treated with a nebulizer four times a day, a procedure that would have made him cough out soggy raisins and potentially spread the coronavirus to people in the same room. But Li found that the visitors who came to the ward when the patient was using the nebulizer were no more likely to get infected than those who came when he wasn't.

Li then considered whether SARS was spreading instead like smoke. He estimated how many droplets the patient released and how many droplet nuclei became airborne. Li tracked their movements with the help of a computer model of Ward 8A's atmosphere stirred up by air conditioners bringing fresh air into the space.

In the model, the virus-laden droplet nuclei spread through much of the ward. The area around the patient had a high concentration of them, because his breath released fresh supplies. As the droplet nuclei spread, they became more dilute. The air conditioners diluted them even more. Li ended up with a map of the concentrations of the virus through the ward. The highest concentrations in his model were located in the places where the most people got sick. In other words, Li's model indicated that SARS was behaving like an airborne disease.

Li came to much the same conclusion when he studied Amoy Gardens, with one key difference. In Ward 8A, the virus had spread only through a cluster of connected hospital rooms. At Amoy Gardens, the virus had formed a vast cloud that spread through a building.

The official investigation had concluded that SARS had spread through Amoy Gardens in large part by direct contact—people either coming close together or leaving the virus on surfaces that others touched. But when Li looked at the patterns of infections, that explanation didn't make sense. According to the official scenario, the Block E staff should have faced the highest risk of infection. After all, they moved around the building all day, turning doorknobs and talking to residents. Yet none of the staff had gotten SARS.

Li and his colleagues made a detailed reconstruction of how droplets would have spread from Wang's bathroom. They even built a mock-up of the Block E drainage system to see how faulty traps released droplets into the air. Then they followed those droplets in a computer simulation of Amoy Gardens. They built it from the same equations that Matthew Meselson and his colleagues had used a decade earlier to reconstruct the anthrax plume at Sverdlovsk.

When Wang had used his brother's toilet, his coronaviruses ended up in the floor drain of the bathroom. Later, the exhaust fan pulled tiny droplets from the drain and into the light shaft, where they rose like a plume of smoke. The plume's tentacles crept into apartment windows on higher floors and then kept moving, wending their way into hallways and other apartments. People may have then inhaled the viruses. The main part of the plume continued to rise up the shaft until it cleared the roof. There it met the winds coming from the northeast. They carried the cloud to other buildings, to Blocks B, C, and D. That airborne spread, Li concluded, fit the pattern of infections best.

○ ○ ○ ○

Lidia Morawska, an Australian physicist, got drawn into the mysteries of SARS in April 2003 with an invitation to go to Hong Kong. The World Health Organization wanted her to help investigate the Amoy Gardens outbreak too.

Morawska was an expert on the spread of particles through the air. Growing up in Poland, she wanted to become a nuclear physicist. But in graduate school, she shifted to studying radon, a radioactive element that escapes as a gas from rocks and soil. It can cause lung cancer as it builds up in unventilated indoor air.

As a postdoctoral researcher at the University of Toronto, Morawska surveyed dust and other particles that carried radon through the air. She learned how to use a sensitive device to detect them in a laboratory, and wondered what she might find if she stepped outside her building. One day she stood on a Toronto street with the probe. She was shocked at all the kinds of particles floating in outdoor air that were absent from her laboratory chamber.

Morawska abandoned radon and dedicated herself to the air entirely. In 1991, she moved to Australia, where she opened a lab at the University of Queensland. There she measured particles of pollution coming from cars, fungal spores growing on ceilings, and smoke rising from cigarettes. Twelve years later, when the Amoy Gardens outbreak occurred, the World Health Organization thought her expertise on microscopic particles might help them figure out how the virus was spreading so fast.

"I was petrified about the idea of going to Hong Kong at the time," Morawska told me. Still, she started getting ready for the trip. She didn't know much about infectious diseases, so she started delving into the scientific literature. As she did her homework, SARS subsided, and WHO abandoned its plans for an Amoy Gardens investigation. But even after her trip to Hong Kong was canceled, Morawska continued to learn more about airborne infections.

Germ-laden droplets were different from the solid particles Morawska had studied up to that point. Like Li, Morawska made her way back into the work of William Firth Wells to understand their physics. She then made her way forward, searching for more recent studies that had built on his work, using the latest methods to study droplet nuclei released from human airways. She found a grand total of three.

"I thought, 'Wow, such an important area, and there's so little science on this,'" Morawska said. SARS might be gone, but there was science left to do. Tuberculosis continued to kill more than a million people a year, and it was conceivable that respiratory diseases like influenza were airborne too. Stopping their spread might keep millions of people out of hospitals and save billions in health care costs.

The confusion that had emerged about how SARS spread made it painfully clear to Morawska how much twenty-first-century scientists still didn't know about airborne transmission. In a 2006 manifesto, she declared that the science of droplet nuclei had barely budged since the Wellses had studied them. She judged the world's understanding of how pathogens spread in the air as "less than basic."

The logical place to start learning about their spread was in people's airways, where droplets formed. But it was impossible to look into those dark realms. "We can only see what comes out," Morawska said. So she invented a means to make precise measurements of droplets as they escaped from people's mouths and noses so that she could infer how they had been created in the first place.

In her lab, Morawska built a giant doughnut-shaped tunnel. Volunteers poked their heads inside through a hole on the underside, and at Morawska's instruction they talked, coughed, or simply breathed. A probe hanging down in front of their faces captured some of the droplets they released and measured the size. A fan halfway around the doughnut blew the air through a filter to trap long-floating droplets, ensuring that the probe caught only ones freshly released from the volunteers.

The tunnel experiments helped Morawska create a clearer picture of the droplets people make. Morawska discovered that the photographs of sneezes that Marshall Jennison had captured with his stroboscope had revealed only a small fraction of the droplets people hurled into the air when they sneezed or coughed. She also found an abundant supply of droplets in the air when her volunteers merely talked. Even their quiet breaths released beads of water.

Morawska's doughnut-tunnel studies also helped illuminate how we generate those droplets. Different parts of the airway produce them in different ways. When air flows over the large tubes of the airway—the throat and bronchi in the lungs—it ruffles the mucus coating the walls and pulls up threads that snap apart into beads. At the end of an exhalation, some smaller branches in our lungs squeeze shut; they then pop open again as we inhale. That snap liberates its own batch of droplets, which can also fly out in the next exhaled breath. As we talk, the vibrations of the larynx unleash droplets too. Even the movements of our mouth that produce consonants and vowels break off saliva from our teeth and lips, then set it loose into the air.

○ ○ ○ ○

When the SARS epidemic hit Canada, Lydia Bourouiba was in Montreal, thinking about turbulence. As a graduate student at McGill University, she was studying the physics underlying both the spirals of water draining out of bathtubs and the wheeling arms of the Milky Way. The agnostic elegance of the equations gave Bourouiba a deep pleasure. When she wrote her dissertation, she introduced it with an epigraph from the mathematician Henri Poincaré: "If nature were not beautiful, it would not be worth knowing, life would not be worth living."

SARS cast a pall over that beauty. The death and confusion that gripped Canada left a mark on Bourouiba after the epidemic ended. She turned her attention to diseases, bringing to them her mathematical expertise. To study the spread of influenza, she transformed virus-carrying geese into variables in an equation. Droplets can also carry pathogens through the air, and she realized she could use her knowledge of turbulence to study how they flew. Like Morawska and Li before her, Bourouiba started paging through scientific journals to see what others had already done. She discovered the work of William Firth Wells and recognized him as a pioneer.

Bourouiba was impressed by the insights that Wells made despite using only rudimentary physics. He treated droplets as if they existed in two crude categories: big ones that automatically fell to the ground and small ones that could float and spread airborne infection. Wells also simplified reality by analyzing each droplet in isolation from the surrounding air. Bourouiba knew from her work on turbulence that reality was far more complex.

Bourouiba assumed that someone must have followed up on Wells's work with more sophisticated physics. "Surely someone has modeled the process of a sneeze, a cough, breathing in indoor space," she thought. "Surely this has been done, right?" It hadn't.

In 2010, Bourouiba took a job at MIT, where Wells had trained a century before. It was also at MIT that Marshall Jennison and Harold Edgerton had photographed sneezes. After Edgerton's death in 1990, the university set up a center in his name to help scientists use high-speed cameras for their own research. Bourouiba went to the center to get help taking new pictures of sneezes.

In her lab, Bourouiba unfurled a sheet of black felt and hung it on a wall. Instead of stroboscopes, she set up high-speed digital video cameras. Bourouiba enlisted students for her experiment and tried to get them to sneeze by having them inhale pepper and dust. But tickling their noses with a swab proved to be the most reliable trigger. When the volunteers sneezed, Bourouiba didn't just film the droplets they produced. A smoke machine also released a plume of smoke that illuminated the drifting air.

When Bourouiba analyzed her movies on a computer, she confirmed her hunch: droplets released from a sneeze do not travel in isolation. They are part of a lung-made cloud composed of warm gas and liquid. Its momentum holds it together and pushes it forward as its warmth lifts it toward the ceiling. Carried along inside the cloud, even large droplets can travel much farther than Wells would have predicted. As they fly inside the cloud, they get extra time to evaporate into droplet nuclei that can then float on their own. Bourouiba suspected that these clouds spread pathogens farther as well, increasing the threat of airborne pathogens to anyone who might inhale them. "I am definitely more aware of sneezers and coughers all around," Bourouiba told a reporter who visited the lab.

By modern standards, Bourouiba would later write, Wells's theory "seems overly simplified." But she was not dismissing Wells as a miasmatist from the Middle Ages. He had developed the science of airborne infection, and it was long past due to start pushing it forward.

fifteen

A STATE OF PREPAREDNESS

On November 1, 2005, President George W. Bush returned to Bethesda, Maryland, to give another major speech at the National Institutes of Health. Less than three years had passed since he had gone there to announce Project BioShield, a $6 billion plan to protect the country from airborne biological attacks. Now Bush stepped up to an NIH podium once more to outline an even more ambitious plan: to prepare for the next influenza pandemic.

"Three years ago, the world had a preview of the disruption an influenza pandemic can cause, when a previously unknown virus called SARS appeared in rural China," Bush said. An influenza pandemic, the president warned, would be far worse. "Our country has been given fair warning of this danger to our homeland," he said.

By the time Bush took to the stage at NIH, no new case of SARS had been recorded for over a year. SARS-CoV had vanished from humans, and no one could even find it in an animal. But viruses were not done with humanity. The H5N1 strain of influenza kept flaring up in one country after another. Millions of chickens and ducks were being slaughtered to slow its spread. In Southeast Asia, dozens of people who handled infected birds got sick. About half of them died. The virus could not spread readily from person to person, at least not yet. The world had enjoyed a bit of good luck when SARS proved controllable. But if H5N1 evolved into a human flu, that luck might turn very bad.

Amid headlines about H5N1 outbreaks, the historian John Barry published a deeply researched account of the 1918 pandemic. It rattled many readers, including President Bush. "You've got to read this," Bush told his top homeland security advisor, handing her a copy. "Look, this happens every hundred years. We need a national strategy."

A few months later, at the National Institutes of Health, Bush unveiled that strategy. His $7.1 billion plan would prepare the United States to fight a new pandemic that might otherwise kill millions of Americans, along with millions more beyond the country's borders.

Bush's plan was a milestone in modern public health. Over the next fifteen years, it set the terms for how the country prepared for influenza, along with any other planetwide outbreak that might occur. Over those next fifteen years, Ebola and other viruses would cause terrifying new waves of infection. But influenza would remain the dominating worry.

In his 2005 speech, Bush singled out one of the traits of the flu that made it so dangerous: its capacity to evolve into a form that could spread over thousands of miles in a matter of weeks. "If the virus were to develop the capacity for sustained human-to-human transmission, it could spread quickly across the globe," he warned. And yet, despite all the billions that would be spent on his plan, precious little attention would be paid to how the virus spreads. Over the next fifteen years, public health experts would offer guidelines based on old assumptions about the transmission of respiratory diseases. And while a small group of scientists spent that time on the work of the Wellses to see if influenza and other diseases could travel by air, authorities would largely ignore their work.

○ ○ ○ ○

Bush's pandemic plan established some important new ways to safeguard public health. The US government sped up vaccine research and added antiviral drugs to the Strategic National Stockpile. It set up a monitoring network to do a better job of tracking new strains. But Bush's plan also suffered some fundamental flaws. For one thing, it failed to prepare for a future pandemic as a public health emergency. Instead, Bush effectively treated influenza as one more form of biological warfare.

The new plan echoed Alexander Langmuir's strategy to prepare for a Soviet attack. To defend itself against an incoming wave of influenza, the United States would gather intelligence in the first days of the outbreak, and it would then launch a military-style response. Influenza pandemics and exploding anthrax bombs were so merged in Bush's policies that he talked about his strategy as if it would defend against them both. "By putting in place and exercising pandemic emergency plans across the nation, we can help our nation prepare for other dangers—such as a terrorist attack using chemical or biological weapons," Bush claimed.

But the bird flu flare-ups that inspired the pandemic plan did not come from terrorist crop dusters spraying influenza viruses on American cities. H5N1 was thriving in the modern food industry. By the early 2000s, farmers were raising chickens, ducks, and pigs—all suitable hosts for influenza—in unprecedented numbers. On many farms, the virus could jump easily among the crowded animals. Industrial agriculture acted like an incubator in which viruses could multiply, mutate, and swap genes. Bush's pandemic plan did not call for any reforms that might lessen the risk of a new strain emerging from factory farms.

Treating influenza like a biological weapon also hurt the public health system in the United States. Bush saw community hospitals and local public health offices as the front lines of defense against a pandemic, but his plan largely ignored the crippling neglect that they had suffered for decades. Even as the Bush administration made plans for pandemics, it continued to inflict more damage. White House budgets regularly slashed spending on public health, including for the Centers for Disease Control—except when it came to money for biodefense.

And while the Bush administration spent a lot of money on scientific research into influenza, it spent most of those funds on antivirals and other technological defenses against a flu attack. It gave far less attention to understanding the basic nature of influenza viruses—including their transmission. That neglect is all the more startling when you read the pandemic plan itself. Its authors acknowledged that the administration had had little idea of how influenza spreads. "The amount of direct scientific information is very limited," they wrote.

In the face of that limited information, the authors of the plan assumed that influenza mostly spread at short range—by direct contact with a sick person, by touching contaminated surfaces, or by getting hit in the face by a heavy coughed-up droplet. In case of a pandemic, the

Bush administration advised people to stay three feet away from anyone with flu symptoms. Hospital workers should protect themselves with surgical masks. There was no need to bother with N95 respirators. There was no need to consider the possibility that influenza might float in airborne droplet nuclei. And yet there was little firm evidence that any of this was true.

When the Bush administration launched its pandemic plan, experts were embroiled in a debate about how the flu spreads. Neither side could point to a definitive experiment that settled the question, so they looked back in history to a few significant outbreaks. When the Livermore Veterans Administration Hospital was hit by Asian flu in 1957, its ultraviolet lights seemed to protect some of its patients—a suggestive clue that the virus was airborne. Two decades later, in 1977, an outbreak that took place on an airplane in Alaska also displayed some hallmarks of airborne spread. One person sick with the flu boarded the plane. Shortly after takeoff, an engine failed and the plane was forced to land. The passengers had to wait cooped up in the plane for hours before they were transferred to another flight. Thanks to the engine failure, the pilots could not run the ventilation system. Within a few days, thirty-eight of the passengers who breathed in that unventilated air came down with the flu.

Raymond Tellier, a Canadian physician and a veteran of Toronto's SARS outbreak, argued that those outbreaks demonstrated that influenza might well be airborne. If flu viruses could be carried in floating droplet nuclei throughout hospitals, then hospital staff should not rely only on surgical masks to stop them. "Recommending the use of N95 respirators, not surgical masks, as part of the protective equipment seems rational," Tellier wrote.

Tellier's opponents questioned his evidence. An airborne outbreak was not the only possible explanation for what had happened aboard the Alaska airplane, for example. As the passengers waited on the tarmac, they did not simply sit in their seats. They got up, stretched their legs, and milled up and down the aisle. Along the way, they might well have grazed their hands across a contaminated seat back. "Effective aerosol transmission is absent in the natural state," a team of Canadian doctors declared.

Bush's plan did not acknowledge the debate. Of the $7.1 billion the government spent to prepare for a pandemic, less than a thousandth went to fund studies on how that pandemic might spread.

In one of those few studies, researchers at the Centers for Disease Control joined forces with scientists in Hong Kong to measure how well handwashing and wearing surgical masks stopped the flu. They identified people who tested positive for influenza and invited their entire households to participate in the study. Some of the volunteers would wash their hands every time they sneezed and wear surgical masks except when eating or sleeping. The study grew to encompass more than two hundred fifty households in Hong Kong. It was the biggest clinical trial of masks ever undertaken. And it proved a disappointment.

Overall, the scientists found that handwashing might protect against the flu if people took up the precaution quickly after an infection. Surgical masks didn't seem to help at all. The researchers couldn't rule out the possibility that N95s, with their virus-trapping fabric and tight seal, would work better. It was also possible the surgical masks worked poorly because the volunteers hadn't used them properly. Only 49 percent of the volunteers who tested positive for the flu wore masks at all. And in their households, only 26 percent of the volunteers masked up. A protection from an airborne disease can't do much good if people don't use it.

Meanwhile, another study on the spread of influenza got underway at Harvard. A doctor named Donald Milton began building a machine to catch viruses in people's breath. Milton was fifty-six at the time, with decades of experience studying the harms that come by air. But now he turned his attention to airborne infections, and the shift felt a lot like coming home.

○ ○ ○ ○

In the late 1950s, when Milton was a boy growing up in Baltimore, his father would take him in the evenings to the city college. An engineer by day, his father hosted a classical music show at night on the college radio station. "You walked through these dark hallways until you found this little oasis of light in the radio studio," Milton told me.

Across the street, a hulking hospital lurked in the night. Milton knew nothing about it at the time, but later he would realize the hospital was Loch Raven. While Milton spent evenings with his father, William Firth Wells was lying in bed across the street dying of cancer, as his tuberculosis experiment took place overhead.

It was Milton's mother who inspired him to pursue a career defending people's health from polluted air. She had grown up in Akron, Ohio, where the laundry on clotheslines turned gray from factory smoke and the sidewalks crunched with particles of coal. She developed chronic lung diseases that would torment her for the rest of her life. "What I remember about Mom was her sitting in bed with a box of tissues, coughing, bringing up a lot of stuff, and reading some huge book," Milton said.

Milton went on to get a medical degree and study cotton lung, a disease caused by inhaling cotton fibers in textile mills. He later went to the Harvard School of Public Health, where he investigated the healing power of ventilation. In a study of workers at the Polaroid Corporation, he found that workers in well-ventilated buildings called in sick less often. He estimated that the company would annually save four hundred dollars per employee by providing them all fresh air.

Perhaps, Milton speculated, the stale air in the Polaroid buildings was making people sick because it contained floating bacteria and viruses. To search for them, he teamed up with Sebastian Johnston, a British gene-sequencing expert, on a machine to pull viral genes out of the air. As other pioneering aerobiologists were starting to grab DNA from the air over oceans and mountains, Milton and Johnston headed into office buildings around Boston. In the buildings with high levels of carbon dioxide—a sign of poor ventilation—they caught more cold viruses in the air.

Milton discovered that his fellow doctors were not much interested in his genetic update on the Wells air centrifuge. They speculated that the genes he gathered came from dead viruses that posed no threat. After all, everyone already knew that few diseases aside from tuberculosis were airborne. The prevailing belief was that respiratory infections didn't spread by droplet nuclei; short-range droplets and fomites were largely responsible. "You grow up with this in medicine," Milton said. "It's more religion than science, I would say."

Like Lidia Morawska and other scientists who were starting to take airborne infection seriously, Milton recognized that Wells's theory needed to be ushered into the twenty-first century. He proposed some updates in an influential essay that he wrote with Chad Roy, an infectious disease expert then at the US Army Medical Research Institute of Infectious Diseases at Fort Detrick. In the piece, published in the *New England Journal of Medicine* in 2004, Milton and Roy argued that the transmission of diseases probably fell along a spectrum. At one end were diseases like tuberculosis that could infect people only by traveling through the air like smoke. At the other end were diseases like HIV that never spread through the air—at least outside of Joshua Lederberg's nightmares. But in between those extremes were many diseases.

Some might spread in small floating droplet nuclei as well as in large short-range droplets. Some might also travel by touch. Depending on the conditions of an outbreak, pathogens might have more success traveling by one route than by another. Milton and Roy used SARS as a case study. It might very well spread short distances in coughs. But that was no reason to ignore the possibility that it could also turn into vast airborne plumes that drifted through a building or even from one building to another. Yuguo Li's research in Amoy Gardens had shown how easily that could have happened.

"As perplexing as it may be," Milton and Roy warned, "the peculiarity of the transmission of the SARS coronavirus in Amoy Gardens may be a harbinger of unorthodox transmission patterns associated with emerging infectious agents in the modern built environment."

A year after Milton and Roy published their 2004 essay, Bush turned America's attention to influenza. Milton saw it as exactly the sort of confusing disease that needed to be pinned to its proper place on the spectrum. He began investigating influenza by trying to answer one basic question: what floated in the breath of people infected with the virus?

Milton and a team of engineers designed a device to collect droplets from sick volunteers. It looked like an enormous scientific gramophone. Volunteers stuck their heads into its horn and exhaled. The droplets they released got swept into a pipe where a probe measured their size. Milton's contraption could sort the droplets, shunting large droplets into one vial and small ones into another. Later, Milton and his colleagues could search each vial for genes from influenza viruses. Inspired by a cone-shaped particle collector built in the 1960s called Gesundheit, Milton called his new creation Gesundheit II.

When Milton was close to finishing the construction of Gesundheit II, he ran into a CDC official. "He said, 'You know why we funded you?'" Milton recalled. "'To put the nail in the coffin of this idea that aerosols are important for infectious diseases.'"

Milton kept going, and by January 2009, he was ready to give Gesundheit II a trial. At the University of Massachusetts Lowell, where Milton was then teaching, his team put up posters around campus: GOT FLU? WE'LL PAY YOU. Woozy students came Milton's way, and those who tested positive for the flu were invited to sit in front of Gesundheit II, lean their faces into the funnel, and breathe.

Milton found that small particles contained far more viruses than the larger ones. He suspected that was the case because the small particles came from deeper in the airway where more viruses were replicating. In his experiments, Milton also had the sick students put on masks and breathe into Gesundheit II. Masks drastically reduced the number of viruses Milton detected. While the Hong Kong trial had not provided strong evidence for masks stopping the flu, Gesundheit II hinted that they could help.

Milton was pleased that his virus gramophone was delivering results, preliminary as they might be. Now he wanted to use the machine in more ambitious experiments. But as he began making plans, nature started to run an experiment of its own.

○ ○ ○ ○

When Barack Obama succeeded George Bush as president in January 2009, the country was in a crisis: the housing bubble dramatized in *The Big Short* had popped and was crashing the economy. But within three months, Obama had a second crisis to manage. In March, two children in Southern California fell ill with an unusual form of influenza. It quickly became clear that the virus was infecting other people and that it represented a new strain that came to be known as 2009 H1N1. The pandemic for which Bush had prepared five years earlier had arrived.

Just how bad it would be, no one could say. This was not H5N1, the bird flu strain scientists had been eyeing anxiously. It was a new virus that had come about through the mingling of genes from older flu viruses. Its closest genetic matches came from viruses that had infected North American pigs, North American birds, Asian pigs, and humans from who knows where.

The first person known to die of 2009 H1N1 was a tax worker in Mexico named Adela María Gutiérrez. She fell ill in early April and went to a series of doctors for help until she had to be hospitalized. Only then did her doctors determine that she had influenza. Soon after her death, it became clear that other people in Mexico were dying of 2009 H1N1 as well.

Experts would later determine that Mexico was where the virus had gotten its start. Pigs imported from other countries had brought their own flu virus strains. On giant farms, the animals passed the viruses around, and the viruses merged into new hybrid forms. Once 2009 H1N1 started infecting people in Mexico, it swiftly spread to other countries.

Now each of those countries faced the same challenge as people had in the 1918 pandemic: they had to slow the virus down. Scientists could still offer only patchy information about how it spread. Governments responded to their vague advice in different ways. Mexico closed schools for weeks and told bars, restaurants, and clubs to shut down as well. At airports, white-coated workers used heat sensors to check for feverish passengers. Soldiers handed out surgical masks at subway stations. President Felipe Calderón called attention to the threat of 2009 H1N1 by refusing to shake hands or even to kiss people on the cheek.

Hong Kong, which had learned hard lessons from the SARS epidemic, was ready with a sophisticated strategy. Its labs quickly launched genetic tests for the new strain. Thousands of medical workers already organized for an emergency response got to work. When Hong Kong registered its first positive test for 2009 H1N1—a traveler from Mexico who had stayed a few hours in a hotel—the hotel was cordoned off and its two hundred guests were quarantined for a week. More than 88 percent of people who felt flu symptoms wore masks in public; 20 percent of symptom-free people did as well.

Meanwhile, in the United States, Obama took the threat of 2009 H1N1 seriously. He had learned about the deadly history of influenza in his days as a senator, when he worked on the country's pandemic preparedness plan. "What I knew scared the hell out of me," Obama later wrote. "It was too early to tell how deadly this new virus would be. But I wasn't interested in taking any chances."

The Centers for Disease Control developed a test for the new strain. The United States also led a global effort to track the flu across the planet, using the Internet to share new findings. Obama opened up the country's pandemic stockpile, distributing antiviral drugs and other supplies, and prepared hospitals for a surge of patients. And the CDC launched an effort to slow the spread of the virus, based on the assumption that direct contact and short-range coughs were largely responsible.

The agency recommended washing hands to clear viruses from the skin. They advised covering coughs and sneezes to protect people nearby. Anyone who felt sick should stay home in order to avoid passing the viruses to others at work or school. If cases rose, people would need to cut down the number of close contacts they made each day. The CDC called this unfamiliar practice social distancing.

In hospitals, the CDC recommended that medical workers going into rooms where patients with H1N1 were being treated put on N95 respirators. Obama authorized the release of 85 million N95s from the Strategic National Stockpile, but the demand for them was so intense that shortages quickly followed. The CDC advised hospitals running out of N95s to let their staff use surgical masks instead. Meanwhile, the World Health Organization caused more confusion by recommending surgical masks rather than N95s.

The mixed messages made Obama's difficult job even more difficult. He needed to persuade Congress to give him billions of dollars to put the pandemic plan into full action. That would require him to make it clear how bad the pandemic could become, without stoking panic. To deliver the government's message, Obama dispatched Vice President Joe Biden to appear on television.

In some ways, Biden was the obvious choice. The lanky former senator had a broad grin and seasoned charm. But Biden also had a long history of blurting out things he later had to take back—either because they were wrong or because they were all too true. He once called himself "a gaffe machine."

The gaffe machine was operating at full power during an appearance on the *Today* show. The host Matt Lauer asked Biden what advice he'd give to his family if they wanted to fly to Mexico for a vacation. Flashing his grandfatherly smile, Biden replied that it wasn't Mexico that people should be worried about. It was the air.

"I wouldn't go anywhere in confined places right now," he said. "It's not that it's going to Mexico. It's that you are in a confined aircraft. When one person sneezes, it goes all the way through the aircraft."

Biden added that he would tell his family not to ride a subway either. "If you're out in the middle of a field and someone sneezes, that's one thing. If you're in a closed aircraft or closed container or a closed car, closed classroom, it's a different thing."

The scientific understanding of influenza in 2009 didn't definitively back Biden up. But he wasn't necessarily wrong. Shortly before the pandemic, the Institute of Medicine had released a report in which it refused to come to any firm conclusion about how the flu spread. "Debate continues about whether influenza transmission is primarily via the airborne or droplet routes and the extent of the contribution of the contact route," the authors wrote.

But when Biden broached the idea that H1N1 spread by air, outrage followed. Even the mere suggestion of such an idea would have been bad for business, his critics complained. "The vice president's advice, if actually followed by other Americans, could seriously imperil the economic health of the US airline industry and of New York's businesses, which greatly rely on employees and customers who travel by subway," the *New York Post* declared.

The Obama administration pulled Biden back from the public eye, setting him to work on the pandemic in private. He talked to governors and mayors about what they needed to manage the pandemic. Meanwhile, cabinet secretaries and White House representatives hastily issued statements countering Biden's remarks. From now on, the Obama administration would leave public updates about the pandemic to Anthony Fauci.

By then, Fauci had become a familiar face on television. Twenty-five years had passed since he was appointed the director of the National Institute of Allergy and Infectious Diseases. He oversaw the development of anti-HIV drugs and a program to distribute those drugs around the world, saving 23 million lives—an effort that earned him the Presidential Medal of Freedom. During the Amerithrax crisis in 2002, Fauci provided the public with updates about how the government was trying to protect the public from bioterror. The following year, he was back to explain what scientists were learning about the threat of SARS. And now, in 2009, Fauci returned to talk about H1N1, yet another frightening new threat.

At the time, Fauci accepted the view of CDC experts that only a few diseases were airborne, carried by aerosols smaller than five microns. "If it's five micrometers or more, it's a droplet that's going to hit the ground," he recalled to me in an interview. "And if you're further away, and it doesn't land on your mucosa or in your eyes or on your nose—or if it's not on your hands when you touch your face—it won't spread. That was the general paradigm."

And so, when Fauci went on television and took calls with reporters, he did not raise Biden's fears about the air people shared. Instead, he talked about a vaccine against 2009 H1N1 that was in development. Because 2009 H1N1 was new to humanity, it would need a new vaccine, and so Fauci couldn't say for sure how long it would take to make it. "We can't hasten the time frame that you need to do clinical trials," he told a reporter. In the meantime, the world would have to dodge the virus based on a fuzzy understanding of how it reached millions of people.

○ ○ ○ ○

When the 2009 H1N1 pandemic hit, Linsey Marr was already used to viruses throwing her life into chaos. For six years she had been working as an environmental engineer at Virginia Tech, where she studied smog. Her husband, Erich Hester, had recently been hired by Virginia Tech as an assistant professor in engineering, and the two young academics juggled their work with raising their young son. On paper, it looked manageable. But their son kept bringing illnesses home from day care, turning balance to chaos.

"My life was dominated by interruptions to my workday because my kid was sick, and I had to drop everything and figure out what to do,"

Marr told me. "He was sick all the time. And I was sick all the time. My husband was sick all the time. It was a rough time."

Marr didn't blame her son's day care teachers. "I felt like they had pretty strict hygiene practices," she said. "Parents and kids, as soon as they came into the room, would have to wash their hands. They had kids washing their hands a lot. And anytime the younger kids put something in their mouth, it would be removed from circulation and go into what they call the Yucky Bucket."

Despite the Yucky Bucket, despite all the other protective measures, the day care center regularly got slammed with waves of infection. Sometimes half the children got sick at once. "It just happened so fast and involved so many of them, that a better explanation seemed to be that the stuff is going through the air," Marr said.

With a mix of exasperation and curiosity, Marr started looking into what scientists knew about how diseases spread. With the 2009 H1N1 pandemic raging, she focused on influenza. "I was really surprised to find out that we didn't really know the mechanisms of how the virus got from one person to another," Marr said. "There were blanket statements: 'Oh, it's in these droplets that people cough onto each other.' But I never was able to find direct evidence of that."

As an environmental engineer, Marr found some claims just daft. Over and over again, she read that airborne infections could be caused only by droplet nuclei smaller than five microns, because only they were small enough to float. Droplets measuring five microns or bigger moved like cannonballs under gravity, Marr read. They could spread infection only if people fired them directly at other people with coughs or sneezes.

"The physics of it was totally wrong," Marr said. Every year, she would turn to the chalkboard in her lecture hall and derive equations to show her students that particles much bigger than five microns can readily stay in the air for a long time. Smog is full of bigger particles, and winds regularly carry grains of sand across oceans.

Wondering where that five-micron myth had come from, Marr found scientists citing the same source: "On Air-Borne Infection," a paper William Firth Wells had published in 1934. Marr was unfamiliar with Wells, so she decided to look up the study. "That paper is really good. He was really right on," Marr said. "But he never said the distinctive cutoff was five microns."

Marr saw a problem in need of solving. She decided to turn some of her efforts away from smog and toward the physics of droplets in the transmission of influenza. As if to confirm she had made the right decision, her whole family got sick with the flu—very likely H1N1—at the end of 2009. "All three of us just could not even get out of bed for twenty-four hours," she remembered. "That is probably the worst I felt in my life. Every cell of your body aches."

o o o o

Marr's cells might have been spared those aches if she had gotten a 2009 H1N1 vaccine. But the push to make the vaccines bogged down in the summer of 2009. As American pharmaceutical companies produced vaccine viruses in chicken eggs, researchers discovered that they grew much slower than expected. By October, the United States had only 11 million doses out of an expected supply of 160 million. Meanwhile, the start of the school year accelerated the spread of 2009 H1N1, as the virus jumped from one unvaccinated child to the next. By the time vaccines went into widespread use at the end of 2009, the pandemic had already passed its peak.

All told, 2009 H1N1 infected about a quarter of all people on Earth, and an estimated 203,000 of them died. Compared to past pandemics, it inflicted only a modest toll. More people died in an average year of seasonal flu. It's possible that 2009 H1N1 was not very deadly because older people had some unexpected immunity to the virus. Whatever the reason, the world's relative luck held out.

Over the next few years, public health officials looked back at 2009 to judge how well Bush's pandemic plan had worked in its first major test. David Michaels, the director of the Occupational Safety and Health Administration, was appalled by how influenza had ripped through hospitals, infecting staff and patients. The optional guidelines that the Centers for Disease Control had offered about stopping the spread of influenza were insufficient, he decided. It was time for clear-cut workplace rules.

OSHA had already reduced HIV and hepatitis B infections at hospitals by drafting rules to prevent accidental needle sticks. Now Michaels wanted to tackle infectious diseases that could spread by other means. "It became obvious this is something we could deal with," Michaels later told me.

In his review of the data, Michaels concluded that influenza could probably spread in heavy, short-range droplets as well as lightweight droplet nuclei. "We recognized from the beginning that this was more than just droplet transmission," Michaels recalled. And if influenza was at least partly airborne, then workplace rules would have to require protections that could withstand that threat. "You can't just use a surgical mask," Michaels said. "You have to use N95s and other protections."

OSHA began drawing up a draft rule that would apply not just to health care facilities, but also to any workplace where highly contagious infectious diseases might spread. Employers would be required to come up with plans to protect workers. The rule was more stringent than the advice offered by the Centers for Disease Control about who might consider wearing a mask at a hospital. And when OSHA shared an early version of the draft with other agencies, the CDC protested.

"We always heard from CDC saying they don't agree with our understanding. The CDC pushed back constantly," said Michaels.

When I asked Michaels thirteen years later why the CDC had resisted, he could only speculate. The agency was led by older infectious disease physicians who didn't see much good evidence for airborne infections. They might still have believed in the five-micron myth. Unpersuaded by the evidence, they might not have wanted to force hospitals and other workplaces to spend extra money on the protections airborne infection would demand. To protect people from pathogens in droplet nuclei, the hospitals might have had to refit their buildings for better ventilation, keep stocks of expensive N95s, and train their staff on how to wear them properly. "We always got pushback from the hospital industry as well," Michaels added.

Michaels decided to negotiate. At first OSHA put influenza in the airborne category, along with diseases such as measles. But in a concession to CDC, they moved influenza to the droplet category—meaning that it spread only a few feet at most and could be blocked by surgical masks.

Even with that downgrade, Michaels believed that the new standard would improve the health of workers. It would cut down on the spread of familiar airborne diseases like measles, and it would create a first line of defense against new diseases. When the next SARS came, the workplace rule would ensure that it couldn't build up in the air and cause avoidable deaths.

Michaels knew that it would take years for his airborne standard to snake its way through the government's regulatory labyrinth. But he guessed that by 2019 it might go into effect. If another pandemic hit in 2020, it might keep workers safe.

○ ○ ○ ○

Although the CDC opposed Michaels's new workplace standards, the agency did recognize how little was known about influenza spread. They could not say how 2009 H1N1 had managed to infect 1.6 billion people in a matter of months. The CDC decided to hold a meeting the following year at its Atlanta headquarters to review the evidence. "It is surprising that, despite numerous studies dating back to 1918, the predominant mode or modes of transmission of influenza between humans remain to be clearly documented," the meeting's organizers later wrote.

Disease transmission experts like Donald Milton came to share their work and to talk about what research should be done next. Some talked about recent experiments on ferrets and guinea pigs in which the animals readily spread the flu from one cage to another—which seemed possible only by airborne infection. But a Canadian doctor named Mark Loeb recounted how he compared infections in workers who wore surgical masks to those who wore N95s in a Montreal hospital during the pandemic. He found no difference. The result called into question how important airborne spread was. And Donald Milton—who had moved to the University of Maryland at the start of the pandemic—described how he was continuing to have sick students breathe into his machine to gather data.

At the meeting, Milton talked about the kind of experiment that might cut through the fog of confusion. He envisioned a study akin to the one that took place in 1957 at the Livermore VA hospital, where ultraviolet lights seemed to block the influenza virus. But that study was not a formal experiment; the hospital staff just switched on lamps that were already in place. Milton wished for a new influenza study that would have the same rigor that William Firth Wells and Richard Riley used in their tuberculosis experiment at Loch Raven when Milton was a boy.

Milton met a scientist at the meeting who was also eager to run such an experiment. Jonathan Van-Tam, an epidemiologist at the University of Nottingham, already had carried out a study in which he sprayed influenza viruses up the noses of volunteers and then had them spend two days with uninfected people to see if they could spread their infection. They ate together and watched television together. When they played Monopoly, they released droplet-laden shouts when they landed on Boardwalk or had to go directly to jail. When they played Twister, they rubbed their clothes and skin together. Through all that contact, the infected volunteers passed on the virus to four out of fifteen uninfected ones.

Now Van-Tam wanted to run a bigger experiment. Milton joined his team, using Gesundheit II to regularly measure how many flu viruses were in the breath of infected volunteers. The team called itself the EMIT Consortium (EMIT standing for Evaluating Modes of Influenza Transmission). Three years would pass after the Atlanta meeting before they were ready to launch the new study.

The scientists took over a hotel in England, where they infected forty-two volunteers with a nasal spray and placed each one in a small group of uninfected people. Out of the seventy-five uninfected volunteers, thirty-five took no precautions to avoid picking up the flu. The other forty rubbed sanitizer on their hands and wore face shields to block any big droplets fired their way from coughs. But if influenza was indeed airborne, it might have been able to infect them despite those safeguards. Virus-laden droplet nuclei could float around the sides of the shields and then get inhaled.

After all the years of preparation, EMIT ended in disappointment. The virus managed to hop only once from an infected volunteer to an uninfected one. The volunteer who got sick did not wear a face shield or use hand sanitizer. But with just one infection to analyze, Milton and

Van-Tam could not draw any firm conclusions about how influenza spread. They were left to guess why their experiment had failed, and to think about a new one that might work.

Experts who had been skeptical that influenza was airborne before the pandemic saw little reason to change their view in its aftermath. John Conly, the Canadian doctor who had argued for short-range transmission of SARS, coauthored a review of acute respiratory infections in 2013 that brushed the possibility away. "It had been assumed in the past that the airborne route of transmission was important," the experts wrote. "Research over the years has provided evidence that this is not the case."

The World Health Organization invited Conly to draw some lessons from the experience with 2009 H1N1 for how to deal with future pandemics. He and his colleagues acknowledged that the scientific understanding of how droplets and droplet nuclei carried pathogens was still evolving. But as they described how diseases spread, they still divided aerosols into two familiar groups: droplets bigger than five microns and droplet nuclei smaller than five microns. The team recommended a wide range of precautions if an entirely new respiratory infection struck. Hospitals should even prepare for the possibility that it was airborne, spreading from room to room like smoke. But such drastic measures were not necessary for influenza. Conly and his colleagues recommended precautions only against fomites and short-range droplets.

○ ○ ○ ○

While Milton and Van-Tam were trying to crack the nature of influenza, a second coronavirus emerged. In June 2012, a sixty-year-old man in Saudi Arabia came down with pneumonia and died after eighteen days. His lungs were loaded not with SARS-CoV but with a different kind of coronavirus. More people in the region began succumbing to the new pneumonia, which came to be known as Middle Eastern Respiratory Syndrome, or MERS for short. The coronavirus, which was named MERS-CoV, jumped from bats to camels, which then sometimes passed it on to people.

MERS could not spread in homes or schools or factories, but it could race through hospital wards, infecting bedridden patients with weak immune systems. In May 2015, a man brought MERS from the Middle East to his home in South Korea. After he developed pneumonia, he spread it through clinics and hospitals as he sought medical help. The outbreaks he sowed raised the disturbing possibility once more that coronaviruses could travel through the air. Scientists who visited hospitals where MERS hit captured MERS-CoV from the air in hallways. Epidemiologists found that the virus was traveling from patient to patient even when they were lying in beds more than six feet apart.

Once hospitals put controls in place, they conquered MERS much as they had SARS. The new coronaviruses subsided, killing fewer than nine hundred people. The world largely forgot about MERS, along with the possibility of its airborne spread. Yet the world remained vulnerable to airborne nightmares.

On Christmas Day in 2013, in the West African country of Guinea, a baby got a fever, vomited, and died. More people came down with similar symptoms in the region. The disease turned out to be Ebola, which had never struck West Africa before. In early 2014, it reached towns and then cities, spreading through Guinea, Sierra Leone, and Liberia. It found victims left vulnerable by poverty, civil war, and broken health care systems. "We talked about the Ebola stare," one doctor later recalled. "You could diagnose this Ebola just by looking at someone. They would get out of an ambulance, and they would just look through you. They were just so overwhelmingly fatigued they could barely speak to you, and they'd stagger in, lie down, and then often die within hours."

As the death toll climbed above those of previous outbreaks, the world worried that Ebola was just one airplane ride away. Thanks to *The Hot Zone* and *Outbreak*, the disease took on an almost mythical horror. Fears of airborne Ebola—the nightmare that Preston and others had toyed with—now resurged in the United States. People sent letters to newspapers calling for a ban on travel from the infected countries to prevent Ebola's airborne spread. Michael Osterholm published an op-ed piece in the *New York Times* in which he warned that Ebola might be on the verge of becoming airborne. Invoking *The Hot Zone*, Osterholm asserted that in the Reston facility, Ebola had spread through the air from monkey to monkey. "We must consider that such transmissions could happen between humans, if the virus mutates," Osterholm warned.

Five years earlier, Biden had been mocked for worrying about airborne influenza. But now airborne Ebola proved all too easy for people to imagine. The CDC felt the need to publish a web page explaining that Ebola demonstrated no ability to spread by air. And Anthony Fauci had to pay a visit to the White House to allay Obama's concerns.

"Tony, explain to me what that's all about," Obama asked Fauci. "Could it go airborne?"

"I sat down with him," Fauci later recalled to me, "and said, 'If you look at the epidemiology of it, the people who are coming into physical personal contact—the morticians, the health care providers in the family—they're getting infected. Other people are not getting infected. Number two, there is no evidence of infection in the preclinical stage. It's only when someone is dramatically ill."

Ebola never became airborne, and the United States did not suffer an outbreak. In West Africa, the outbreak peaked in 2015 and dwindled over the next few months. A total of 11,325 people died—all but fifteen of them in Guinea, Liberia, and Sierra Leone. The United States helped stop the outbreak by supplying protective equipment like gloves, suits, and goggles. Public health workers, some of whom sacrificed their lives to fight Ebola, helped communities thwart the virus by modifying some of their most cherished practices, such as washing the bodies of the dead before burial. Ebola was stopped by reckoning with how it spread on Earth, not how it might float like a cloud in a movie.

○ ○ ○ ○

After the Ebola epidemic ended in January 2016, Obama entered his last year as president. With a healthy respect for infectious diseases, he ordered an update to the government's plan for a pandemic. The National Security Council put together a pandemic playbook complete with a flow chart of actions to take in the face of different pathogens—the worst of which would be capable of "rapid airborne transmission."

The playbook would become part of Obama's legacy to help future presidents respond to a medical catastrophe. But that legacy had a number of flaws. Obama's administration responded to the influenza outbreak in 2009 by burning through much of the Strategic National

Stockpile. In the years that followed, they failed to secure enough money from Congress to replenish the masks. The Obama administration also came up with an ambitious plan for a factory that would crank out 1.5 million masks a day. But the plan was still on the drawing board when Obama left office.

When Donald Trump came to the White House, Obama's team tried to walk the new president's staffers through the steps they should take in an influenza pandemic. But Trump's people seemed to consider the exercise "really stupid," in the words of one member of Obama's team. The pandemic playbook was promptly forgotten. "It just sat as a document that people worked on that was thrown onto a shelf," a former official who served in both the Obama and Trump administrations told reporters for *Politico*. "It's hard to tell how much senior leaders at agencies were even aware that this existed."

As the Trump administration settled into power, it demonstrated a hostility to preparations for a pandemic. OSHA had finally finished its draft of an infectious disease standard for workplaces after seven years of effort. The White House promptly abandoned it. It fired the pandemic czar. And it made no effort to rebuild the country's supply of masks and scrapped the plans for the mask factory.

○ ○ ○ ○

In January 2017, as Trump took office, Linsey Marr flew to Singapore. She had been working for over seven years on airborne infection, and she was finally going to her first scientific workshop on the subject. It attracted most of the scientists in the world working on the problem—a motley crew with a motley set of results to share on viruses, building design, and the flow of air. "That was the first time we had really come together and put these different things together," Marr said. Marr got to meet scientists she regularly cited in her own papers but had never met in person—people like Yuguo Li, Donald Milton, and Lidia Morawska. Altogether, they came to twenty people.

Marr told the little band about some of the work on influenza she had been doing in Virginia. She had been carefully observing how long the virus could survive in the air. After an hour, Marr found, the viruses remained viable. Milton talked about moving to the University of Maryland, where he got hundreds of students to breathe into Gesundheit II. He estimated that in half an hour, people sick with the flu could exhale a thousand infectious viruses. They did not have to sneeze to release them. Breathing was enough.

At the end of the meeting, the scientists drew up a list of questions they should address in the years to come. But they would not be able to answer those questions unless they could get support from their colleagues.

"How does this community break through current dogma?" they asked.

The community tried as best it could. In 2018, Lydia Bourouiba recorded a talk at TEDMED, an annual conference on medicine. Wearing a bright red blazer and with a tiny microphone taped to her cheek, Bourouiba paced a stage as she introduced her audience to William Firth Wells and then showed her own pictures of turbulent clouds bursting from people's mouths. She warned that the clouds could waft across entire rooms and spread through buildings. If we could better understand the basic principles of airborne infection, we would be better able to deal with new pathogens. "We may even be able to prevent the next pandemic," Bourouiba promised.

Word of Donald Milton's work also spread, with the help of Anthony Fauci himself. Milton had just published the results of the Gesundheit II experiment in the *Proceedings of the National Academy of Sciences*. His finding that droplet nuclei could carry influenza viruses drew attention from the press. A television reporter asked Fauci what Milton's work meant for the spread of influenza.

"What a machine like the Gesundheit tells you is that it actually can spread by aerosol," Fauci declared. And aerosols could travel beyond two or three feet. "So what we've learned from that is that a person can go into a room and not necessarily sneeze on you to get you infected, which is worrisome," Fauci said.

It was an astonishing thing to hear the country's top infectious disease expert say on television. Yet Fauci's comment, like Bourouiba's TEDMED talk, failed to dispel the confused neglect of airborne infection. The dogma remained as strong as ever. Public health authorities continued to claim that only droplets smaller than five microns across could cause airborne diseases. When the US government updated its guidelines for preventing pandemic influenza, it still assumed the flu was spread mainly by large droplets falling to the floor at short range. It recommended washing off surfaces, covering sneezes, and canceling school if needed. The guidelines made no mention of stopping airborne spread. They did not bring up N95s, better ventilation, or ultraviolet light.

The full threat of an airborne disease seems to have eluded many experts on infectious disease—even when they were trying to imagine a worst-case scenario. That failure of imagination was put on display in October 2019 in a hotel near Central Park. A group of scientists, public health experts, and politicians gathered there for a new pandemic game.

Twenty years had passed since the first tabletop exercise drove home the catastrophic consequences of an airborne bioterror attack. The early games had used familiar scourges like anthrax and smallpox, but their designers were now turning their attention to emerging diseases. The World Health Organization was warning that the next pandemic might be caused by a pathogen new to science—what they referred to as Disease X. Epidemiologists would not be able to rely on previous pandemics to understand how Disease X spread, and vaccine makers would have to create new formulas from scratch and test them on the fly. The players who gathered to play the 2019 game now faced their own Disease X: a hypothetical coronavirus that had spilled over from pigs in Brazil. The disease it caused, coronavirus acute pulmonary syndrome (CAPS for short), moved as fast as the flu and killed as efficiently as SARS. It jumped from Brazil to Portugal and then to the United States.

The players didn't have a chance against CAPS. What made it particularly hard to fight was its ability to infect many people without causing symptoms. As carriers silently spread the disease, governments got locked in time-wasting conflicts about how to allocate money and expensive antiviral drugs. Vaccines remained far out of reach as the pandemic raged. Meanwhile, social media platforms were overrun with trolls who fueled distrust of public health authorities. As people stayed home to avoid infection, the global economy froze. By the end of the exercise, 65 million people had died.

The exercise was designed to get governments and corporations to think about the unthinkable. And yet CAPS should have been worse. "Transmission is via the respiratory route, mostly by respiratory droplets," the center's team wrote, "with some proportion being airborne during aerosol-generating medical procedures." If the game designers drew on the work of scientists like Yuguo Li and Donald Milton, they could have allowed CAPS to spread like smoke through schools, offices, planes, and bars.

"We could never get any traction with the medical community, because we weren't MDs, I guess," Marr recalled. She came to accept that her little community would work in obscurity for the foreseeable future. At some point the neglect would have to end, because physics could not be denied forever. "But I seriously thought it was going to take thirty years," Marr said.

° ° ° °

While CAPS was wreaking havoc in New York, Kim Prather was working in San Diego on a new theory about airborne infections. William and Mildred Wells had dismissed the outdoors as a place where droplet nuclei could spread diseases. But as Prather looked out at the ocean, she wondered if the waves might set loose microbes that could cause diseases—perhaps even cholera. Prather was imagining, in effect, a twenty-first-century cholera miasma.

Prather knew that the vast majority of microbes that rise from the ocean are marine species, adapted for life at sea. Some bob in the water capturing sunlight. Others act like single-celled vultures, scavenging organic carbon released by dead organisms. But Prather knew that the ocean off the coast of San Diego regularly carried dangerous human pathogens. When heavy winter rains overwhelmed a sewage plant several miles to the south in Tijuana, its sewage flowed out an estuary into the ocean and was then carried north. As the waste got pushed toward the shore by pounding waves, San Diego beaches were sometimes closed to prevent swimmers from ingesting bacteria that could make them sick.

When Prather looked out at the contaminated waters, she watched the waves rise and break. In her mind's eye, she could see microbes getting captured inside droplets that soared into the air. She knew from her research up and down the Pacific Coast that winds blowing eastward could carry some of those living droplets onshore. Could people who lived miles inland inhale bacteria from the sewage?

In 2015, Prather had led an experiment to see whether it was even a question worth asking. Her students dumped a fluorescent pink nontoxic dye into the estuary. For days, they flew an airplane over the coast and watched the dye flowing into the ocean and then drifting north for miles. Prather's team also set up air samplers almost half a mile inland to catch droplets blowing off the waves. Some of the droplets glowed pink. In 2019, Prather followed up on that experiment by trying to catch microbes instead of pink droplets. After heavy rains fell in January, her team captured sewage bacteria in the winds blowing over Imperial Beach, two miles north of the Mexico border.

Prather kept her results quiet at first, knowing that they might create a huge controversy that could spread far beyond Southern California. "It's a global issue," she told me. "You can just think of all the regions where the water is contaminated on the coastline." Half of Bangladesh's population lives along the Indian Ocean, she observed. Cholera was a menace to the people who lived on its coastline. Perhaps some of the bacteria were infecting people not through their drinking water or tainted food but by air.

Before getting ahead of herself, Prather wanted to see if people in Imperial Beach really were inhaling bacteria and viruses from Tijuana sewage, and then to see if pathogens made anyone sick. The best people to study, Prather decided, were lifeguards. For hours a day, they scanned the waves from their beach towers, breathing in the ocean wind and anything it carried.

Prather was an expert at building wave machines and air samplers, but examining lifeguards was something new for her. She puzzled over how to run her experiment until she met a San Diego entrepreneur named Dawn Verdugo. Not long before, Verdugo had founded a company called yuFlu to develop an at-home influenza test. Customers would put a swab in their noses, rub it around a few times, and then send it to yuFlu to analyze. The company would fish out genetic material from the viruses, bacteria, or fungi multiplying in their nostrils and let their customers know if they found the signature of influenza.

Prather found the notion of pulling genes out of people's nostrils verging on science fiction. The technology would now make her experiment possible. She would wait for the next heavy rain to send sewage into the Pacific, and then she would set up air samplers to catch droplets flying in from the ocean. At the same time, her team would use yuFlu tests to swab the lifeguards and discover what was living in their noses. If her hypothesis was right, she would find some matches.

In February 2020, Verdugo delivered a box of swabs to Prather. She set it on her desk, with a plan to start organizing the experiment over the next few weeks. And then one day in March, Prather didn't come to work. On that day, the entire campus of the University of California San Diego stood empty. The box of swabs sat on Prather's desk for months in the silent building. Disease X had arrived.

PART 5

Wuhan and Beyond

sixteen

DISEASE X

Something was not right with the old couple, Zhang Jixian thought.

Zhang was a doctor in Wuhan, a Chinese city of 12 million residents situated where the Han River flows into the Yangtze. In 2003, she had served at the front lines of the SARS epidemic, and she had gone on to become the director of the respiratory and critical care department at the Hubei Provincial Hospital of Integrated Chinese and Western Medicine. On December 26, 2019, Zhang puzzled over an elderly married couple who came to the hospital with high fevers and hacking coughs. Even with her deep experience with pneumonia, she was not sure what was making the old couple sick.

At first she thought it might be the flu. After all, a wave of influenza was roaring through Wuhan that winter. But a flu test came back negative. So did tests for other common viruses. When Zhang looked at the CT scans of the couple's lungs, she noticed they both had hazy curtains of ground-glass opacities—a sign of inflammation and damaged tissue. That was not something she saw in typical pneumonia cases. But it was something she might have expected from SARS, a coronavirus that no one had seen for over a decade.

The couple's son had brought them to the hospital, and Zhang told them to call him in from the waiting room. She observed that he was in good health. Nevertheless, she demanded that he get a CT scan on the spot. The son refused, protesting that he was not sick. He accused Zhang of trying to fleece his family. Zhang persisted, and eventually she won out.

When the son's scan arrived, Zhang saw ground-glass opacities a third time. Seeing opacities in someone in good health was particularly odd—even for SARS. Zhang suspected she was dealing with something new.

Three people infected with an unknown disease, one that could reach deep into the lungs: two old people dangerously ill and the third oblivious to the pathogens multiplying inside him. In that one family in Wuhan, the future of the world was inscribed.

○ ○ ○ ○

Over the next few days, Zhang saw more patients with the same symptoms. "Something is definitely wrong!" she told a hospital superior. She suspected the pathogen moved easily between people. Zhang didn't know exactly how, but that uncertainty made her only more cautious. She isolated infected patients and ordered her staff to put on N95 respirators and canvas suits. Zhang did not know it, but other doctors around Wuhan were seeing patients with the same symptoms, and some of them had also become alarmed. "At the end of December, the signs of human-to-human transmission were already very obvious," Zhang Li of Jinyintan Hospital later recalled. "Anyone with a little common sense could reach that assessment."

A number of the cases had a link to the same place: the Huanan Seafood Market, a football-field-sized complex of stalls where food—including live wild mammals—was sold. On December 31, the Wuhan Health Commission released the first public notice of the outbreak, identifying twenty-seven cases. The announcement offered a murky mix of speculation and contradictory guidance. "So far the investigation has not found any obvious human-to-human transmission," the commission said—ignoring the opinion of Zhang Li and other Wuhan doctors. If the disease was instead spreading from animals at the market to people, as the notice implied, then stopping the outbreak would require eliminating that exposure. But the commission recommended instead to avoid virus-laden air. "The disease can be prevented and controlled by maintaining indoor air circulation, avoiding closed, airless public places and crowded places, and wearing a mask when going out," it said.

The next day—the first day of 2020—Wuhan police shut down the market. Doctors across Wuhan collected sputum from patients to test for pathogens. Their mucus shared a common inhabitant: a new coronavirus. It was not SARS-CoV, nor was it MERS-CoV. It was a new coronavirus that needed a name of its own. Virologists decided to call it SARS-CoV-2.

Although the outbreak was now burning across Wuhan, the city officials reported only a few new cases. Their rules for counting cases masked the true scale of the disaster: patients had to be running a fever, have some link to the Huanan market, and test positive for SARS-CoV-2. Even after a sixty-one-year-old woman from Wuhan showed up in a Bangkok airport on January 8 with the new disease, officials remained quiet. It was not until January 20 that Chinese experts publicly acknowledged that the virus was spreading from person to person. An observation that a single patient managed to infect fourteen people made that conclusion inescapable.

By then, much of China was on the move. People were traveling across the country to visit relatives for the Lunar New Year, and residents of Wuhan took the virus with them. They infected many others in late January 2020. A few of them got sick in revealing ways.

On January 22, a twenty-four-year-old man boarded a bus in the city of Changsha, more than two hundred miles from Wuhan. The man—scientists would later refer to him as Mr. X—got on the bus feeling tired but otherwise fine. Forty-five other passengers boarded the bus, which drove over three hours to another city. There Mr. X changed to a minibus to get to his home village, traveling with fifteen other passengers. When Mr. X got home, he developed a fever and tested positive for SARS-CoV-2. Contact tracers discovered that seven of his fellow passengers on the bus became infected after riding with him. On the minibus, two people got sick as well.

Mr. X did not cough or sneeze on his journey. He talked to no one. And yet he managed to infect people rows away from him. All he did was breathe.

On January 23, a family boarded a train in Wuhan and headed six hundred miles south to Guangzhou, the Chinese city where the first major outbreak of SARS had taken place seventeen years earlier. In Guangzhou, the family went out for a Chinese New Year lunch, choosing a crowded restaurant on the third floor of a five-story building. The five of them sat down around a table next to a big window along the west wall. One of them, a sixty-three-year-old woman, was carrying SARS-CoV-2.

To counter the muggy Guangzhou weather, the restaurant ran five air conditioners that were bolted to the top of the north wall. They sucked in stale air from the room, cooled it, then blew it out at the south wall. Each air conditioner created its own swirl of recirculating air that enveloped a set of tables. The air kept swirling all afternoon, like tiny stalled hurricanes.

After lunch, the sixty-three-year-old woman fell ill and ended up in a Guangzhou hospital, where she tested positive for SARS-CoV-2. In the following days, four other members of her family also got sick. The virus had not limited itself to their table: diners at the tables on either side of the family also got sick. The woman from Wuhan managed to infect nine people in all, some as far as ten feet away. But they had all sat inside the same stalled hurricane of recirculated air. Outside their viral vortex, there were no other infections.

No one at the time recognized the lunch in Guangzhou as a superspreading event. No one jumped off the bus from Changsha, shouting that it carried passengers freshly infected with SARS-CoV-2. Weeks would pass before those two telling incidents came into sharp focus, only after public health workers and scientists pieced together hospital records, closed-circuit TV recordings, and seating plans. In late January 2020, when Mr. X took his ride and the Wuhan family had their New Year's lunch, SARS-CoV-2 remained mysterious. Beyond China, the mystery ran even deeper.

o o o o

American health officials saw the December 31 warning from Wuhan, but they didn't know what to make of it. "We were getting an evolution of information that seemed to change from week to week," Anthony Fauci told me later. At first the Wuhan pneumonia seemed like bird flu:

a dangerous disease that could spread from animals to people but not between people. "And then it's 'Oops, wait a minute, it can spread from person to person,'" Fauci said. "And then a week or so later, 'Well, wait a minute, it transmits really, really well from person to person.'"

It wasn't until January 18 that Health and Human Services Secretary Alex Azar first informed President Trump about the disease. He warned Trump "that the virus could potentially be a serious public health threat," as he later recalled. Trump reportedly brushed off the warning. He was apparently more interested in scolding Azar for angering his supporters by supporting a ban on flavored e-cigarettes.

The threat of a pandemic did not move Trump to action, despite his well-known germophobia. "I'm not a big fan of the handshake," he said in a 1999 television interview. "I think it's barbaric. I mean, they have medical reports all the time. Shaking hands, you catch colds, you catch the flu, you catch this. You catch all sorts of things. Who knows what you don't catch?" After Trump announced he was running for president in 2015, new campaign staffers were informed that they were not allowed to cough or sneeze in the same room as Trump. When Trump became president, his germophobia did not abate. He was accompanied one day aboard Air Force One by his White House communications director Anthony Scaramucci and noticed he was hoarse. Trump would not let Scaramucci sit next to him on the plane until his doctor gave the communications director a shot of penicillin in the rear end.

The day after Azar warned Trump, the first case of SARS-CoV-2 came to light in the United States: a man who had returned from Wuhan to Washington State. By then, Trump was in Switzerland to attend the World Economic Forum. A reporter at the meeting asked the president about the news of the first American case. "We have it totally under control," Trump said. "It's one person coming in from China. We have it under control. It's going to be just fine."

Another week passed before any of Trump's team dared to confront him with the fact that it would not be just fine. At a January 28 intelligence briefing, his national security advisor Robert O'Brien warned that the coronavirus would become the Trump administration's greatest national security threat. O'Brien's deputy, Matthew Pottinger, backed up that dire prediction. Pottinger had worked for the *Wall Street Journal* in

China before joining the White House, and he had been talking to his old sources there. One doctor told Pottinger that half of the people who tested positive for the SARS-CoV-2 had no symptoms. It had the ability to spread silently, which would accelerate its dissemination. "Forget SARS," the doctor told Pottinger. "Think 1918."

Pottinger recommended that Trump ban flights from China. Anthony Fauci and other advisors agreed. Trump took the action on January 31 and assured Americans nothing more had to be done. "We pretty much shut it down coming in from China," Trump declared in a February 2 interview.

But off camera, he had something else to say. On the evening of February 7, 2020, Trump called the journalist Bob Woodward to chat about a book Woodward was writing about his presidency. Trump said he was confident that China's president Xi Jinping would rein in the epidemic. But he also worried that the situation was "very tricky."

Woodward asked Trump what he meant by tricky.

The president said that the new disease was "more deadly than even your strenuous flus."

Months would pass before Woodward reported what Trump said next: that the coronavirus spread in a dangerous manner. "It goes through the air," Trump said. "That's always tougher than the touch. You don't have to touch things. Right? But the air, you just breathe the air and that's how it's passed."

By the end of January, some countries had begun taking aggressive steps to stop the virus. South Korea, haunted by the SARS and MERS outbreaks, carried out large-scale testing and distributed protective gear from a medical stockpile. But the United States, supposedly the world's leader in pandemic preparedness, failed spectacularly. The test that the CDC created for SARS-CoV-2 proved defective, and weeks would pass before they fixed it. All the while, evidence was emerging that SARS-CoV-2 was indeed spreading silently. On January 30, researchers in Germany published a case study on a symptomless traveler from Shanghai who had spread the virus at a business meeting. On February 17, Chinese researchers reported substantial undocumented infections in Wuhan. Two more papers that came to similar conclusions appeared in the next few days. And yet, even in late February, the CDC did not consider those studies significant enough to change its policies. At least not yet.

° ° ° °

Yuguo Li, the University of Hong Kong engineer who had tracked the 2003 SARS outbreak in Amoy Gardens, didn't know what to think at first. "At the beginning, I didn't want to say it was airborne," he told me. "There was no evidence. You only knew it spread very fast."

It was possible that SARS-CoV-2 was spreading in fomites—droplets that dried on surfaces and then transferred the viruses to people's hands and then to their mouths and noses. To figure out what was happening, Li sent WeChat messages to ten of his former students who were working across China. He asked for details about any outbreaks they heard of, and for them to forward his request. Once Li had a database of municipal outbreak reports, he began studying it, hoping to see a pattern.

In Virginia, Linsey Marr could rely only on international news to keep up with the outbreak. "I was worried about it in January of 2020, based on what I was seeing in China," Marr later recalled, "the speed at which it was spreading, and photographs of health care workers coming out of China in full respiratory protection." The effort to put on full bodysuits, goggles, and masks suggested the Chinese knew the virus was especially dangerous. "It's time and cost intensive. And a pain," Marr said. "I don't think they would use that unless they really needed it."

In Australia, Lidia Morawska was getting ready to travel to Bonn for a meeting. At the end of January, Germany still seemed safe enough to visit. But when Morawska boarded her flight in Brisbane, it was almost empty. "That was something out of this world," she said. When Morawska changed planes in Dubai, her connecting flight was full. She strapped on her surgical mask. As far as she could tell, the only other masked passengers were all Asians.

After Morawska finished her meeting in Bonn, she made her way back to the Frankfurt airport on February 4. She was headed for Japan for a week of skiing. Waiting for her flight, Morawska noticed an email message from Junji Cao, a colleague in China with whom Morawska had recently worked on a survey of bacteria in city air.

Cao said he was worried that public health authorities were ignoring the possibility that SARS-CoV-2 was airborne. He had written a commentary and asked Morawska if she would collaborate with him on it. Morawska spent her flight to Japan editing Cao's draft. Once she got back to Australia, she traded emails with Cao as they finished the piece. Looking back at the SARS epidemic in 2003 and at their own subsequent studies, they worried that SARS-CoV-2 might travel long distances in droplet nuclei floating in the air. They urged the world to take the idea seriously.

One high-profile journal rejected their piece, and then a second turned them down as well. One editor assured Morawska and Cao that it didn't need to be published. Public health authorities were already considering the possibility of long-range airborne transmission. Morawska found that claim to be bizarre. In February 2020, public health authorities were flatly rejecting that possibility. They were even telling the entire world to ignore it.

○ ○ ○ ○

On February 11, 2020, the World Health Organization held a live press conference at the agency's headquarters in Geneva. A group of top infectious disease experts crowded together around three rectangular tables arranged in a horseshoe. Not one wore a mask. At the center sat the director of the World Health Organization, Tedros Adhanom Ghebreyesus.

Tedros, a fifty-four-year-old doctor who had served as Ethiopia's minister of health, looked into the camera and began delivering an update. "First of all, we now have a name for the disease, and it is Covid-19," he said. Next, Tedros turned to the numbers. "One thousand and seventeen people in China have lost their lives to this virus," he said. "Outside China there are 393 cases in twenty-four countries and one death."

The WHO was coordinating a global response to the disaster, Tedros said. That included developing vaccines. "The first vaccine could be ready in eighteen months," he cautioned, "so we have to do everything today using the available weapons to fight this virus."

The officials then started answering questions from the reporters tuning into the press conference via their computers around the world.

A journalist with CNN asked Tedros to comment on a Hong Kong doctor's claim that the coronavirus might ultimately infect 60 percent of the world. Tedros warned against speculations. He pointed out that some people thought in 2015 that Ebola would sweep the planet too. But public health experts extinguished that epidemic, even without a vaccine.

"Of course Ebola and this are not the same," Tedros acknowledged. "Ebola is lousy. This is airborne—corona is airborne."

Tedros went on to explain what he meant. "It's more contagious," he said, "and you have seen how it went into twenty-four countries, although it's a small number of cases. In terms of potential to wreak havoc, the corona is very different from Ebola. Corona has more potency, virulence. We take it more seriously."

The next reporter asked about the disease's new name. As her voice piped over the speakers, the man sitting next to Tedros leaned over and spoke to him off mic. It was Mike Ryan, a burly Irish doctor who served as executive director of WHO's Health Emergencies Programme. Ryan scribbled something on a piece of paper and drew a line under it. Tedros looked at it and smiled. A few moments later, at the end of another question, Tedros jumped in.

"Okay," he said. "Sorry. I used the military word *airborne*. It meant to spread via droplets or respiratory transmission. Please take it that way, not the military language," he said, laughing. "Thank you."

It was a hard comment to decode. What was military about *airborne*? Was Tedros trying to say that Covid-19 was not a form of biological warfare? Was he trying to keep people from panicking that the virus could spread like smoke? Why had he laughed? What had Ryan written? The reporters that day did not ask such questions. They moved on, and Tedros never circled back.

Instead, the WHO continued to downplay the possibility that Covid-19 might be airborne. "Covid-19 is transmitted via droplets and fomites during close, unprotected contact between an infector and infectee," a joint WHO–China mission announced on February 28. "Airborne spread has not been reported for Covid-19 and it is not believed to be a major driver of transmission based on available evidence."

To reduce the spread of Covid-19, WHO offered up the same advice that had been offered for years, including the need to stay clear of soggy raisins. "Maintain at least 1 metre (3 feet) distance between yourself and other people, particularly those who are coughing, sneezing and have a fever," the agency advised.

That advice frustrated Donald Milton. He had had a queasy feeling about Covid-19 ever since January, when he read a report from China. He was startled to see an X-ray of ground-glass opacities in a ten-year-old, which suggested that the virus was multiplying in the deepest recesses of the child's lungs. If that were true, then the child might have been able to spread the virus simply by breathing. The narrow passageways of the lungs would have stuck shut and opened, producing tiny droplets that could easily stay airborne. If that were true, Milton realized, then the WHO's advice wouldn't keep people safe.

"I would think about more than just washing my hands," Milton recalled telling people. "I would think about the risk of airborne infection."

And yet China and the World Health Organization were brushing away that possibility based on old assumptions rather than looking at new evidence. "I don't think they know," Milton told National Public Radio. "And I think they are talking out of their hats."

○ ○ ○ ○

In Boulder, Colorado, Jose-Luis Jimenez was getting very confused. Jimenez worked at the University of Colorado as an atmospheric chemist, studying how pollution from cars and refineries spreads through the air. He was not an expert on infectious diseases, but he knew enough about the air to know that what the World Health Organization was saying about Covid-19 did not make sense. He was particularly baffled by the claim that droplets bigger than five microns would fall to the ground within a few feet.

"This is so mind-boggling to me," Jimenez said. "It just made it obvious that they were full of shit."

Jimenez shared his befuddlement with Linsey Marr, who had worked with him on pollution research. Marr, who now had a decade of research on influenza behind her, assured him he wasn't crazy. Marr had moved from worry to exasperation as news outlets parroted the claim that soggy raisins bearing SARS-CoV-2 would travel only a few feet from the infected. "This just infuriated me," Marr told me. "I was just frustrated because I knew the history of incorrect thinking about it."

A colleague of Jimenez at the University of Colorado was getting confused too. Shelly Miller was an expert on the dangers that indoor air can harbor, from tuberculosis to wildfire smoke creeping into homes. She was aware of the CDC's guidelines about disinfecting surfaces and such, but in early March a news item made her suspicious that Covid-19 was airborne. "It occurred to me that something was off," she said.

Miller's attention was caught by a story about a New York lawyer named Lawrence Garbuz, who had been rushed to the hospital with a SARS-CoV-2 infection. He had not traveled to China, nor had he had close contact with someone who had. After being taken to the hospital, Garbuz fell into a coma that would last for weeks. New York governor Andrew Cuomo established a one-mile "containment zone" in Garbuz's town, as if he could snuff out the virus in that little patch of New York State.

A detail leaped out at Miller, one that most people likely missed. "There was this little statement that his neighbor drove him to the hospital, and he got Covid," Miller said. "And I thought, 'Okay, how the hell does a neighbor dragging you to the hospital in fifteen minutes get Covid?' And that's when it occurred to me that something's wrong."

Miller doubted that Garbuz could have infected his friend with fomites he had left behind in the car. It seemed more likely to Miller that Garbuz's friend had inhaled some of his viral breath. If Covid-19 was airborne, Miller worried, it would take a lot more than wiping down school cafeteria tables to keep the disease from spreading.

As Miller grew worried, Yuguo Li was growing less confused. He had been working his way through the reports from his ex-students, and he could see that virtually all the outbreaks in China were taking place indoors—in apartments, restaurants, supermarkets, hotel rooms. That pattern was what he would have expected from viruses floating in poorly ventilated indoor spaces. Li then partnered with Chinese colleagues to investigate some of the better-documented outbreaks. They pumped tracer gas into the Guangzhou restaurant and discovered the bubbles of air that had enshrouded diners. They found the bus Mr. X had traveled on, climbed inside it, and measured the flow of air from its vents in the back to the driver's seat at the front.

By March, Li was confident that fomites were not spreading the virus. "We knew that airborne was possible," he said.

° ° ° °

Even in January, Rick Bright could see that American hospitals were headed for trouble. Bright at the time was the director of the Biomedical Advanced Research and Development Authority, and part of his job was preparing for pandemics. He grew worried about the Strategic National Stockpile, which included N95 respirators and other protective gear to meet the demand that pandemics would put on hospitals, where doctors and nurses would need to put them on to visit patients in isolation rooms or perform aerosol-generating procedures. But the government had been neglecting the stockpile for years. When Bright took a close look at the inventory, he was horrified to find it was down to only 12.5 million respirators. If the United States was hit with a full-blown pandemic, he estimated, it would need 3.5 billion.

That shortage would only get worse if everyday Americans got scared of the virus and snatched up N95s and surgical masks. In Hong Kong, ten thousand people stood in line overnight for a chance to buy masks from a vendor. The few American companies still making masks couldn't meet that kind of demand. Nor could the United States count on Chinese manufacturers for help. In Wuhan and other cities, Covid-19 patients were filling up hospitals, which were already running out of protective gear. To deal with its own crisis, China halted the export of masks.

Bright wanted the US government to meet with domestic mask factories to figure out how to drastically ramp up production. But his superiors waved him off. At one meeting, they assured Bright that if the United States experienced a shortage, they could just ask the CDC to update its guidelines to tell people who didn't need masks not to buy them.

"I can't believe that you can sit there and say that with a straight face," Bright replied. "Do you really believe that changing a CDC guideline to tell people not to wear masks would reduce the panic people would feel once this virus spreads?"

Soon enough, Americans were buying up the country's meager supply of masks. And as masks ran out, the government did exactly what

Bright had feared. Public health officials responded to the crisis by telling Americans not to wear them because they were not effective. On February 27, CDC director Robert Redfield told Representative Chrissy Houlahan in a congressional hearing that masks had no place in the community.

"Should you wear a mask if you're healthy?" she asked.

"No," Dr. Redfield replied.

Two days later, Surgeon General Jerome Adams repeated the message on Twitter. "Seriously people—STOP BUYING MASKS!" he tweeted. "They are NOT effective in preventing general public from catching #Coronavirus, but if health care providers can't get them to care for sick patients, it puts them and our communities at risk!"

A week later, Anthony Fauci appeared on the March 8 edition of *60 Minutes* to speak to the country about Covid-19. As a scientist, Fauci was helping fight the pandemic mainly by overseeing the development of Covid-19 vaccines. But he was also appointed to the White House Coronavirus Task Force in January, and he soon emerged as the public face of the government's campaign against the pandemic. Now seventy-eight, he was comfortable in that role, having spent the previous two decades guiding Americans through a series of crises, from Amerithrax to Ebola. Now Fauci joined press conferences at the White House to help explain the unfolding crisis, and he appeared regularly on television news.

On *60 Minutes* Fauci delivered the same message as Redfield and Adams. He tried to protect the limited supply of masks for hospitals without addressing how the United States had mismanaged its stockpile so badly over the years.

"Right now in the United States, people should not be walking around with masks," Fauci said. "When you're in the middle of an outbreak, wearing a mask might make people feel a little bit better, and it might even block a droplet, but it's not providing the perfect protection that people think that it is. And often, there are unintended consequences. People keep fiddling with the mask and they keep touching their face."

People outside the United States might have been wearing masks in public, but Fauci advised American viewers not to follow their example. "When you look at the films of foreign countries, and you see 85 percent of the people wearing masks, that's fine. If you want to do it, that's fine," he said. "But it can lead to a shortage of masks for the people who really need it."

In the back of his mind, Fauci would later tell me, he recalled Gesundheit II. Two years before the pandemic, Donald Milton and his colleagues had demonstrated that airborne droplet nuclei could carry the influenza virus. And when that study came out, Fauci recognized the possibility that flu could spread across a room like smoke.

But at the start of March 2020, he did not consider that kind of evidence strong enough to advise people to wear masks outside hospitals, especially during a mask shortage. The standard precautions for influenza should have been enough for Covid-19: people who felt sick should stay home so they wouldn't spread the virus.

The Centers for Disease Control recycled its influenza guidelines as well. In March, the agency stated on its website that SARS-CoV-2 was thought to spread mainly in close contact, traveling short distances in respiratory droplets produced by coughs and sneezes that landed on mouths and noses. The CDC called for healthy people to stay six feet away from those who were sick rather than wear a mask. The agency also recommended washing down frequently touched objects and surfaces. It cautioned against touching one's eyes, nose, and mouth. It reminded people to practice etiquette and cover coughs and sneezes.

That guidance was a choice, not an inescapable conclusion from the evidence available at the time. The day after Fauci appeared on *60 Minutes*, Japan issued its own policy. Its public health officials advised people to avoid the three conditions where clusters of infections occurred: closed spaces with poor ventilation, crowded places with many people nearby, and close-contact situations such as conversations. Places where all three conditions overlapped were the most likely places for people to get infected. That was because they were the places where the air was most dangerous. The advice was simple enough to put on a poster with room to spare: "Avoid the 'Three Cs'!"

○ ○ ○ ○

On March 10, the day after Japan unveiled its Three Cs, the Skagit Valley Chorale met for a rehearsal. Yvette Burdick drove north from Seattle to direct the singers. The evening had gotten chilly by the time she reached the Presbyterian church. At the front door, Burdick pumped disinfectant from a bottle into her hands and then walked into Fellowship Hall. A few dozen people were assembling in the high-ceilinged room. The singers took their seats, and Burdick gave them the cue to sing.

Four days later, Burdick woke up with a raging fever. She suspected that she had come down with Covid-19; she had just heard from Ruth Backlund and Carolynn Comstock that they were sick with something as well. Burdick's wife, Lorraine, called around to hospitals and clinics to find one that would test Yvette for the coronavirus. Tests were still in such short supply that she could not get one. Carolynn Comstock had no luck either. But the Backlunds did. It helped that they were in their seventies, and it also helped that their son was an emergency room physician in Seattle and had contacts he could work. The Backlunds drove to a clinic and had swabs stuck up their noses. It would take days to get the results, and while they waited, they followed the CDC's guidelines, isolating at home.

Burdick got emails and calls from more sick singers. Debbie Amos, a retired doctor, was among them. When she started coughing on Saturday night, she thought it was just a case of allergies. But Amos's husband suggested she go to their other bedroom just in case she was infected. She left the room, but it was too late: he got sick as well.

Amos was pretty sure she had Covid-19, but she didn't know how exactly she was supposed to feel. It seemed as if her sense of smell had vanished. She felt extraordinarily drained of energy. "I would wake up, do my bathroom routine, and then get dressed and go downstairs," Amos said. "And by that time, I would be so tired again that I would come and lay on the couch. I'd either just lay there until I felt a little not so tired, or I'd take a nap that might be for another couple of hours."

Ruth Backlund felt her own strange constellation of symptoms. She would have fits of yawning, and aches spread across her midriff as if she had been gardening all day. "Odd stuff," Ruth recalled. Carolynn Comstock experienced a milder bout. "I've had colds that were worse," she said, shrugging. Still, Comstock stayed in isolation, getting food delivered to the house and doing crossword puzzles. Like Amos, Comstock lost her sense of smell. She whiled away the time using Google Translate to read papers by German scientists about anosmia.

On March 15, Burdick sent an email to the choir: "We have some news of concern to relate," she wrote. Six singers in total had reported getting sick. Debbie Amos and Mark Backlund—the two doctors in the choir—answered questions and created a communication system so that everyone in the group had another member checking in on them. And they did all that even as they suffered their own cases of Covid-19. "I just felt like I was in a tornado," Amos said.

○ ○ ○ ○

As the Skagit Valley Chorale hunkered down, Covid-19 cases soared across the United States. In early March, Deborah Birx, the White House coronavirus response coordinator, developed a plan to shut down the country for fifteen days or so. Other White House officials battled against it, believing the pandemic would be no worse than ordinary influenza. Birx's original concept ended up as a watered-down set of suggestions that sick people stay at home and old people practice social distancing. On March 16, Trump announced the plan, confident that the country was close to crushing the pandemic. "I'd love to have it open by Easter, okay?" he said.

Instead, the pandemic kept blowing up. In New York City, the sound of ambulance sirens began rippling through the air both day and night. A sense of economic doom pervaded the country. The stock market crashed. Restaurants, stores, and companies closed down. In a matter of weeks, 700,000 Americans lost their jobs. US schools and colleges closed down too, sending tens of millions of students home to struggle through the rest of the academic year on Zoom or to simply vanish from their teachers' radar.

Hospitals were swamped by victims of Covid-19, and doctors scrambled to find ways to help them. Very few could spare the time to investigate the virus itself. The University of Nebraska Medical Center in Omaha was an exception. For years, researchers there had worked on defenses against biological warfare, and they had built one of the country's few biocontainment units. John Lowe, a pathologist at the center, oversaw a study on the spread of SARS-CoV-2 from the Covid-19 patients there. His team detected genes from the virus in hospital rooms—even in out-of-the-way nooks that they probably could reach only by floating through the air. Lowe and his colleagues also captured viral genes from the hallway air.

When Lowe's team released their preliminary results on March 26, they cautioned that they hadn't proven that viable viruses were floating in the air. But they also believed it was time to take the possibility of airborne Covid-19 seriously. In an April 1 letter to the White House, the National Academy of Sciences flagged the Nebraska study, along with three similar ones from China. The findings, the academy warned, were "consistent with aerosolization of virus from normal breathing."

o o o o

As the experts on airborne droplets sounded the alarm, they mostly went ignored. And when they did to grab some attention, their message often got garbled in the growing panic.

In San Diego, Kim Prather closed down her wave machine lab. Packing up her office, she took one last look at the box of swabs on her desk. She would not be testing lifeguards anytime soon. The sight of the box made her think about how much had changed in a matter of weeks. Not long ago, she had thought it strange to perform a genetic test on a snot-covered swab. Now thousands of tests were being done every day. The future had arrived sooner than she had expected.

Prather worried about the mysteries that still cloaked SARS-CoV-2. If it could travel through the air, how far exactly could it go? After all, in her own work she was used to measuring particles traveling for miles. But how far a coronavirus could go and still pose a risk remained unknown. Prather was contacted by a *Los Angeles Times* reporter, asking about the crowds of people then packing California's beaches. When the reporter asked whether the virus would fall to the sand before it traveled six feet through the air, Prather just laughed. The conversation then shifted to the research Prather had been doing at Imperial Beach on pathogens from sewage.

"I wouldn't go in the water if you paid me a million dollars right now," she said. She was thinking about the assortment of pathogens she was finding in the Pacific, not SARS-CoV-2. It was true that the coronavirus was turning up in wastewater. But Prather suspected that those viruses were so badly damaged by then that they could no longer infect people.

That distinction vanished when the *Los Angeles Times* published its story. "Beachgoers Beware: Virus May Be Lurking," the headline read. In the story, Prather sounded as if she thought she could get Covid-19 from a swim. It wasn't Prather's first time talking to reporters, but it was her first time riding the pandemic news cycle. Her Twitter feed and email inbox filled with insults, obscenities, and death threats.

Lydia Bourouiba, the MIT professor who had taken pictures of sneezes, published a warning of her own on March 26 in the *Journal of the American Medical Association*. She had started on the piece long before Covid-19, after editors had seen her 2018 TEDMED talk and invited her to contribute an essay. "In January, I essentially went back to them and said, 'Look, this thing in China clearly doesn't look good,'" Bourouiba recalled. "'We should wrap this up.'"

In her essay, Bourouiba introduced readers to William Firth Wells, fifty-six years after his death. She recounted his experiments and shared her videos of sneezes racing from people's mouths in ghostly plumes as far as twenty-seven feet. Bourouiba warned public health officials that their arbitrary six-foot rule might not be enough, especially for health care workers. They needed to appreciate "the distance, timescale, and persistence over which the cloud and its pathogenic payload travel," Bourouiba wrote.

USA Today reported Bourouiba's warning, and later that day it came up at the White House, where President Trump joined the Coronavirus Task Force for a press conference.

John Roberts, a national correspondent for Fox News, asked Anthony Fauci about the story. "There is a professor from MIT, Dr. Fauci, who suggests that the coronavirus can be carried on droplets a distance of twenty-seven feet," Roberts said. "Do you buy into that? If that might be the case, does that suggest that current social distancing guidelines may need to be extended?"

Trump stepped back and Fauci took to the podium. He was visibly irritated. "This could really be terribly misleading, John," Fauci said. He proceeded to downplay Bourouiba's warning by gripping the sides of the podium, leaning back, and delivering a big mock sneeze.

"If you go *way* back, and go *ACHOO*, and go like that, you might get twenty-seven feet," Fauci said. "When you see somebody do that, get out of the way. That's not practical, people. That is not practical, John. I'm sorry, but I was disturbed by that report, because that's misleading. That means that all of a sudden the six-foot thing doesn't work."

Bourouiba's friends let her know by email about Fauci's performance. She watched it online in disappointment. "I suspect that Fauci had not actually read the paper," she later said. "I think he tried to do a good job, but he maybe had his own echo chamber. Or he was just not aware of that research, and he just needed to give an answer."

Other researchers were having even worse luck than Bourouiba. Linsey Marr and Joseph Allen, an expert on indoor air at the Harvard School of Public Health, submitted an essay to the *New England Journal of Medicine*. The journal rejected it without comment. Meanwhile, Lidia Morawska still couldn't find a journal that would publish her piece with Junji Cao. "No one seemed to listen," Morawska recalled.

Morawska grew even more frustrated as the World Health Organization shut aerobiologists out of the advisory groups it organized for Covid-19. Those groups were mostly made up of infectious disease experts like John Conly, the Canadian doctor who had been skeptical about airborne infections before the pandemic. Meanwhile, Morawska was getting horrifying reports from colleagues in Italy, where Covid-19 had established itself and was killing hundreds of people every day. Doctors and nurses were getting infected even after taking precautions against large short-range droplets.

Despite those reports, the World Health Organization was vigorously dismissing airborne infection. In late March, it responded to a meme circulating on the Internet claiming that "COVID -19 is confirmed as airborne and remains 8 hrs in air!" WHO rebutted the meme with its March 28 tweet: *"FACT: #COVID19 is NOT airborne."*

"This was so, so utterly nonsensical," Morawska recalled. "How can you react to nonsense like this?"

Morawska ran out of patience. WHO's nonsense on Twitter drove her to draft a petition. It asked Tedros to stop denying the risk of airborne spread.

Shelly Miller felt an immense sense of relief when she got an email from Morawska asking her to sign it. "Why are we the only ones that get this?" Miller recalled thinking. She signed. So did Jose-Luis Jimenez. So did Yuguo Li, Linsey Marr, and Donald Milton. Morawska's list of signatories grew to thirty-six experts. Group 36, as they came to call themselves, sent the petition to WHO on April 1. They kept the correspondence private, hoping that behind closed doors they could persuade WHO to change.

"Dear Dr. Tedros Adhanom Ghebreyesus," they wrote, "We, scientists from around the world, who have worked for many years on the characteristics and mechanisms behind the transport of droplets expired by humans, and on airflow patterns in buildings, appeal to you to recognize the significance of the airborne spread of SARS-CoV-2 (COVID-19) and advocate for preventive measures to mitigate this."

Group 36 called for measures to keep the public safe from Covid-19 by reducing the number of viruses floating in the air. Buildings should be better ventilated. Indoor air should not be recirculated. People should avoid the flow of air coming from other people. And big groups in indoor spaces should be discouraged.

"We fear that the lack of recognition of the risk of airborne transmission of COVID-19 has grave consequences," they wrote. "People think that they are protected if they adhere to the current recommendations, but they are not."

Morawska got a quick reply from WHO, and an hour later she was on the phone with Mike Ryan and Maria Van Kerkhove, the Covid-19 technical lead. To Morawska, they sounded defensive on the call, but it ended on a hopeful note: they agreed to set up another call on April 3, which other members of Group 36 could join.

In Colorado, Jimenez woke up before sunrise for the call. He took his laptop from his nightstand and logged into Zoom. His wife woke up as well and asked if she could watch. Jimenez switched on the computer's speakers, and the two of them watched groggily as the faces of dozens of people from Group 36 and WHO popped up on the screen and the call began.

"I have to describe this call as traumatizing," Jimenez recalled.

Kerkhove started the conversation off politely, but it quickly devolved. Some WHO representatives rejected Group 36's concerns outright, dismissing their evidence as weak. Julian Tang, a British virologist with Group 36, pushed back, pointing out that WHO was claiming that fomites and large droplets spread SARS-CoV-2 while it had no rigorous proof of its own. Before long, Jimenez recalled, the temperature hit the boiling point. He watched John Conly scream at Morawska.

"He kept yelling, 'But, Lidia, where is your evidence? Where is your evidence?'" Jimenez recalled. "I was thinking, 'Jesus Christ, what is this?'"

The Group 36 scientists would later publish a collective account of the shocking experience. "After the call was finished, disappointed and frustrated, we wondered, Why are they acting like this?" they wrote. "Why are they so bluntly rejecting our arguments?"

I later asked Conly about Group 36's account. He stressed that he was not directly involved in the conflict over WHO's guidelines. With over thirty advisors from around the world helping to develop the guidelines, he wasn't that high on the WHO ladder. Conly thought that scientists like Li and Morawska did great work, but he saw many gaps in the evidence about how Covid-19 spread. They all deserved careful scientific inquiry. With so much complexity and uncertainty to contend with, disputes were inevitable. "The flattest pancake in the world has two sides," Conly told me.

Reeling in frustration after the April 3 call, Group 36 conferred about what to do next. "We were like, 'Okay, we need to find more evidence,'" Jimenez said. They had to find an outbreak of Covid-19 that they could document, taking proper account of the role that air played in the infections.

It just so happened that Richard Read, the Seattle bureau chief for the *Los Angeles Times*, had just broken a story that seemed to fit their needs exactly. Its headline read, "A Choir Decided to Go Ahead with Rehearsal. Now Dozens of Members Have COVID-19 and Two Are Dead."

○ ○ ○ ○

Six days after the March 10 rehearsal, the Backlunds officially learned they had Covid-19. Mark Backlund called the Skagit County Health

Department, which launched an investigation. Five days later, on March 21, it issued a press release that was a mix of urgency and haziness: "Skagit County Public Health is investigating a cluster of recently confirmed COVID-19 cases that has been traced to a group meeting of approximately 60 people in early March," the department announced. "With the strongest sense of urgency, the Skagit County Health Officer urges the community to abide by all health officer recommendations. People should postpone or cancel any gathering outside of immediate household members."

By then a lot of the singers were recovering, but some continued to decline. Mark Backlund slept most of the day, and he babbled sometimes when he awoke. He had to force himself to eat. The smell of coffee became so disgusting to him that Ruth could not even brew it in the morning. Three other singers fared even worse and ended up in the hospital. One was Nancy Ann Hamilton, an eighty-three-year-old retired postal worker known to her friends as Nicki.

Nicki had spent much of her life traveling the world on a shoestring budget; in Skagit Valley, she was active with the Democratic Party. After she came down with what felt like a cold, it took her husband, Victor, a few days to persuade her it might be something much worse. On March 19, he took Nicki to the Skagit Valley Hospital, where she went into isolation. When he called her the next morning, she said she was feeling better already, but when he called again that evening, she said she felt terrible. That was the last thing Victor ever heard Nicki say. She died the next day, on March 21. Victor spent her last hours alone at home.

Word of the outbreak finally started to spread. On March 25, a friend of Nicki's named Ken Stern mourned her loss in a letter to the *La Conner Weekly News*. "Nicki Hamilton loved to see her name in print. She will not read this, of course. How many more Skagit County names will be printed as fatalities, victims of this virus? Working to minimize the spread is the work of every citizen."

By then a second singer was also spiraling downward. Carole Rae Woodmansee talked to her grandchildren from the hospital by phone. She assured them that she was not scared and said she was praying for them. She quoted a hymn:

This is my story, this is my song,
Praising my Savior all the day long.

Woodmansee died on March 27, her eighty-first birthday.

"The hardest thing is that there was no goodbye," Carole's youngest child, Wendy Jensen, told a reporter. "It was like she just disappeared."

Richard Read got wind of the outbreak and contacted Burdick, Comstock, and the Backlunds. He also emailed Linsey Marr to get her scientific opinion. Marr couldn't say whether the Skagit Valley Chorale had suffered an airborne outbreak, because too many questions were unanswered. How exactly had people been arranged around the room? Was there a pattern to those who got sick and those who didn't? But even without answers to such questions, the outbreak should serve as a stark warning, Marr said. "This may help people realize that, hey, we really need to be careful," she told Read.

Read's March 29 article led to more press attention, including an interview on CNN for Burdick. The news anchor Erin Burnett asked how the choir was coping with the trauma. "One of the ways that we would have dealt with it is by being together and singing together," Burdick said. "And we can't do any of that."

While many people sent the choir their sympathies, anonymous accusations also flooded in. "One can only imagine how many people are infected because of their stupidity," one commenter said on Facebook. In fact, the choir earned the praise of the county health department for taking their isolation seriously. By staying at home, they prevented the outbreak from extending past their ranks. But the trolls who skulked online had an appetite only for shame, not for epidemiology. Their attacks sent the singers into hiding, and so Comstock was suspicious when Read passed along a message from a scientist named Jose-Luis Jimenez on April 3.

Comstock asked her daughter, a tech recruiter, to check Jimenez out. Her daughter looked into his online presence and saw his long list of scientific publications. "She goes, 'Yeah, he's legit,'" Comstock recalled. She emailed Read and told him to let Jimenez get in touch.

Jimenez emailed her with Group 36's proposal. "I thought this was wonderful," Comstock recalled. "Let's use this so we can learn something about this thing."

By then, the Skagit County Health Department was deep into their own investigation. They discovered that another singer in the rehearsal had started feeling symptoms on March 7. She tested positive for SARS-CoV-2 six days later. The Skagit County Health Department concluded that the woman, whose identity was never revealed, had brought the virus to Fellowship Hall. They called her the index case.

The county health investigators concluded that the index case had probably spread the virus to fifty-two people the night of March 10. They emphasized how often the singers got within six feet of one another—gathering around a table to grab oranges, then stacking chairs together at the end of practice. "These factors likely contributed to making this a high-risk event," the investigators later wrote. They acknowledged that long-range droplets might have helped spread the coronavirus. But they did what so many epidemiologists had done before them and focused on fomites and close-range transmission.

The health department's final report was heavy on medical detail and light on physics. The investigators did not explain how short-range large droplets could have traveled the distance from the index case to all the infected singers. They did not examine the air in Fellowship Hall or consider how it could have allowed droplet nuclei to float. But those were exactly the kind of questions Group 36 wanted to answer with the choir's help.

After the chorale board agreed to let Group 36 investigate the outbreak for themselves, Comstock got a call from the health department. "You should not work with this person," Comstock recalled them saying. The department followed up with a letter cautioning the choir not to work with any researchers who lacked approval from an institutional review board. "Skagit County Public Health is willing to support the group to screen researcher proposals for appropriate approvals and documentation," they wrote.

Comstock bridled. "That's like telling me I can't read a book because it's banned," she said. And with that, the Group 36 study began.

seventeen

HISTORY SET US UP

By April 2020, Rick Bright's fears had been fulfilled. The distribution of masks in the United States had devolved into chaos. Hospitals were largely left to themselves to decide who needed to wear a surgical mask and who should put on an N95. Some hospitals allowed face coverings only during aerosol-generating procedures, such as using a nebulizer to deliver drugs into a patient's lungs. Otherwise, they discouraged their staffs from protecting themselves from ordinary hospital air. After one Texas doctor walked down a corridor while wearing a mask, he got a furious text from his supervisor: "UR WEARING IT DOWN A PUBLIC HALL. THERES NO MORE WUHAN VIRUS IN THE HALLS AT THE HOSPITAL THAN WALMART. MAYBE LESS."

In hospitals where workers were allowed to put on masks, they sometimes discovered that the inventory had been moldering for years. Some masks fell off people's faces as the brittle rubber bands snapped. Soon hospital supplies started running out, and the Strategic National Stockpile failed to meet the demand. The Trump administration did not follow Bright's advice and get American factories to ramp up the production of masks. State governments were left to compete for contracts from price-gouging vendors overseas. The shortages became so dire that some health care workers regularly reused N95s for days. "I feel like we're all just being sent to slaughter," one New York nurse told a reporter.

In Washington, Matthew Pottinger began arguing for the public to wear masks as well. He pointed to Hong Kong and Taiwan, where many people wore masks and where Covid-19 rates were low. That was no coincidence, he believed. Many people in the White House bristled at the idea, not because they were concerned about the mask shortage but because they didn't think Covid-19 was all that worrisome. When Pottinger began wearing a mask to work at the White House in mid-March, he felt as if he had dressed up like a clown.

By the end of March, however, Pottinger had an unexpected new ally in the fight for masks: the Centers for Disease Control. Its epidemiologists had finished up an investigation of an early nursing home outbreak in Kirkland, Washington. Instead of testing only the symptomatic residents, the scientists tested them all. Of the twenty-three residents who tested positive, they found, only ten showed symptoms. Combined with studies coming from other countries, the new study suggested that silent spread might be common.

Now the agency began to give serious thought to Pottinger's idea. If everyone wore masks in public, people who were unwittingly infected could trap some of their exhaled viruses. Masks might also afford some protection to the uninfected as they inhaled the shared air. But if the CDC simply issued new guidance that the public should start wearing masks, it could worsen the ongoing shortage in hospitals. At the end of March, the agency decided on a compromise. A few preliminary studies on cloth masks carried out before the pandemic suggested they could block some exhaled viruses. The CDC advised the White House to tell people to make cloth masks at home.

Some of Trump's advisors tried to block the guidelines, but they lost. On April 3, President Trump announced the new policy. "The CDC is recommending that Americans wear a basic cloth or a fabric mask that can be either purchased online or simply made at home," he said. Fauci joined in urging people to put on masks in an interview with PBS. "The thinking is really now influenced by information that's coming in," he said, less than a month after dismissing masks on *60 Minutes*.

Birx would later look back at the switch on masks as a huge public health blunder. "Americans wanted and needed clear, evidence-based guidance on masks, and they weren't getting it," she later wrote. Instead of clear, evidence-based guidance, Americans got to watch Surgeon General Adams demonstrate how to make a mask at home. Standing at a table in front of a row of military flags, Adams folded a black T-shirt and then tied rubber bands to the sides. Somehow, when he pulled it over his mouth, it fit like a snug mask. "It's that easy," he said. It was not.

Trump made matters worse by revealing his contempt for masks. Fauci would later offer his own interpretation of the president's antipathy to covering his face: Trump "equates wearing a mask with weakness." When the president made the April 3 announcement, he sullenly recited the new policy. And when he was done, he noted that he wouldn't follow it.

"I just don't want to be doing—I don't know, somehow sitting in the Oval Office behind that beautiful Resolute Desk, the great Resolute Desk," he rambled. "I think wearing a face mask as I greet presidents, prime ministers, dictators, kings, queens—I don't know. Somehow I don't see it for myself. I just, I just don't."

◦ ◦ ◦ ◦

On April 10, Lidia Morawska and Junji Cao finally published their warning. "Based on the trend in the increase of infections, and understanding the basic science of viral infection spread, we strongly believe that the virus is likely to be spreading through the air," Morawska and Cao wrote. "All possible precautions against airborne transmission in indoor scenarios should be taken."

Unfortunately, those strong words appeared in *Environment International*, a niche journal that focused on topics such as pesticides and wildfires. They went mostly unnoticed, and the world did not start to treat SARS-CoV-2 as if it could drift like smoke across a room. Supermarkets installed plexiglass walls to shield cashiers from soggy raisins, ignoring the possibility that droplet nuclei might drift around them. Meanwhile CNN's chief medical correspondent Sanjay Gupta demonstrated on television how to wipe down groceries with disinfectant wipes.

Morawska was disappointed to find that the World Health Organization's reaction to Group 36 had not budged. The day after the *Environment*

International piece appeared, the WHO Health Operations, Infection Prevention and Control Technical Team emailed Morawska to rebut her arguments. "The role of airborne transmission for SARS-CoV-2 is predominantly opportunistic and mainly limited to aerosol generating procedures," they told her. The WHO team refused to tell poor countries with small public health budgets to divert money "into a costly and potentially unfeasible direction."

Morawska thanked WHO on behalf of Group 36 for its response. At the end of her note, she added a warning. "We believe that the matter is so important and urgent that we will have to consider any avenues available to bring it to the attention of the general public, the medical community and authorities in charge of public health," she wrote.

In San Diego, Kim Prather was looking for those avenues as well. After her media fiasco over the risk of Covid-19 on California beaches, she started posting regularly on Twitter about airborne transmission. Working with the virologist Robert Schooley and Chia Wang, an atmospheric chemist, she wrote a commentary that they sent to *Science*, one of the world's biggest journals. While they waited for it to be accepted, the death toll accelerated. On May 24, 2020, the *New York Times* filled the newspaper's front page with names of the dead. The United States had suffered 100,000 deaths from Covid-19. "The immensity of such a sudden toll taxes our ability to comprehend," the newspaper wrote.

Three days after that dark milestone, the commentary from Prather and her colleagues came out in *Science*. They laid out the current understanding of how droplets spread viruses over short and long ranges. The scientists argued that social distancing alone could not keep people safe. "For society to resume, measures designed to reduce aerosol transmission must be implemented, including universal masking and regular, widespread testing to identify and isolate infected asymptomatic individuals," they warned.

○ ○ ○ ○

Group 36 decided to turn their petition into an open letter. They reached out to more of their colleagues to sign it. Prather signed on, along with more than two hundred other scientists. Morawska then started looking for a venue where they could share the petition and stir up a public outcry.

She approached top journals once more. And once more she was rebuffed. It seemed to Morawska that journal editors were not paying attention to the evidence, both old and new, that she was presenting them. One editor accused her of fostering panic "by warning against a mode of SARS-CoV-2 transmission for which the evidence so far is very weak."

Jimenez meanwhile worked with Comstock on collecting evidence he hoped would counter that opinion. Comstock and the rest of the choir no longer had to cope with hate mail, but they still carried a psychological trauma. With the pandemic only growing worse, they realized that they would not be getting together to sing anytime soon. "To have that taken away, it was a big hole in our lives," Debbie Amos told me.

Comstock filled the silence by working with Group 36. "I went from having a very busy life to having nothing," she said. "So this was a project." Comstock worked her way through the roster of people who had been at the March 10 rehearsal.

"Everybody needed to say what had been happening—'this is what happened to me,'" she recalled. "It was exhausting." Once she and the singers worked through their pain, they turned to Jimenez's questions. Comstock asked if her fellow singers had eaten a snack at the rehearsal or if they had used the bathroom. She asked how close they had gotten to the index case. On a chart, Comstock marked the location where each person had sat on the evening of March 10. She was struck by just how far away some of them had been from the index case. "Even the people forty feet away got it," Comstock said.

Shelly Miller volunteered to lead the analysis of the outbreak. "I said, 'This is my area. I got this,'" she recalled. Miller added her own research to the effort, making calls to reconstruct the layout of the Mount Vernon Presbyterian Church. She chatted with the company that serviced its furnace to gather clues about how air moved inside the building. Then Miller started evaluating the routes SARS-CoV-2 might have taken the night of March 10 from the index patient to fifty-two other singers.

First, Miller and her colleagues considered fomites. When the index case went to the church bathroom, she might have left viruses on its doorknob. But only six other singers used the bathroom that night, making it unlikely that it had been important for the spread. Miller and her colleagues also ruled out the possibility that fomites smeared on the chairs spread the virus. Carolynn Comstock didn't even put her own chair away, and yet she still got infected.

Having discounted fomites, the Group 36 team considered whether the index case might have spread the virus in large droplets that fell to the floor within a few feet of her. Other singers at Fellowship Hall that night did not recall her sneezing or coughing, which suggested she did not expel many soggy raisins. Even if she had, the seating chart argued against that route of spread. Because so many singers had stayed home on March 10, the chairs directly in front of the index case—the ones most likely to be hit by soggy raisins—had been empty.

The only other plausible route was droplet nuclei. The index case presumably produced an abundant supply of them as she sang and made her larynx quiver. To see if her droplet nuclei could infect fifty-two people, Miller and her colleagues used an updated version of the Wells-Riley equation—the model that originated from Wells's work in the 1930s and was improved by his prodigal protégé, Edward Riley, in the 1970s. Miller ran calculations to estimate how many viruses one person would have to release to infect fifty-two other people in a space the size of Fellowship Hall. She then compared her number to studies of how many viruses Covid-19 patients exhaled. The numbers lined up. It looked eminently plausible that the index case could have spread the infection by air.

Two months after Group 36 started their search for evidence, they had it. On June 17, Shelly Miller and nine coauthors released their study, "Transmission of SARS-CoV-2 by Inhalation of Respiratory Aerosol in the Skagit Valley Chorale Superspreading Event." They concluded that inhaling shared air was likely the leading route of transmission that night in Fellowship Hall, and they discouraged any further group singing during the pandemic.

Comstock still couldn't join people in song, but she felt gratified to see the silencing of the Skagit Valley Chorale become part of the scientific literature. "It was a very good catharsis for a lot of us to say, 'Oh, well, something good might come from this crappy thing,'" she told me.

In Australia, Lidia Morawska added the study of the Skagit Valley

Chorale to the open letter. She also added other studies pointing to airborne infection, including Yuguo Li's research on the Guangzhou restaurant. She sent the open letter to the journal *Clinical Infectious Diseases,* which accepted it on June 30. Morawska had warned the World Health Organization that her team would make the most of the letter, and they followed through. They sent advance copies to journalists and gave interviews to drive home their conviction that Covid-19 was airborne. "We are 100 percent sure about this," Morawska told Read at the *Los Angeles Times.*

WHO officials could no longer quietly dismiss Group 36's concerns in back-channel exchanges. At first, they brushed off questions from reporters, downplaying the importance of the air. "Especially in the last couple of months, we have been stating several times that we consider airborne transmission as possible but certainly not supported by solid or even clear evidence," Benedetta Allegranzi, the WHO's technical lead for infection prevention, told the *New York Times.*

But WHO officials also quietly tweaked scientific briefing documents. "Short-range aerosol transmission, particularly in specific indoor locations, such as crowded and inadequately ventilated spaces over a prolonged period of time with infected persons cannot be ruled out," the agency now said. Despite that shift, WHO still recommended that hospital staff wear only surgical masks, and suggested N95 respirators only for aerosol-generating procedures.

○ ○ ○ ○

The conflict between Group 36 and WHO echoed William Firth Wells's struggles eight decades earlier. Wells had looked at diseases with an engineer's eye, using physics and statistics to track movements of pathogens inside droplets. He discovered that none of that mattered much to doctors, who put more stock in epidemiology. When Covid-19 arrived in 2020, WHO set its guidelines in much the same way. And when critics began to speak out against Group 36, they displayed a similar philosophy. In an August 2020 commentary, John Conly and his colleagues at the WHO Infection Prevention and Control Research and Development Expert Group for COVID-19 dismissed the Group 36 studies as "opinion pieces" rather than rigorous science. They declared that "SARS-CoV-2 is not spread by the airborne route to any significant extent." Conly also signed a protest against Morawska's open letter, along with more than three hundred other doctors. "The concerns raised by the authors are not borne out in clinical experience," they said.

Morawska and her colleagues weren't just making a mistake, Conly and his cosigners declared. They were harming public health with their hijinks. "This has resulted in confusion and fear in the general public, mistrust in healthcare workers, and a risk of a deepening divide between experimental scientists and healthcare epidemiologists," they wrote.

Group 36 responded to the critics by reminding them about tuberculosis. Epidemiologists did not recognize how it spread by looking only at case counts. It took William Wirth Wells obsessively pushing for the Loch Raven experiment at the end of his life to make it clear that tuberculosis spread only by air. Exalting epidemiology above every other line of evidence ensured that crucial facts would go overlooked.

The open letter clearly did not settle the debate about Covid-19 and airborne infection. But Group 36 succeeded in grabbing the attention they needed to make their case. "That's when we became respectable," Jimenez said.

○ ○ ○ ○

By the summer of 2020, some patterns of the pandemic were emerging. Old people were far more likely to get severely ill than younger adults or children. Diabetes and other conditions could also raise the risk of severe Covid-19. But biology did not act alone. Economic inequality helped determine who stayed healthy and who got ill. Computers and Wi-Fi allowed white-collar workers to stay home and away from potential infection. But workers in so-called essential industries had to mingle. They drove buses, they did laundry in hospitals, they stocked supermarkets.

Meatpacking workers proved especially vulnerable to Covid-19 as they worked long shifts side by side in crowded plants. It's possible that the cold air in the plants kept the virus viable longer, giving it more of a chance to infect new hosts. Doctors who treated sick meatpackers sounded an alarm, but industry leaders persuaded Robert Redfield to keep the plants running. Covid-19 cases soared ten times higher in counties with meatpacking plants than in counties without them. In South Dakota, OSHA found that Smithfield Foods failed to protect its workers. At least 1,294 Smithfield meatpackers got Covid-19. Four died. And for that failure, the company paid a fine of just $13,494.

Covid-19's blows fell particularly hard on Americans of color. They were overrepresented in the essential industries that exposed people to a greater risk of infection. They were more likely to take subways and buses to work rather than drive alone. They then went home to apartments and houses that tended to be crowded, enabling the virus to spread. The disparity was driven further by the American prison system. US jails had long been disproportionately filled with people of color, and now they proved particularly dangerous places in a pandemic. Prisoners lived in cramped quarters with poor ventilation, with new detainees bringing in more viruses from the surrounding communities. Everyone had to breathe the air, but some people had to breathe more viruses than others.

Richard Corsi, an environmental engineer who had signed Morawska's open letter, spent the summer of 2020 thinking about what could be done to reduce the risk of airborne infections. Ever since Corsi had been a boy growing up in Los Angeles, he had dreamed of ways to fix the air. His asthma, aggravated by the city's air pollution, forced him to sit out many Little League baseball games. He once doodled a jet flying over his house carrying a giant vacuum that sucked up all the smog from the sky.

Corsi went on to become a professor at the University of Texas, where he studied indoor air pollution—the mix of chemicals, particles, microbes, and mold that floats inside homes. During a bad wildfire season, Corsi assigned some students to try to build cheap air purifiers that could pull smoke out of indoor air. They taped a filter on a box fan, but the contraption didn't work very well. The air flowing through the filter created resistance, which slowed the fan motor down.

When Covid-19 emerged, Corsi didn't know what to make of it at first. "I was still washing my hands a lot," he told me. Corsi was not an expert on airborne infection, but he was familiar with the struggles of his friend Donald Milton. "He went through years of frustration," Corsi said. Yuguo Li's report about the Guangzhou restaurant startled Corsi. He followed up on it with some calculations of his own and concluded that Covid-19 was probably airborne. He was appalled to see that possibility fail to make its way into Covid-19 guidelines.

"There certainly was pushback from the public health community. They just didn't want to go there," Corsi said. "When you look at the pathways for possible exposure, we were told to protect ourselves against all of them, except for the airborne route. Why? Why drop the one that, to so many people, looks like the obvious one?"

After signing the open letter, Corsi wanted to do more. If Covid-19 was spreading by the airborne route, how could that route be blocked? Corsi turned his attention to schools, which had closed for the last three months of the school year and didn't seem to know what to do about the fall. Some people called for them to open up again, while others warned that they would become hot spots for new infections. Few people involved in the decisions paid attention to what Corsi and other experts believed mattered most: keeping the air clean.

Writing with Joseph Allen in the *Washington Post*, Corsi urged schools to make a plan right away. "We have limited time and funds to get students and teachers back to school safely, but we can—and must—do it," they wrote. Cleaning the air was crucial to making schools safe. In some schools, that might simply mean opening windows. Schools that couldn't take that option—because the weather wouldn't permit it or because their windows had been permanently sealed—could use air purifiers.

The government would need to step in to make sure that schools stayed safe, Corsi and Allen wrote. In 2019, manufacturers made a total of 2.9 million purifiers, but American classrooms would need 3.4 million purifiers—at a cost of a billion dollars. "With reopening schools, we're mostly hearing about roadblocks and 'that can't happen' thinking," they wrote. "Let's get creative and live by what we teach."

Corsi and Allen were, in effect, resurrecting the crusade of William and Mildred Wells to make the air a common good, like water or food. In 2020, it would have been a scandal for schools to let sewage flow out of their water fountains or to serve children rancid turkey sandwiches for lunch. Corsi and Allen wanted schools to give the air the same respect.

No one answered their call. Instead of rushing millions of air purifiers into schools that summer, Congress drafted a relief bill. When schools opened in September 2020, the bill was still a long way from a vote. Like hospitals before them, schools were largely left to make their own decisions about how to handle the pandemic. Many stayed closed that fall. Some of the schools that opened required students to wear masks while staying six feet apart from one another. A number of private schools and public schools in rich neighborhoods were equipped with modern ventilation systems that could exchange the air in classrooms quickly. Some schools managed to shake money out of state and federal coffers to upgrade their systems. But many schools with bad ventilation could do nothing.

A few teachers decided to do what their schools would not and put cheap homemade air purifiers in their classrooms. They built them according to plans created by Richard Corsi.

One day, Corsi was on Twitter when he noticed a conversation about the high price of commercial air purifiers. Someone asked Jose-Luis Jimenez if putting a filter on a box fan would be able to remove SARS-CoV-2 from the air. Jimenez tagged Corsi to see what he thought. Corsi recalled his old classroom assignment to make filters for wildfires.

Adam Rogers, a senior correspondent for *Wired*, followed up on the tweet with a call. As Rogers asked about the physics of fans and filters, Corsi's mind clicked on an idea. He knew that a filter taped to a single box fan would not work. But what if someone constructed a cube, with a box fan as one of the walls and filters as the other five? If the fan blew air out of the cube, more air would flow into it through the filters, trapping virus-bearing droplets. But the filters would not slow down the motor, Corsi reasoned, because they were not directly attached to it.

"You've got me really pumped up on this right now," Corsi told Rogers. "I don't have a lot of free time, but this would be something to build a prototype of."

Rogers next called Jim Rosenthal, an air filter manufacturer who had been a part of the Twitter conversation with Corsi and Jimenez. When Rogers described the imaginary cube, Rosenthal thought it might work. After the call, Rosenthal tried out different combinations of fans and filters. He ended up with a contraption that cleaned the air at an impressive rate. It cost less than a hundred dollars in parts.

"As usual, Rich is right," Rosenthal announced on Twitter.

Donald Milton, impressed by the invention, dubbed it the Corsi-Rosenthal box. He tried to get the organizers of the October 7 vice presidential debate between Kamala Harris and Mike Pence to put them onstage—not only to protect the candidates but to promote clean air as a protection against Covid-19. The organizers refused, opting instead for plexiglass. The transparent barriers might have looked impressive, but Milton considered them nothing but hygiene theater.

Even without a TV appearance, Corsi-Rosenthal boxes gained a cult following in late 2020. Universities put Corsi-Rosenthal boxes into classrooms. High school clubs put school colors on their creations. Some people fashioned miniature boxes. Others lit them from inside so that they glowed like paper lanterns. Corsi cheered on those efforts and established a foundation to share plans for the boxes. He held contests for the best new design and judged hundreds of entries.

Corsi had no way to know exactly how many people were using the boxes, but he guessed they ran into the tens of thousands. "You see school kids in Wales, Australia and New Zealand making them," he told a reporter in 2022. Corsi refused to patent the design and even talked to lawyers about keeping anyone else from doing so. "This is a movement," he said.

○ ○ ○ ○

By the fall of 2020, Covid-19 had become the biggest pandemic crisis since the 1918 flu. It had reached almost every country on Earth; only a few remote Pacific islands lay untouched. The confirmed cases pushed past 40 million, the official deaths rose over a million, and the true totals of both were higher—likely much higher.

Each country contended with Covid-19 in its own way, from rigid lockdowns in China to voluntary suggestions in Sweden. The United States contained multitudes. Masks were required in some places and scoffed at elsewhere. Within the federal government, the battles grew more intense. Trump grew impatient with Fauci, who would not say that the pandemic was almost over. The White House barred Fauci from most television appearances. Trump supporters flooded Fauci's office with hate mail. In August, he opened a letter from which a white powder wafted onto his face and hands. A hazmat team rushed in to decontaminate him. "I thought, this is insane," Fauci later wrote. "There I was standing naked, being sprayed down by guys in space suits."

Tests revealed that the powder was not anthrax but just a harmless chemical. But Fauci remained at odds with the White House—especially after the addition of a new advisor to the task force: the Stanford radiologist Scott Atlas. Atlas lobbied to drop any recommendations for masks, considering them useless. He pushed to stop testing people without symptoms. Atlas even predicted the pandemic would likely grind to a halt if 20 or 25 percent of Americans got infected—an idea that one infectious disease expert called "the most amazing combination of pixie dust and pseudoscience I've ever seen."

As the presidential campaign ramped up, it added more chaos to the country's response to the pandemic. In 2016, Trump had gotten elected by casting the government as hopelessly corrupt. "Nobody knows the system better than me, which is why I alone can fix it," he declared at the Republican convention that year. Now, as he ran against former vice president Joe Biden, Trump distanced himself from many government policies. When thousands of people parked their cars in front of Michigan's state capitol building in a protest of the state's lockdown, Trump tweeted, LIBERATE MICHIGAN. When the Centers for Disease Control prepared to require masks on buses, airplanes, and other public transportation, the Trump administration informed the agency that would not happen. When public health authorities advised against large indoor gatherings, Trump held rallies of thousands of supporters. Soon enough, Trump got Covid-19.

The virus entered the presidential airway in late September 2020. According to Mark Meadows, the White House chief of staff, Trump tested positive on September 26. He had just thrown a party at the White House to celebrate Amy Coney Barrett's appointment to the Supreme Court. In the days that followed, a number of people who attended the ceremony also developed Covid-19. Although Trump began feeling run down, Meadows wrote, he ignored his positive test and flew off to a campaign event in Pennsylvania. When he tested negative on a different test, Trump chose to believe that one. He herded campaign staffers into a small room to help him rehearse for his debate against Biden. It was Group 36's nightmare. And as they would have predicted, most of the people in the debate prep room got sick. Former New Jersey governor Chris Christie ended up in an intensive care unit near death.

Two days after the debate, Trump began to have trouble breathing. He was given a third test, which came up positive again. Trump's blood oxygen level was reportedly so low that his doctors thought he might have to go on a ventilator, and they rushed him to Walter Reed hospital for treatment. The *Washington Post* later estimated that the president might have exposed more than five hundred people before being hospitalized.

Trump's doctors gave him oxygen, along with a cocktail of monoclonal antibodies and other drugs. From his hospital bed, the president called Fauci to tell him he "felt like fucking shit." But after three nights at Walter Reed, Trump recovered enough to return to the White House. Camera crews filmed him as he climbed a staircase to a second-floor balcony. He dramatically pulled off his mask, stuffed it in his pocket, and entered the White House, where he would likely fill the air with coronaviruses for days.

On the other side of the country, the Skagit Valley Chorale was getting used to its new status as a textbook case of airborne transmission. Anthony Fauci gave a lecture for Harvard Medical School in which he described "the now-famous Skagit, Washington, choir outbreak where one symptomatic person infected 87 percent of the group." For Fauci, their experience was one of the crucial pieces of evidence that opened his mind about the airborne spread of Covid-19.

While the choir lived on in the scientific literature, Burdick wasn't sure if it would survive as a singing group. It had gone months without a rehearsal, and a concert in person was out of the question for the foreseeable future. Burdick persuaded the singers to rehearse on Zoom, with an eye to putting on a virtual performance at the end of the year.

Only fifty singers were willing to try. After practicing their songs, they each recorded their parts and emailed them to Burdick. Some of the recordings included dogs barking and dishwashers running. Some of the singers were so unsettled by the sound of their solo voices that they dropped out of the concert.

With the help of audio-editing software, Burdick spent November stitching what few recordings she received into a digital chorus of voices. On December 11, the Skagit Valley Chorale streamed its concert. Burdick kicked off the event by standing in front of a camera to welcome the audience watching on their phones and computer screens.

Carolynn Comstock and Ruth Backlund then appeared together in another square onscreen, standing in front of a fireplace. "We dedicate this performance to Nicki Hamilton and Carole Ann Woodmansee," Backlund said. The songs then began, as photographs of beach sunsets and farm fields filled the screen. At the end of the hour, Burdick wished the viewers a safe holiday. "We don't know yet what we're going to do for the spring," she said.

On the day of the concert, the US Food and Drug Administration gave the green light for a Covid-19 vaccine made by Pfizer and BioNTech. As a doctor, Debbie Amos was stunned by the results. "I remember saying in a board meeting, 'If we get a vaccine before the end of 2020, it will be a miracle,'" she said. "And we did, and I'm still to this day shocked at how fast that vaccine was ready. That made it much more safe for us to consider getting back together again."

But early December also brought an ominous surprise. The southeast corner of England experienced a baffling surge of Covid-19 cases, and virologists discovered that people were getting infected with a new variant carrying a dozen mutations. The variant, which was later named Alpha, was able to spread faster than other forms of the virus. In a matter of weeks, it created a global surge of cases, lifting the global toll of Covid-19 to more than sixteen thousand deaths a day in January 2021.

In Maryland, Donald Milton investigated Alpha with Gesundheit II. He had students infected with the new variant breathe into his device. He then compared the results to the ones he had gotten earlier in the pandemic. Milton found that people infected with Alpha had up to a hundred times more viruses in their droplets. It looked as if the variant might be dominating the pandemic because it did a better job of traveling through the air.

○ ○ ○ ○

Vice president Joe Biden beat Trump in the November election, and the following month he traveled to a hospital near his home in Delaware. Biden wore a black turtleneck instead of his usual suit. Over his mouth he wore a pale blue surgical mask, and over that, he wore another mask made of black cloth.

With cameras filming him, Biden sat down and looked over at a nurse fitted with a N95 mask and a face shield.

The nurse, Tabe Mase, prepared a syringe. "Are you ready?" she asked the president-elect.

"I'm ready," he replied.

Biden rolled up his sleeve and received his first Covid-19 vaccine shot. "I'm doing this to demonstrate that people should be prepared when it's available to take the vaccine," he told the assembled reporters.

Even before he was sworn in, Biden was signaling that vaccines would become a top priority of his administration in its fight against the pandemic. Trump also got vaccinated the following month, but no cameras captured that moment. The development of Covid-19 vaccines had been one of the few bright spots in Trump's handling of the pandemic. But he also knew that many of his followers rejected them.

Biden's inauguration took place on January 20, 2021, nearly a year after the first positive test for Covid-19 in the United States. At his swearing in, he did not look out at a crowd on the National Mall. Instead, the ground was planted with nearly 200,000 flags in memory of the dead. More than 415,000 people had actually died of Covid-19 by then, but there wasn't enough room on the mall for even half that number of flags.

After he took office, Biden brought in some of the Obama-era officials who had helped him respond to the 2009 H1N1 pandemic and the West Africa Ebola outbreak. As he had promised, they worked hard to get vaccines into arms. Fewer than two months after Biden's inauguration, he celebrated the first 100 million doses delivered. He promised even better news to come. If the country kept up the pace of vaccination, it would finally enjoy a summer that felt closer to normal. By Independence Day,

Biden promised, the country would begin to declare its independence from the virus.

But critics worried that the Biden administration was declaring victory before the fight was over. Assuming that vaccines would quickly end the pandemic, the president was not taking enough other actions to stop the virus's spread. In February, Linsey Marr and other experts drafted a letter calling on the White House to stop the airborne spread of Covid-19. The following month, in testimony Marr gave to Congress, she kept pushing for the government to reckon with its transmission. "Calling the virus 'airborne' is the clearest way to convey how it is transmitted," Marr said. "Airborne, meaning 'borne by air,' is directly analogous to the terms waterborne, food-borne, blood-borne, and vector-borne for describing how pathogens are transmitted."

With the Alpha variant on the loose, and the possibility of even more dangerous variants evolving soon, Marr urged Congress to bring more masks to Americans. According to some studies, surgical masks were not holding up well against SARS-CoV-2. In Israel, a hospital reported that a child spread it to six health care workers, all wearing surgical masks, who visited him during grand rounds. Marr called on Congress to help people get high-quality N95s, which were finally being manufactured in abundance. It was time to make N95 masks available not just to medical workers carrying out risky procedures, Marr said, but to everyone.

Marr discovered that opposition to that idea remained strong. At the same congressional hearing where she spoke, the American Hospital Association submitted a statement urging the government to spare its members N95 requirements. "A rigid new standard has real potential to add for hospitals and health systems a new layer of conflicting and impractical regulatory burden at precisely the wrong time," the association said. To back that request, it pointed to the CDC's current recommendations to control Covid-19. "The CDC continues to hold that COVID-19 is primarily spread through close contact, not airborne transmission," the association declared. And it was true: the CDC had yet to officially change its guidance.

And then, two weeks after the hearing, another attack came. John Conly and a team of researchers released a systematic review of the studies that might determine if SARS-CoV-2 was airborne or not. They were not impressed.

Conly and his colleagues argued that studies on the spread of Covid-19 had to follow Koch's postulates properly. Researchers needed to use the best tools at their disposal to confirm each step in the transmission of the virus by any route. While Robert Koch had been limited to Petri dishes and light microscopes, twenty-first-century scientists could conduct other studies, such as reading the genetic sequence of pathogens they captured. Based on those standards, Conly and his colleagues concluded that the eighty-nine studies of Covid-19 in the air so far—including the Skagit Valley Chorale analysis—were all "low quality."

Conly's team picked apart one study after another, noting how they fell short. The Skagit Valley Chorale study, for example, was based mostly on interviews, building plans, and computer models. Group 36 had not captured viruses from the air on the night of March 10. They could not point to genetic tests confirming the viruses were SARS-CoV-2. Conly's team also raised questions about studies in which scientists did manage to capture genetic material from SARS-CoV-2 floating in the air, such as at the University of Nebraska Medical Center. The Nebraska researcher had not made a strong case that the coronaviruses to which those genes belonged could replicate. Even if SARS-CoV-2 did move in the air, the studies didn't prove that it could cause Covid-19 in people who inhaled it.

"The lack of recoverable viral culture samples of SARS-CoV-2 prevents firm conclusions to be drawn about airborne transmission," the scientists concluded.

Conly and his colleagues declared that this weakness was not unique to new studies of Covid-19. In earlier studies on influenza and other respiratory diseases, scientists sometimes caught viral genes from the air but failed to demonstrate that airborne viruses were still viable. Conly and his colleagues pointed to a single disease that had clear evidence of airborne spread: tuberculosis. The evidence had come from the experiments at Loch Raven carried out in the late 1950s by William Firth Wells, Richard Riley, and Cretyl Mills.

Group 36 responded swiftly with a series of rebuttals. Raymond Tellier and Julian Tang accused Conly and his colleagues of arbitrarily ignoring studies without good reason. "It strongly suggests a lack of familiarity with the methods of clinical virology," they wrote.

And Jose Luis Jimenez and Linsey Marr caught Conly's team in a contradiction. Why did they accept the Loch Raven experiments as valid evidence while rejecting the newer studies on Covid-19? Jimenez and Marr pointed out that Wells, Riley, and Mills never captured a single free-floating bacterium from the air. Their only evidence came from their sick guinea pigs. According to the standards that Conly and his colleagues set, tuberculosis had not been proven to be airborne. By those standards, chicken pox or measles had yet to be proven to be airborne as well. "This would clearly be nonsensical," Jimenez and Marr wrote.

These debates brought William Firth Wells back to the world's attention after seven decades in obscurity. Some of the Group 36 scientists published paeans to Wells in scientific journals. They praised him in a feature about the five-micron myth published in *Wired* and recounted his work on National Public Radio. They cast the deadly confusion over Covid-19 as the enduring misunderstanding of Wells's message.

"My impression is that history set us up," Jimenez told me.

Jimenez and his colleagues painted a portrait of Wells tailored for the Covid-19 pandemic. For them, he was an uncomplicated visionary. Lost were the years that Wells had spent as an oyster wizard, his struggles to be taken seriously as an expert on human diseases, and his knack for becoming his own worst enemy. No mention was made of how the germ warfare industrial complex snatched his work or of his ravings about his book as he thrashed and died in Loch Raven. And Mildred Weeks Wells, who had done so much on airborne infection both with and without her husband, remained largely in the shadows in 2021, as she had in life.

○ ○ ○ ○

Throughout the spring, the Biden administration continued to issue cheerful pronouncements about the pandemic. Early epidemiological studies were already confirming that vaccines lowered the risk of both infection and death from Covid-19. In May, Rochelle Walensky, Biden's new director of the CDC, declared that vaccinated people did not need to wear masks. "If you are fully vaccinated, you can start doing the things that you had stopped doing because of the pandemic," she said. "We have all longed for this moment when we can get back to some sense of normalcy."

A month later, Biden celebrated the end of the Alpha surge. "America is headed into the summer dramatically different from last year's summer," he said, "a summer of freedom, a summer of joy, a summer of get-togethers and celebrations—an all-American summer that this country deserves after a long, long, dark winter that we've all endured."

That summer did not deliver freedom from Covid-19. Instead, it delivered Delta, a new variant that evaded some of the antibodies produced by vaccines. While vaccinations continued to reduce the risk of death, they became less effective at stopping infections. On Independence Day, when the country was supposed to declare independence from Covid-19, an outbreak occurred in Provincetown, Massachusetts. Rain drove the revelers, mostly vaccinated, off the beaches and into bars and restaurants. A thousand people got infected over the holiday weekend. Very few of them ended up in hospitals, but the lesson was inescapable: Walensky reversed the CDC's policy and called for vaccinated people to wear masks indoors again.

Delta brought its own record-breaking surge over the next few months. In unvaccinated people, it proved more prone to cause severe Covid-19 than Alpha, filling hospitals with patients again, and sending more bodies to mortuaries. America had yet to escape the darkness.

By then, the Skagit Valley Chorale had gone over a year without singing together. They now knew that singing was one of the riskiest things they could do. They were no longer the only case study in singing outbreaks. Scientists had documented similar ones that struck choruses in other countries. Many members of the Skagit Valley Chorale said they would only gather again if the group took every possible step to stay safe.

"A cat that's burned fears warm water," Ruth Backlund said.

Several members told Backlund they would not come back unless every member had a vaccination card in hand. Others did not believe that should be a requirement. Carolynn Comstock was one of them. While she had given her children every required vaccine when they were growing up, she had not gotten vaccinated against Covid-19. She thought her March 2020 infection provided her with immunity.

After a fierce debate, the board voted in favor of a vaccine mandate. Comstock and her husband left the choir rather than get the shot. “We are the lepers,” she told me.

Now the choir had to decide where to sing. They would not go back to the Mount Vernon Presbyterian Church, because some members felt unsafe there. “I had a post-traumatic feeling,” Ruth Backlund said. “I didn’t want to go back into that room.”

As the search for a new home continued, the choir gathered one chilly day in October outside a local elementary school. A metal overhang would be their rehearsal space for the time being. “We choose not to be defined as the infamous choir from Washington State, even though that’s how you can find us on the internet,” Burdick told a reporter who came to watch. The masks muffled their voices. It started to rain as they rehearsed, and the drops drummed so loudly on the overhang that Burdick could barely hear their singing.

The Skagit Valley Chorale finally found a new home at the Salem Lutheran Church. When the singers gathered there, they stayed six feet apart from one another and kept a set of double doors to the outside opened. Mark Backlund taught himself how to make Corsi-Rosenthal boxes and brought two of them to every rehearsal. A carbon dioxide monitor let the singers make sure that fresh air was moving around them as they exhaled. When winter set in, the choir kept the doors open and put on hats and gloves. Debbie Amos had no complaints. “It felt jubilant; it felt wonderful,” she said. “To hear our voices again, it just felt like a release.”

As the Skagit Valley Chorale began to sing again, the World Health Organization finally issued a clear public statement that the virus was airborne. On December 23, 2021, the agency updated an existing web page with information about Covid-19. “Current evidence suggests that the virus spreads mainly between people who are in close contact with each other, for example at a conversational distance,” it wrote. “The virus can also spread in poorly ventilated and/or crowded indoor settings, where people tend to spend longer periods of time. This is because aerosols can remain suspended in the air or travel farther than conversational distance (this is often called long-range aerosol or long-range airborne transmission).”

Jimenez was glad to see the change, but he considered the long delay one of the biggest mistakes in the history of public health. "They made an enormous error that allowed the disease early on—when we could have stopped it—to spread everywhere, to get to every nook and cranny of the world," he said.

Indeed, by the time the World Health Organization felt comfortable using the word *airborne*, an even more potent variant was sweeping the planet. WHO dubbed it Omicron. It proved far better than Delta at infecting people who had immunity from vaccines or previous infections. When Donald Milton studied Omicron with Gesundheit II, he found that people who were vaccinated and boosted against Covid-19 could still exhale the variant in their breath. In the face of an even more contagious variant, WHO changed its advice about protection. It advised health care workers to switch from surgical masks to N95s.

Omicron proceeded to create the biggest spike of infections of the entire pandemic. While vaccines reduced the risk of disease, hospitals still saw a new wave of seriously ill patients. But the Biden administration made few new efforts to slow the spread of Omicron. "Two years into this crisis and a year into Biden's presidency, we seem to be even worse off than we were under Trump in the most lethal metric," Steven Thrasher, a Northwestern University professor, complained. "More deaths are taking place under the Democrat than occurred under his predecessor."

The Skagit Valley Chorale crept slowly back toward their previous life. For their 2021 Christmas concert they performed in person—not in front of a live audience, but in front of a live-streaming camera at the Lutheran church. Wearing a black N95 mask and furry reindeer antlers, Ruth Backlund stood in front of the seated singers. "We are filled with joy to be able to sing together in person," she told the viewers at home.

The following month, when Omicron swept across Washington State, it struck Backlund again. With her immunity from her first bout with Covid-19, along with her vaccines and a booster shot, she had a far less harrowing time with her second infection. "It's an entirely different ailment," she said. "It was just like a horrible, horrible head cold where you have to sit up in bed. But it only lasted five days. None of the odd other things that went on the initial time."

After Ruth recovered, she rejoined the choir to rehearse for their spring concert. In April 2022, they returned to McIntyre Hall. They were ready to sing before an audience, but they also required ticket holders to show proof of vaccination and brought carbon dioxide monitors onstage. They took up where they had left off two years before, although they had lost two singers along the way.

"I chose pieces in the repertoire that were emotionally powerful," Burdick said. One piece, "Sing You Home," addresses someone who has just died.

Know that you will live
On the lips of those who knew
What it was you had to give
And what it was they learned from you

"It became kind of an anthem for us," Burdick said.

eighteen

A MARK ON THE AIR

The Covid-19 pandemic made the ocean of gases surrounding us visible. The transparent currents of air turned to streams of gleaming smoke suffused with droplet nuclei launched from people's mouths and noses like swarms of airborne stars.

Over the course of 2022, Group 36's manifesto—once too radioactive to publish—became widely accepted. A panel of 386 scientists convened to review research on Covid-19 and come up with a consensus that, in their words, "can serve as a strong basis for decision-making to end Covid-19 as a public health threat." In November 2022, the panel published their findings in the journal *Nature*. At the top of the list they offered their conclusion about how the pathogen spread. "SARS-CoV-2 is an airborne virus that presents the highest risk of transmission in indoor areas with poor ventilation," they wrote.

And, as if to underline that message, Omicron surged back into China that month, overwhelming the country where Covid-19 had gotten its start three years before.

◦ ◦ ◦ ◦

When Covid-19 burst out of Wuhan in January 2020, President Xi Jinping imposed some of the harshest lockdowns in the world. At one point, he confined 760 million people almost completely to their homes. Xi's Zero Covid policy effectively drove down the outbreak in just a few months. But Xi barely loosened the lockdown. China continued to test aggressively, and people who tested positive were forced to stay with their families in their apartments. Sometimes police would seal apartment doors from the outside.

By the end of 2020, Xi was boasting that Zero Covid was a spectacular success. "The trend of the times is in favor of us," he declared. The times continued to favor Xi through 2021 as cases remained low. Omicron began causing flare-ups in China in early 2022, and Xi simply cracked back down.

But Omicron was profoundly different from the virus that Xi had contended with when it emerged out of Wuhan. It easily slipped past his measures. All Xi could do was impose even harsher restrictions. He put Shanghai—a city of 25 million—in lockdown for two months. Barbed-wire fences sprang up around apartment buildings where residents tested positive. Children were separated from their parents.

Xi's harshest edicts failed to hold back Omicron. Instead, they triggered protests on a scale China had not witnessed since the Tiananmen Square demonstrations in 1989. It looked to Xi as if Zero Covid might become his political doom. In November 2022, Xi abruptly dropped Covid-19 enforcements, allowing Omicron to rage through the country.

It was impossible to know precisely how many people were getting sick, because Xi also scrapped the country's testing program. When people died of Covid-19, the government ignored the cause. In the month after Xi lifted Covid restrictions, China officially reported a grand total of seven Covid-19 deaths. Reporters for foreign outlets did their best to find out what was actually happening. They listened to frantic families describe bringing their loved ones to hospitals, only to be turned away from overflowing wards. They watched ambulances wandering cities, searching for a place to unload their patients. Funeral homes and crematoria were overwhelmed. "There's been so many people dying," a worker at a funeral goods shop told an Associated Press reporter. "They work day and night, but they can't burn them all."

Scientists outside China tried to guess the size of the Omicron wave. Yuguo Li tapped his network of contacts across the country, sending out a thirty-two-question survey on WeChat on December 31, 2022. He got back 4,421 responses, from twenty-nine of China's thirty-one provinces.

Among the answers to Li's questions were descriptions of the experiences families had had with Covid-19. Based on his respondents' replies, Li and his colleagues estimated that 2.2 percent of people in China had had Covid-19 before December 2022. Over the course of December, that number leaped to 79 percent. "That's a lot," Li said. He repeated himself, his voice trailing off, as he contemplated the scale of what he had documented. "That's a lot . . . that's a lot . . ."

If Li's survey accurately reflects China's experience as a whole, a billion people across the country got Covid-19 in a single month. Other researchers ended up with similar numbers. And out of those billion people, Li estimated that a million might have died. Other scientists have pinned the number at two million.

It is hard to find an outbreak in modern history of comparable speed and scale. During the 1918 influenza pandemic, an estimated 500 million people got infected, but it took the virus two years to reach them all. Once Omicron had free rein in China, it needed only a few weeks to infect over twice as many people. "I guess it's really telling us a story," Li said. "How come so many people could get infected within such a short time?"

Li started the pandemic unsure of how SARS-CoV-2 spread. By 2021, he had concluded it was predominantly airborne, but he suspected it could also be spread by other means. And at the end of 2022, examining the evolved coronavirus, Li changed his mind again.

His survey of the China outbreak revealed that people who kept their living room windows open were 50 percent less likely to get Covid-19. If the hallways in their apartment buildings were well ventilated, they enjoyed a similar reduction in their risk of infection. People who lived on the twentieth floor or higher in a building were 80 percent more likely to get sick compared to those living on the ground floor—presumably because updrafts were spreading the virus.

Omicron, Li concluded, had become even more adept at spreading through the air. "I now believe it is not predominant—it is likely stronger than predominant," he said. "For Omicron, it is probably nearly 100 percent. I cannot find another way to explain how one billion people got infected in December."

○ ○ ○ ○

Despite the staggering carnage that Omicron caused in China, the world largely turned its attention away from the coronavirus. In January 2023, Quinnipiac University conducted a survey about the most urgent issues facing the United States. Inflation was the top response. In a distant second place was immigration. At the bottom of a long list of issues, mentioned by fewer than 1 percent of the people surveyed, was Covid-19.

People weren't thinking much about Covid-19, but they were still getting it. And they were still dying of it. In 2023 alone, the *Economist* estimated that SARS-CoV-2 killed 3.19 million people worldwide. That toll raised the total number of lives lost in the pandemic to 28.43 million. The world was now ready to accept three million deaths a year as ordinary.

It's understandable that people would crave some normalcy. But it's no reason to let the atmosphere go dark again. The aerobiome still enfolds us, still reaches inside us with every breath, whether we think about it or not.

While Covid-19 continued to claim lives, other airborne pathogens continued to spread like smoke. Measles floated from person to person, despite the existence of a powerful vaccine that can provide lifelong immunity. As vaccination rates sagged in many nations, the world saw a sudden post-pandemic spike: from 941 cases in 2022 to more than thirty thousand in 2023. Tuberculosis, the centuries-old scourge, continued to ride on human breath. Before the pandemic, more than a million people died of TB each year. After the pandemic subsided, its death toll remained about as high.

Yuguo Li and his colleagues suspected that measles and tuberculosis had a lot of company. In a 2022 paper, they argued that all respiratory viruses are airborne, at least to some degree. Epidemiologists had long dismissed this idea based on the fact that people tended to get sick from many respiratory diseases if they were close to their index cases. That seemed to argue in favor of soggy raisins, not droplet nuclei, as the route the pathogens took to new hosts. But Li and his colleagues questioned that reasoning. If a pathogen is carried in droplet nuclei, it may also be easier to inhale at close range. At longer distances, the pathogen becomes more dilute as the droplet nuclei drift apart from one another. But even at close range, the protections required to stop an airborne disease are different from the ones that will work on large, short-range droplets. If Li and his colleagues were right, then droplet nuclei may carry a long list of diseases, including influenza, parainfluenza, respiratory syncytial virus, and even smallpox.

Li and his colleagues took a bold position with that hypothesis, but they still limited their vision to the world of indoor air. They considered only the threats that pathogens could pose as they floated in unventilated spaces. But the Covid-19 pandemic spurred a number of scientists to take a closer look at the possibility that diseases of humans or animals can spread outdoors, perhaps for hundreds of miles.

Some of the first evidence for long-range infection came from veterinarians who studied foot-and-mouth disease. The viruses multiply to staggering numbers inside an animal, and the sores they cause in its mouth lead to virus-saturated drool that can spread to other animals nearby. But an infected animal will also spew droplet nuclei that can take to the wind. If the weather cooperates, an outbreak on one farm will be followed by dozens of others, which may ignite miles downwind. In a few cases, veterinary scientists have found evidence of clouds of foot-and-mouth virus crossing national boundaries, traveling hundreds of miles.

Dust storms may also spread diseases over vast distances. One of those pathogens, a fungus called *Coccidioides*, lurks in the soil across arid regions of North and South America. It has likely lived there for millions of years, sprouting threads that break down organic matter. Kangaroo rats and other rodents that dig burrows in spore-infested ground breathe in *Coccidioides* spores, which grow harmlessly in their lungs. When the animals die, the fungus may feast on their decaying carcasses and extend its tentacles back into the soil.

In 1889, a thirty-three-year-old soldier stationed on the plains of northern Argentina developed the first-known human infection of *Coccidioides*. Domingo Escurra thought at first a spider had bitten his cheek. He tried stopping the growth with tobacco; when that failed, he tried chopping out the diseased flesh with a penknife. But the lesion kept growing, and then new ones appeared across his face and neck. After Escurra was hospitalized, his doctors realized he had a fungal infection. It reached his lungs and slowly started spreading elsewhere in his body, even turning up on his feet. In 1898, after nine agonizing years, Escurra finally died. His case so unsettled his doctors that they decapitated him for posterity. His head floats in a jar that remains on display at the National Institute of Parasitology in Buenos Aires.

Soon after Escurra's case came to light, doctors in California discovered the same disease plaguing migrant farmworkers in the San Joaquin Valley. Their cases led to the ailment's common name: valley fever. Healthy people who inhale airborne spores of *Coccidioides* can usually destroy them. At worst, they experience a nagging cough for a few weeks. Valley fever is far worse for people who inhale a heavy dose, as well as for those with suppressed immune systems.

Domingo Escurra did not have to dig a burrow to inhale *Coccidioides* spores. Instead, the fungus came to him. In aerobiological surveys, scientists have found *Coccidioides* in dust storms. One particularly big storm near Bakersfield, California, traveled four hundred miles to Sacramento in 1977. In the weeks that followed, more than a hundred people there were diagnosed with valley fever.

Other human pathogens may also travel far on air currents. In western China, measles outbreaks often occur after dust storms blow through. It's possible that the dust grains people inhale are studded with viruses. Winds may also be the solution to a number of medical mysteries—diseases for which scientists have yet to find causes.

In 1961 a Japanese doctor named Tomisaku Kawasaki examined a boy who had suffered a fever for two weeks and had bloodshot eyes, peeling skin, and a strawberry-colored tongue. "I had never experienced this kind of unique symptom complex in my 10-year pediatric career," he later wrote. But over the next decade he saw dozens of children who had the same symptoms. After he published reports on the cases, they came to be known as Kawasaki disease.

While some children recovered from the symptoms, others died of heart attacks. Some seemed to get better but suffered hidden damage that would kill them years later. In Japan, one in every hundred children suffers from Kawasaki disease, but the disease is not unique to the country. Every year, three thousand children in the United States are diagnosed with it, and the disease has been documented in more countries as doctors learn what to look for. Experts on Kawasaki disease suspect that it comes about when some kind of infection triggers an oversized response from the immune system. But no one has been able to follow Koch's postulates and isolate a pathogen that causes its symptoms.

One clue to the cause of Kawasaki disease is its timing. It occurs around the world, but in each place—be it South Korea, Italy, or New Zealand—clusters of children tend to get sick together. The clusters rise and fall in a rhythm that spans hundreds or thousands of miles. The rhythm matches up fairly well with shifts in the winds that blow east across the farms of China. It's possible that some unknown pathogen in those fields rises into the air, bringing invisible waves of Kawasaki disease across the planet.

○ ○ ○ ○

As deadly as airborne pathogens may be, they make up only a sliver of the aerobiome's diversity. It would be wrong to think of the atmosphere as merely a stockpile of nature-made biological weapons. Every day we inhale millions of organisms, the vast majority of which do no harm to us at all. If that day happens to be in spring, we may take in thousands of pollen grains and tens of thousands of fungal spores. We breathe in hundreds of species of bacteria and viruses. These visitors to our airways may include a rust spore from a wheat field hundreds of miles away, algae thrown up by waves pounding the Pacific Coast, or bacteria just returned from a trip to the stratosphere.

When these organisms enter the human airway, they arrive in a fortress of immunity. Over millions of years, our ancestors evolved a series of defenses to limit what gets inside us. They begin in the nose, where mucus-coated hairs trap incoming particles and droplets. Hairlike cilia that line the airway down into the lungs ripple like seagrass. Together, they ferry foreign objects out of danger, to be coughed out or sent down to the stomach for destruction. The mucus in which the cilia flutter is rich with microbe-killing compounds. A special division of immune cells crawls over it, spewing out antibodies, triggering inflammation, killing infected cells, and swallowing up viruses and other intruders.

We need these defenses to stop pathogens adapted to parasitize us. But a lot of the things we inhale cannot cause infectious diseases in humans anyway. A pollen grain may contain genetic instructions for building a daffodil, but it can't make your lungs bloom with flowers. On a beach you may inhale viruses adapted to infecting marine bacteria. They lack the molecular machinery to infect your human cells.

Other species can thrive in our airways without making their presence felt. The nose alone is typically home to more than a dozen species, many of which get in there in the air we breathe. Some airborne organisms float all the way through the nasal cavity and down the airway into the lungs. The fungus *Pneumocystis jirovecii* can survive only in the warm, moist habitat of the human lung. There it scavenges nutrients floating around the lung. Every human being will inhale *Pneumocystis* at some point in their lives and exhale new spores that can infect others. But almost none of them will be aware of harboring a fungus.

That doesn't mean *Pneumocystis* is our friend. It does us no favors, and the only reason it doesn't make most people sick is because their immune systems hold the fungus in check. An HIV infection that weakens the immune system allows *Pneumocystis* to explode, sometimes causing a lethal case of pneumonia. It might seem strange to tolerate a potential killer in our bodies, but our immune system has evolved to use a sophisticated mix of tolerance and vigilance. There are simply too many things floating in our breath to kill them all, because the immune system would kill us in the process. Its defenses against intruders are effective in small doses, but too much inflammation and too many toxic chemicals can damage our tissues. The lungs, with their delicate passageways for oxygen, are especially vulnerable to friendly fire.

A healthy immune system can expel pollen without incident. It allows *Pneumocystis* to lodge in the lungs but keeps it under control. And it annihilates measles viruses and other pathogens once it learns how to make antibodies against them. This balance is too complex to be encoded simply in our genes. Our immune systems learn throughout our childhood, figuring out what to attack and what to ignore.

Our ancestors probably started adapting to the aerobiome when they first began gulping air in Devonian swamps 360 million years ago. As they adapted to life on land, the aerobiome itself changed. When seed plants came to dominate the land, our reptilian ancestors began inhaling pollen grains, for example. The aerobiome altered slowly enough for our ancestors to evolve new adaptations to it. But in just the past few thousand years, the aerobiome has changed drastically. We are responsible for that jolt.

The agricultural revolution triggered a series of changes that let a host of new airborne diseases thrive. Before farming, hunter-gatherers stayed on the move and built temporary shelters on the go. During the agricultural revolution, farmers became anchored to one place and built themselves permanent homes. Families slept together in these new buildings. The walls kept the droplet nuclei they exhaled from dispersing, making it more likely that someone in their household would inhale them.

The surplus of crops and meat that farmers produced meant that other people could buy their food and do different kinds of work. Instead of living on farms, they built houses in villages; eventually towns emerged and then cities. For airborne pathogens, those settlements provided more indoor space in which to float successfully to a new host. New spaces emerged—temples, jails, markets—where people could exhale pathogens and others could inhale them. As populations increased, the pathogens were more likely to find new victims, allowing them to sustain chains of infections rather than burning out. And the emergence of long-distance trade between towns and cities linked those growing populations into a huge network of potential hosts.

Agriculture also spurred the evolution of new pathogens. Each species of livestock carried its own set of bacteria and viruses. Farmers lived in close contact with those germs, some of which took the opportunity to jump the species barrier. A few managed to adapt inside their new hosts. Measles viruses likely evolved from ones that infected cattle. Influenza viruses started as an intestinal infection in chickens and ducks, which then shifted to the airways of people. Coronaviruses likely jumped from time to time from mammals—including domesticated pigs as well as the rodents that made human settlements their new homes.

Not every microbe that turned into an airborne human pathogen made a jump from animals, however. The ancestors of *Mycobacterium tuberculosis* appear to have started out as bacteria that lived in the soil. Studies on the living relatives of the microbe suggest that it arose in sub-Saharan Africa sometime between two thousand and six thousand years ago—the period during which Africans were establishing the continent's first farms and towns. The forerunners of *Mycobacterium tuberculosis* already had some biochemical tricks that prepared them well for life in our lungs.

Living in the soil, the bacteria were at perpetual risk of getting devoured by amoebae. That threat led them to evolve the ability to survive getting swallowed up by these single-celled predators. They turned the tables, becoming parasites that fed on the amoebae from the inside. They multiplied and then burst out of their hosts to return to life in the soil. When *Mycobacterium tuberculosis* became a human pathogen, it used this same strategy in our lungs. Instead of predatory amoebae, the bacteria are attacked by macrophages. They withstand getting swallowed by the immune cells and then start growing inside them.

But the success of tuberculosis is also due to some new adaptations that made *Mycobacterium tuberculosis* even better at spreading between people through the air. The surface of the bacteria is coated with a protein that tickles the nerve endings in the human airway. That trigger causes people to cough, sending out plumes of droplet nuclei. The sound that made tuberculosis different from other diseases—that led some to call the disease the graveyard cough—was actually the sound of bacteria launching their personal airships.

Over the last few centuries, the explosion of our population has accelerated the aerobiome's transformation. In the eighteenth century, cities boomed with people crowded into tight quarters. Some airborne pathogens—both old and new—thrived in this new environment. Tuberculosis spread easily in the poorly ventilated air of Europe's packed houses and factories. Ships took passengers—both free and enslaved—to new continents, bringing bacteria and viruses with them. Industrial-scale livestock operations allowed influenza viruses to spread among chickens; they could also sometimes hop into pigs, where they could evolve further. As people cut down forests for farmland, they came into more contact with wild animals and their microbes, affording coronaviruses and other pathogens the chance to spill over.

Industrial farming may also be spurring the long-distance spread of diseases such as foot-and-mouth. In 1961, there were a billion cows, and forty years later in 2021, there were 1.53 billion. In the same period, pigs boomed from 400 million to 975 million. As more animals get packed together on farms, they may be able to produce bigger clouds of foot-and-mouth viruses that may be able to travel for longer distances. There's also more dust in the air thanks to modern farming. Since 1750, humans have doubled the emission of dust, largely by clearing land for agriculture. The rise in dust in some parts of the world may be helping to drive the rise in certain diseases such as valley fever.

Our influence on the aerobiome now extends even to the clouds. Pierre Amato and his colleagues documented our mark on the sky by collecting droplets from clouds passing over Puy de Dôme. They extracted microbial genes from the cloud water and searched for ones that endowed the bacteria with resistance to antibiotics. In 2023, the scientists announced that they had found twenty-nine different kinds of resistance genes in the cloud-residing bacteria. Every cubic meter of cloud, they estimated, held about fifty-four hundred resistance genes. A typical cloud floating overhead may hold more than a trillion of them.

The discovery of antibiotics such as penicillin in the twentieth century was one of the greatest triumphs in medicine. When humans started manufacturing antibiotics, they could unleash the compounds at concentrations that the microbes had never experienced before. At first, the antibiotics worked like silver bullets, quickly clearing infections. But then evolution erased much of their power.

As bacteria replicated, they mutated. Some mutations helped them withstand antibiotics, rendering many of the drugs effectively useless. In the 1960s, scientists like Selman Waksman predicted that antibiotics would eradicate many bacterial diseases. But in 2014 alone, 700,000 people died worldwide from infections of antibiotic-resistant bacteria. Five years later, the toll rose to 1.27 million.

That lethal evolution took place not just within the bodies of people who took antibiotics. Farmers discovered that the drugs made chickens and other livestock grow bigger by protecting them from infections, and they came to dispense the majority of antibiotics. Once resistance genes evolved in bacteria in farm animals, those genes could then spread to other microbes. Bacteria can pass along resistance genes from one species to another.

In 2009, a microbiologist named David Graham and his colleagues discovered that the evolution of antibiotic resistance was starting to leave a mark on the Earth. They discovered it in an archive of dirt. Starting in 1979, scientists in the Netherlands regularly scooped soil from Dutch farms. Once the soil went into storage, most of the microbes it contained died. But fragments of their DNA survived. Every gram of dirt preserved billions of bacterial genes.

Graham fashioned molecular hooks to pull genes with resistance mutations out of that genetic soup. When he and his colleagues looked in soil samples dating back to World War II, they struggled to find any resistance genes. But in younger soils, Graham found the genes in swiftly growing abundance. Antibiotic resistance has become a geological marker of humanity.

When the threat of resistance came to light in the 1950s, scientists began tracking the spread of the genes that allowed bacteria to evade antibiotics. They found resistance genes in soil microbes as well as in drinking water. They found microbes harboring these genes in hospital sinks and on countertops. And in recent years, scientists have looked for them in the air. They found them floating in hospitals and on pig farms. Antibiotic resistance genes float through cities as well. An international team of scientists spread out across nineteen cities took the filters out of automobile air conditioners and inspected them for bacteria. The trapped microbes carried thirty different kinds of resistance genes.

The scientists who carried out those studies couldn't track how far the resistant microbes had traveled or how high they had floated into the air. But when Pierre Amato and his colleagues looked in clouds, they discovered resistance genes in vast numbers. Bacteria in clouds seem to be especially well equipped to fight antibiotics. A single airborne bacterium may carry as many as nine resistance genes, each providing a different defense against the drugs.

It's possible that antibiotic-resistant microbes are especially good at thriving in the clouds. Some genes provide antibiotic resistance by allowing bacteria to pump the drugs out of their interiors quickly, getting rid of them before they can cause damage. The stress of life in a cloud may cause bacteria to produce toxic waste that they need to pump out quickly as well.

Clouds may be able to spread these resistance genes farther than contaminated meat and water. Once in a cloud, bacteria can travel hundreds of miles in a matter of days before seeding a raindrop and falling back to Earth. When they reach the ground, the microbes may then pass along their resistance genes to other microbes they encounter. Every year, Amato and his colleagues estimate, 2.2 trillion trillion resistance genes shower down from the clouds. When it rains, we walk through downpours of DNA of our own making.

o o o o

The air, Theodor Rosebury said, is free in evil ways as well as in good ones. And when we have tried to protect ourselves from its evil ways, we have sometimes brought down more evil on our heads.

Rusts had been growing on plants for more than 300 million years when people began to farm. But the new fields of wheat and other crops offered the fungi easy new targets. If a spore of rust landed on a stalk of wheat, its offspring had to float only a short distance before finding another plant they could infect. And as farming expanded, rust spores carried off by winds had better odds of landing on a distant wheat field. Just as towns and other dense settlements changed the evolutionary equation for human pathogens, farms did the same for pathogens of plants. Bible-era farmers in the Near East blamed the arrival of rust on God's wrath. They did not know that they were the ones who had created the evolutionary conditions that shaped their divine punishment.

The large-scale farms of the nineteenth century provided even greater opportunities for rust to thrive. The international trade in seeds led to more farmers growing the same strains of wheat. Rust that adapted to them could spread across vast expanses of identical crops. There was no way to shield wheat plants from the rain of rust spores, so scientists like Elvin Stakman developed resistant strains of wheat. Stakman's student Norman Borlaug made even more improvements. His new wheat strains not only resisted rust, but also produced bigger harvests. In some countries, farmers tripled their production with his crops. In 1970, Borlaug won the Nobel Peace Prize for what came to be known as the Green Revolution.

But Borlaug's victory over the aerobiome was not complete. Rust spores still managed to find enough plants to infect, and they continued to evolve. In 1998, agricultural scientists in Uganda discovered rust growing on wheat that should have been able to fend it off. When South African scientists took a closer look at the rust, they discovered that it was a new race, mutations allowing it to overcome the resistance found in 80 percent of the world's wheat.

The new race was named Ug99. Although it was named for Uganda, the place where it was discovered, Ug99 originally emerged in Kenya in the early 2000s. Once Ug99 could fly, it proved unstoppable. By 2003, long-range winds had carried it to Ethiopia. In 2005, Borlaug—then ninety-one—held a press conference in Nairobi to warn reporters of the danger. "Nobody's seen an epidemic for fifty years, nobody in this room except myself," he said. "Maybe we got too complacent."

Borlaug died in 2009, and Ug99 outlived him. It moved across the Red Sea, attacking wheat fields in Yemen and Iran. Soon, other resistant rusts evolved and spread. One, called TKTTF, sailed across Europe and Asia, as well as Africa. When TKTTF reached Ethiopia, it wiped out almost all the country's harvest in 2013.

The relentless winds are expected to spread these rusts farther. It is just a matter of time, experts suspect, before the spores land in the United States, bringing back the scourge that Borlaug had fought as a young student.

○ ○ ○ ○

Wheat is not the only crop that has to survive in a relentlessly evolving aerobiome. Coffee crops are also under threat from a rust of their own. New forms of late potato blight are menacing fields once more. Some aerobiologists suspect that we would be better off giving up on trying to wall ourselves off from the aerobiome. We might instead change the layout of evolution's arena.

Farmers might benefit from abandoning the search for the single best strain to plant. If they planted a variety of strains, it would become harder for a single strain of pathogen to vanquish all of them. Another possibility is to tolerate some damage to our crops. If the current version of rust can thrive even modestly on farms, mutants will have a harder time gaining an advantage.

Our own health might also benefit if we could find an amicable peace with the aerobiome. In our struggle with airborne pathogens, we must resist the dream of total victory. We know this in large part thanks to the microbes that dwell not in our lungs but in our guts. It is hard for scientists to survey the life in our airways because they can't easily dip a swab into the trachea or below. Scientists who study the microbes in the gut don't have to probe people's intestines for samples. They just have to wait for the next bowel movement.

Feces have demonstrated that some of the microbes that live in a mother's birth canal slip into a baby's mouth during delivery. After birth, a mother's breast milk delivers more microbes into the gut, and babies can then add even more of them by sticking their hands into their mouths.

Those microbes teach our immune cells tolerance so that they don't overreact to harmless distractions and harm our bodies instead. They help break down our food and render toxins harmless. They may even influence our moods by putting mind-altering compounds into our bloodstreams or communicating directly through the nerve endings in our guts.

Some of the most compelling evidence for the help our gut microbes provide us comes from our ongoing disruption of our inner ecosystem. In recent decades, children have taken a growing amount of antibiotics. While the drugs are essential for stopping deadly infections, they also kill off many harmless strains in the gut. The diversity of the children's microbiome has been getting simpler. And as they spend more time indoors, they are exposed to fewer species they might encounter on the ground outside. As they eat more sugar and processed foods, some species thrive while many others struggle. All this disruption may make it harder for the microbiome to carry out the jobs we need it for, such as training the immune system. It might help explain why so many immune disorders, such as asthma and allergies, are on the rise in Westernized countries.

We know much less about the microbes of the airway than the ones that reside in our gut. It appears that some microbes enter it when we touch our mouths and noses with contaminated fingers. But scientists suspect that we breathe in other microbes—airborne fungal spores, bacteria riding on dust, viruses embedded in droplet nuclei. It's possible that even pollen grains deliver microbes on their surfaces.

With their first gasp of air, babies start inoculating themselves with the aerobiome. They live for years close to the ground, inhaling denser clouds of microbes than the adults breathing the air a few feet overhead. It's possible that the diversity of species children breathe may help their immune systems become tolerant. But the environment in which children grow up determines what sort of microbes they inhale. The aerobiome of a suburb full of monoculture yards is less diverse than that of a forest, and a blacktop-covered city contains even fewer species. Inhaling these different menageries may have effects that last long after childhood and that extend far beyond the immune system.

In one intriguing study, Martin Breed and his colleagues at the University of Adelaide tested different aerobiomes on mice. They reared the animals in sealed cages equipped with a fan blowing across a tray of soil to make the air dusty. In some cages, the soil came from a heavily farmed field. In other cages, it came from a eucalyptus woodland.

After letting the mice breathe the dust for seven weeks, Breed and his colleagues observed how the animals behaved. They were particularly curious about the anxiety the mice experienced. In one test, they put the mice on a large black square arena. Anxious mice tend to scurry to the edges, while calmer ones will spend more time investigating the center. Breed found that the mice breathing woodland dust were measurably less anxious than the ones breathing bacteria from farmland. The microbes that the mice inhaled might have spread into their gut as well as into their airways. Once in the intestines, they might have made compounds that alleviated the anxiety that the mice would otherwise have felt.

Experiments like these are mostly good for tantalizing. Their results may well dissolve as scientists carry out more research. But they at least feed the mind with new ways to think about the aerobiome. We don't have to look at it as an incoming rain of biological weapons. We can try instead to treat it as an atmospheric garden. Perhaps we can even consciously tend to the species that fill the air and ultimately make their way into our lungs. Breed and his colleagues envision a place they call the Probiotic City, where parks are infused with beneficial bacteria for people to inhale and where green walls—vertical nurseries of plants growing inside buildings—release microbes from their leaves.

In an age in which we are heating the atmosphere with greenhouse gases, polluting it with smog, and fostering new pathogens, it may be hard to envision this kind of harmony. But it is worth trying, because the aerobiome is not going away even if we stop thinking about it. As long as there is life on Earth, it will fly, and as long as we are here, we will breathe.

epilogue

HAPPY BIRTHDAY, CHITA RIVERA

One winter night in Boston, not long after the Omicron wave had crested, my wife, Grace, and I made our way down Columbus Avenue in a light snow. Far overhead in the dark, bacteria and fungi were seeding snowflakes. The crystals of ice built themselves a sixfold symmetry and then tumbled a mile. Just before they met the pavement, the snowflakes melted away, smearing a cold slick below our feet and making the wind at our faces damp and raw. The snow melted into drops on the glass screen of my phone, which was guiding us to Club Café.

Founded in 1983, Club Café is one of the oldest gay nightspots in Boston still in operation. It opened in the midst of the intense homophobia of the Reagan era, but the owners did not hide the venue behind windowless brick. Instead, they built a tall, curved glass wall to look out at the street front and to let passersby look in. In the late 1980s, when AIDS swept through Boston's gay community, the café became a place for activists to organize.

Four decades later, Grace and I entered the club on a Wednesday evening that was mellow and slow. In the main room, young men talked at high tops. In a side room a drag queen sang karaoke. Some office friends sat at the bar sharing drinks before catching the T back home. Grace and I passed through a glass door set in a glass wall. We entered a space called the Napoleon Room. A bust of Bonaparte rested on a stand with three spears for legs. A full statue of Joséphine, in a Greek outfit, stood nearby. At the far end of the room, cherubs looked down from a purple-lit arch at a baby grand piano on a dais. People were filling up the tables in the Napoleon Room, many of them old and gray. None wore a mask.

Bob and Eliza, friends of ours from Boston, arrived soon afterward. A few months beforehand, they both had gotten Covid-19—twice in the course of eight weeks, in fact—but they had still agreed to join us in a small room full of unmasked people for a night of singing. I had promised a night almost certainly free of Covid-19.

We caught up on news about kids, work, and Covid-19—the troika of our lives—until we were joined by Ed Nardell, the scientific grandchild of Willam and Mildred Wells. He was accompanied by his partner, Douglas, an urban forester. When I had first visited Nardell months before in Boston, he encouraged me to come to Club Café to hear him and his friends sing.

Nardell grew up listening to Italian crooners in the 1950s in Wilkes-Barre, Pennsylvania. It wasn't until his late sixties that he started to sing as well. After Nardell's wife died of cancer, he came out of the closet and joined the Boston Gay Men's Chorus. One night at a piano bar, he struck up a conversation with the house musician, Brian Patton. Patton invited him to join the crowd at the Napoleon Room. Nardell had been a regular there for five years by the time we joined him.

As people passed by, going to their tables, they leaned down, unmasked, to hug Nardell's shoulders. Patton soon arrived too. Sandy haired and middle-aged, he wore a blue plaid blazer. Stepping onto the dais, he inflated balloons and let them float off. They landed on the piano and on a drooping potted plant. Patton laid a sheet cake on top of the piano as well. The numbers 9 and 0 were anchored in the vanilla frosting. That night, we were celebrating Chita Rivera becoming a nonagenarian.

Patton sat down at the piano and honored Rivera by playing one of her songs, "When You're Good to Mama." He segued with shifting chords to "You Must Have Been a Beautiful Baby" and then to a favorite of his as a teenager, "After the Lovin'" ("I had no business singing that when I was seventeen," he observed between verses). Patton began inviting people up to sing.

A white-haired woman in a long blue sweater and bright red glasses quavered through "You're an Old Smoothie":

I'm an old softie, I'm just like putty in the hands of a boy like you.

In March 2020, the Napoleon Room went silent when Governor Charlie Baker closed down bars throughout Massachusetts. "This is not a sprint," he warned. "This is going to be a marathon."

Soon after the club closed, Nardell came down with Covid-19. He suspects he became infected at a rehearsal of the Gay Men's Chorus. It took Nardell two weeks to recuperate, and as soon as he felt well enough, he attacked Covid-19 as a scientist.

He began by volunteering to help the National Academy of Sciences make sense of the early studies of how the coronavirus spread. To him, the transmission of Covid-19 looked suspiciously like that of tuberculosis. He worried that it was traveling in droplet nuclei through the air. Later, Nardell signed Morawska's open letter to public health authorities to take the airborne transmission of SARS-CoV-2 seriously.

Nadell then published a letter in the *Journal of the American Medical Association* with another Harvard tuberculosis expert, Ruvandhi Nathavitharana. Nathavitharana and Nardell pointed to the Skagit Valley Chorale and other outbreaks of Covid-19 as evidence that SARS-CoV-2 might readily spread through the air. They urged that protections go beyond just wearing masks. "Should not air disinfection be deployed in intensive care units, emergency departments, waiting rooms, and ambulatory clinics?" they asked.

Nardell and his colleagues realized that they could run an experiment to see just how similar Covid-19 was to tuberculosis. They converted their experimental TB ward in South Africa to test SARS-CoV-2 instead. They had the patients wear surgical masks, which they then tested for viruses. Only people who were actively exhaling SARS-CoV-2 would be invited to the ward for a day.

The air from the patient rooms flowed to an animal facility. Over half of the hamsters that breathed in air from the rooms where Covid-19 patients were staying got infected. On the days when the rooms were empty, none of them picked up the virus.

The experiment finally yielded results long after a consensus had emerged that Covid-19 was probably mainly airborne. But it made up for its slow pace with its startling results. The animals in the facility were kept more than fifty meters away from the rooms where the patients stayed. And yet the coronaviruses could travel that long distance and infect the hamsters.

○ ○ ○ ○

Club Café reopened in late 2020 and struggled to stay afloat. For months the tables remained widely distanced. The club required masks, which people could take off to eat. In 2021, doormen started checking for proof of vaccination at the door. But the Napoleon Room, sealed off behind its glass wall, stayed closed.

Nardell proposed that the club make the space safe again with ultraviolet light. Eight decades had passed since the Wellses had installed mercury-vapor lamps in the Germantown Friends School, and ultraviolet light technology had evolved dramatically in that time. Mercury-vapor lamps produce ultraviolet light at a wavelength of 254 nanometers. Looking directly at that light can irritate the eyes. The Wellses protected children in their studies by pointing the lamps up to the ceiling so that they sterilized only droplet nuclei that wafted to the upper layer of air in classrooms.

But in the early 2000s, engineers discovered that a mixture of krypton and chloride produces UV light at 222 nanometers. Lamps that produce that light, known as far-UVC, can be safely pointed downward. The Columbia University radiation biologist David Brenner discovered that far-UVC cannot penetrate the skin or the eyes. Yet he and his colleagues also found that it can still protect people from infection. In 2013, they performed a tabletop experiment in which they killed microbes with far-UVC. In 2021, they killed bacteria floating in the air of a lab with far-UVC lamps installed on the ceiling.

The next singer rose in the Napoleon Room: a bearded man who introduced himself as Nabil. He strode between the tables as he belted out *"La Vie en rose."* As I tried to make out the French lyrics, I looked up at the tin-stamped ceiling. I saw six far-UVC lamps that Nardell had arranged to have installed in 2021. They looked to me like smoke detectors, about the size and shape of a hardback novel. Each lamp emitted a faint white gleam.

Now it was Doug Still's turn at the microphone. "Growing up, I thought this was elevator music," he said, "but now I know it's a good song."

Say you're in love,
In love with this guy
If not, I'll just die.

A dance studio owner announced he would sing a piece he had performed for Chita Rivera in person. They had both been at Liza Minnelli's birthday party.

In the roaring traffic's boom,
In the silence of my lonely room
I think of you.

Silverware and keys clinked on glasses. I found myself playing a gourd and shouting back choruses later in the evening. A woman introduced herself as Meg and admitted that she had "become a Covid introvert, so this is a stretch." Meg then began to sing.

Maybe it's best to love a stranger.
Well, that's what I've done.
Heaven help my heart.

Nardell had gotten the UV lights installed and working in time to celebrate his seventy-fifth birthday in February 2022 with a one-man show. The monitors on the walls notified customers of the change. "Club Café is now equipped with the latest in Far UV air disinfection technology to make gathering, talking, eating, drinking, and even singing as safe as possible," they announced.

The Napoleon Room is a far cry from the guinea pig penthouse that sat atop the Loch Raven hospital. But once Nardell installed the UV lights, he tried to collect as much data as he could. He sometimes asked customers and waiters to let him clip a piece of photographic film on their shoulder. As ultraviolet rays fell down through the room, the ones that hit the film left black spots. After collecting twelve film badges, Nardell and his colleagues counted the spots on them to estimate the exposure to far-UVC experienced in the room.

Radiation biologists measure exposure to ultraviolet light with a unit of energy called millijoules per square centimeter. Over an eight-hour period, they recommend that people receive no more than 479 millijoules of far-UVC on their skin. The film badges at the Napoleon Room registered only fifty-three. And yet that tiny dose of UV is deadly for SARS-CoV-2. Nardell estimates that it's the equivalent of exchanging all the air in the Napoleon Room thirty-four times every hour. Hospital isolation rooms are required to provide just twelve exchanges.

It is the curse of experiments on airborne infections that they may work without any way to prove their success. Nardell did not attempt to collect official data about Covid cases among customers or waiters. But the clientele is a close-knit group of piano bar loyalists, and their evenings at the Napoleon Room are the heart of their social lives. Nardell knows of no outbreak of Covid-19 in their ranks since the lights were turned on.

Nardell got up at one point in the evening to help a frail couple onto the stage, and he brought them a pair of chairs to sit on. Like the Skagit Valley Chorale three years before, these were Covid's most vulnerable targets: old people trying to enjoy each other's company by singing together. The wife laid her head on her husband's shoulder as they sang.

Two sleepy people
By dawn's early light,
Too much in love to say good night.

Eliza and Bob both found much to talk about with Nardell. Eliza designed affordable housing and wanted to hear about Nardell's experiences installing ultraviolet lights in homeless shelters to stop tuberculosis. Bob, a neuroscientist, was fond of gadgets. Early in the pandemic, he had bought himself a carbon dioxide sensor to play the part of Max von Pettenkofer. Nardell advised him that a room reading under a thousand parts per million was probably safe. Bob wondered if he should buy a far-UVC lamp for his house.

Nardell shrugged. "Who are you protecting yourself from?" he asked. Nardell does not believe that any one gadget will protect people from airborne infections. In some spaces where strangers often mingle, ultraviolet lights may be a safe, effective choice. But doctors may still need to wear N95s to treat infected patients. Ultraviolet light may not be able to stop droplet nuclei in the short time they need to travel the distance between them. In other places, Corsi-Rosenthal boxes or some equivalent may clear the air well enough. Scientists still need to do more research to know what is effective in each setting—and Nardell hopes they will get it done before Disease X is followed by Disease Y.

"We ought to do what works, not what theory or modeling dictates," Nardell said. "If simple masks and distancing seem to control an infection, that is what we do."

Nardell's argument sounded eminently sensible as we chatted in the Napoleon Room. But beyond those purple-lit walls, he and like-minded researchers were not having much success improving the air. The Biden administration provided money to schools through the American Rescue Plan to improve ventilation in classrooms. Just over half of American schools followed through. The government was developing a plan to install improved filters in the ventilation systems in all federal buildings. But they would treat the project only as an exemplar of what could be done—not as a government-enforced standard for air quality. As for ultraviolet light, Nardell was getting queries from crypto tycoons who liked to prep for a viral apocalypse, but he heard little from public health officials. William and Mildred Wells had dreamed of a society that ensured clean air along with clean drinking water. Even after the arrival of Covid-19, that vision remained mostly out of reach.

Patton finally called up Nardell. He made his way to the piano and began to sing "My One and Only Love." Frank Sinatra first made it famous when Nardell was seven years old, but he preferred the Johnny Hartman arrangement.

The very thought of you makes my heart sing
Like an April breeze on the wings of spring.

On a wet winter night in Boston, an April breeze on the wings of spring sounded like a far-off dream. We enjoyed the thought. We enjoyed Nardell crooning in the purple light. We enjoyed the protection of the lamps overhead. We enjoyed the indoors, where we could stay warm and dry, where we could use our lungs to create our own breezes carrying vibrations that tickled the tiny bones in other people's ears, transmitting our choruses, our memories, our happiness.

acknowledgments

I would like to thank Donald Milton for introducing me to the Wellses, and Ed Nardell for sharing Richard Riley's papers and his own memories of working with him. They ushered me into the universe of aerobiology. My thanks also go to Eric Simonoff, who helped me think my way from an idea to a book proposal.

As I worked further on my book, many scientists and doctors welcomed me on visits, chatted with me on Zoom calls, and answered a stream of emails from me. They include: Cathie Aime, Joseph Allen, Rommie Amaro and her students, Pierre Amato, Lydia Bourouiba, Martin Breed, Wilbur Chen, Kristen Coleman, John Conly, Richard Corsi, Grant Deane, Anthony Fauci, Edward Holmes, Jose-Luis Jimenez, Leda Kobziar, Yuguo Li, Linsey Marr, David Michaels, Shelly Miller, Lidia Morawska, Caitlin Pepperell, Chris Ponting, Kim Prather, Anne Pringle, Andrew Read, Jake Robinson, Greg Sandstrom, Diane Saunders, David Schmale, and Lisa Zeigler.

I am deeply grateful to the Skagit Valley Chorale for their willingness to revisit a hard chapter of their lives, especially Debbie Amos, Ruth and Mark Backlund, Yvette Burdick, and Carolynn Comstock. Thanks also go to Richard Riley Jr. for sharing memories of his father.

Along with interviews, I built this book on archives. I'd like to thank Bethany Antos at the Rockefeller Archive Center, Daniel Barbiero at the National Academy of Sciences, Mary Crauderueff at Haverford College, Michael Gallagher (for work in the Delaware historical archives), Jackie Graziano at the Westchester County Archives, Vicki Killian (who found documents for me at the National Archives), Mary Mann at Cooper Union, Jessica Murphy at the Countway Library at Harvard University, and Tim Wood at the Germantown Friends School archives. Thanks also to the staffs at the American Philosophical Society, the John Hay Library at Brown University, the Rhode Island Historical Society, the University of Pennsylvania, and Yale's Manuscripts and Archives.

I was fortunate to work with some gifted researchers who helped me with fact-checking, translation, and other essential work. Thanks to Oliver Guinan, Lori Jia, Sandra Klos (who also translated German sources), Rafaela Kottou, Matt Kristoffersen, and Sophia Li.

I'm grateful to the Sloan Foundation for a fellowship that supported this book. Stephen Morrow and John Parsley provided expert editing, and I'm grateful for their patience as we went through rounds of edits. Thanks to my mother, Marfy Goodspeed, and brother, Ben Zimmer, for their eerie skills at reading old documents. Thanks to my daughters, Charlotte and Veronica, for putting up with their virus-obsessed father during the pandemic. And most of all, thanks to my wife, Grace, who always keeps the pilot light burning in my heart.

notes

Uncited quotations in this book are from interviews conducted by the author. A longer version of this end matter—including an expanded bibliography, Internet links, and digital object identifiers—can be found online at http://carlzimmer.com/airborne. A copy of that webpage was also archived on the Wayback Machine at http://web.archive.org on February 1, 2025.

ABBREVIATIONS OF PEOPLE

BC: Beulah Chase
CC: Charles Chapin
CW: Charles-Edward Amory Winslow
CL: Charles Lindbergh
ER: Edward Riley
ES: Elvin Stakman
EW: Edwin Wilson
FM: Fred Meier
JS: Joseph Stokes
MW: Mildred Weeks Wells
OR: Oswald Robertson
RC: Roger Crane
RH: Roderick Heffron
RR: Richard Riley
SM: Stuart Mudd
TR: Theodor Rosebury
WW: William Firth Wells

ARCHIVAL SOURCES AND ABBREVIATIONS

APS/JS: Joseph Stokes Jr. Papers. Mss.B.St65p. American Philosophical Society, Philadelphia, PA.

APS/OR: O. H. Robertson Papers. Mss.B.R546. American Philosophical Society, Philadelphia, PA.

CAL: Charles Augustus Lindbergh Papers (MS 325). Selected Correspondence, 1911–1974. Manuscripts and Archives, Yale University Library.

CAL/FM: Correspondence between CL and FM. Charles Augustus Lindbergh Papers. Box 21, Folder Meier, Fred Campbell.

CC/BU: Charles V. Chapin Papers. Correspondence, 1901–1939. John Hay Library, Brown University.

CC/RI: Charles V. Chapin Papers. MSS 343, Series 6. Rhode Island Historical Society.

CF: Commonwealth Fund records. Rockefeller Archive Center, Sleepy Hollow, NY.

CWP: Charles-Edward Amory Winslow Papers. General Correspondence, 1897–1957. Box 31. Manuscripts and Archives, Yale University Library.
EWP: Edwin Wilson Papers. Box 2, Folder 34. Countway Library, Center for the History of Medicine, Harvard School of Public Health.
NAS: National Academies of Sciences–National Research Council Archives. B&A: DNRC: Committee on Aerobiology.
NRC: National Research Council Archives. Committee on Aerobiology. General: 1937–1939.
OHECS: Oral History of E. C. Stakman. Box 1, Volume 9. Elvin C. Stakman papers, ua-01017. University of Minnesota Archives. https://archives.lib.umn.edu/repositories/14/archival_objects/348190.
RRP: Richard L. Riley Papers. HMS c363 Accession #2011-131. Countway Library, Center for the History of Medicine, Harvard School of Public Health.
TRP: Theodor Rosebury Papers. US National Library of Medicine.
USDA/BPI: US Department of Agriculture, Records of the Bureau of Plant Industry, Soils, and Agricultural Engineering. National Archives, College Park, MD.
USDA/CSA: USDA, Records of the Commissioner and Secretary of Agriculture. National Archives, College Park, MD.
USDA/ES: USDA, Records of the Extension Service. National Archives, College Park, MD.
USDA/WB: USDA, Records of the Weather Bureau. National Archives, College Park, MD.
USGPO: US Government Printing Office, Washington, DC.
WCA: Westchester County Archives, White Plains, New York.

EPIGRAPH

ix **W. Szymborska,** *Miracle Fair: Selected Poems of Wisława Szymborska* (New York: W. W. Norton & Co., 2002), 70.

PROLOGUE: THAT'S WHERE IT IS

xv **"This virus is here":** "WHO Director-General's Opening Remarks at the Media Briefing—5 May 2023."
xvii **"Suddenly my annoying allergy symptoms":** K. Cauvel, "Locals Speak Out About Symptoms, Seriousness of Coping with COVID-19," *Skagit Valley Herald*, April 2, 2020.
xviii **"The community should postpone":** Skagit County, "Declaration of Emergency and Health Officer Recommendations," March 10, 2020.
xix **"Individuals should try":** J. Inslee, "Inslee Issues Emergency Proclamation That Limits Large Events to Minimize Public Health Risk During COVID-19," *Medium*, March 11, 2020.
xx **"FACT: #COVID19 is NOT airborne":** https://twitter.com/WHO/status/1243972193169616898.

CHAPTER I: THE FLOATING GERMS

3 **"The entire facade":** "Obsequies of Louis Pasteur," *New York Times*, October 2, 1895.
4 **"Everything gets complicated":** L. Pasteur to A. Chappuis, August 10, 1860, *Correspondance de Pasteur 1840–1895, Réunie et Annotée par Pasteur Vallery-Radot* (Paris: Flammarion, 1940), 73.
5 **"The immensity of these aerial summits":** M. W. Shelley and P. B. Shelley, *History of a Six Weeks' Tour Through a Part of France, Switzerland, Germany, and Holland* (London: T. Hookham, 1817), 151–52.
5 **"I perceived, as the shape":** M. W. Shelley, *Frankenstein, or the Modern Prometheus* (London: G. and W. B. Whittaker, 1823), 203.

5 **"the germs that float":** L. Pasteur, *"Des Générations Spontanées: Conférence de Fait au 'Soirée Scientifique de la Sorbonne' le 7 Avril 1864,"* *Revue des Cours Scientifiques* 1 (1864): 257–65.

5 **"Life and death":** Aristotle, *"De Respiratione,"* in *The Works of Aristotle*, Vol. III: *Meteorologica* (Oxford: Clarendon Press, 1908), 472.

6 **"It attacks everyone":** J. Jouanna, "Air, Miasma and Contagion in the Time of Hippocrates and the Survival of Miasmas in Post-Hippocratic Medicine," in *Greek Medicine from Hippocrates to Galen*, ed. P. van der Eijk (Leiden: Brill, 2012), 127.

6 **"When the air is full":** Jouanna, "Air, Miasma, and Contagion," 125.

7 **"Take your rough hands":** M. E. Palmer, "A Blight on the *Pax Augusta*: The Robigalia in Ovid's *Fasti*," *Classical World* 111 (2018): 503–23.

7 **"For every plant":** J. D. Hughes, "Theophrastus as Ecologist," *Environ Rev* 9 (1985): 296–306.

7 **"Lands which are exposed":** Theophrastus, *Enquiry into Plants and Minor Works on Odours and Weather Signs*, trans. A. Hort (London: William Heinemann, 1916), 203.

8 **"This caused a severe epidemic":** M. W. Dols, *The Black Death in the Middle East* (Princeton, NJ: Princeton University Press, 1977).

8 **"Since it is civil":** G. Geltner, "The Path to Pistoia: Urban Hygiene Before the Black Death," *Past & Present* 246 (2020): 3–33.

8 **"The art of Hippocrates":** S. K. Cohn, "The Black Death and the Burning of Jews," *Past & Present* 196 (2007): 3–36.

9 **"Infection is a spark":** J. K. Stearns, *Infectious Ideas: Contagion in Premodern Islamic and Christian Thought in the Western Mediterranean* (Baltimore, MD: Johns Hopkins University Press, 2011), 71.

10 **"born amid squalor":** G. Fracastoro, *Fracastoro's Syphilis*, trans. G. Eatough (Liverpool, UK: Francis Cairns, 1984), 55.

10 **"It was necessary":** M. Greenwood, "Some Epidemiological Observations on Foot-and-Mouth Disease, with Special Reference to the Recent Experience of Holland," *J Hyg* 26 (1927): 465–89.

11 **"There are many seeds":** Lucretius, *De Rerum Natura*, trans. W. D. Rouse (Cambridge, MA: Harvard University Press, 1924), 575.

11 **"so very varied":** Fracastoro, *Fracastoro's Syphilis*, 53.

11 **"The air is the most suitable":** G. Fracastoro, *Hieronymi Fracastorii de Contagione et Contagiosis Morbis et Eorum Curatione, Libri III*, trans. W. C. Wright (New York: Putnam, 1930), 57.

12 **"Mankind, Quadrupeds and Plants":** R. Bradley, *The Plague at Marseilles Consider'd* (London: W. Mears, 1721), 47–48.

13 **"the wholesome breath":** R. Evans, *The Fabrication of Virtue: English Prison Architecture, 1750–1840* (Cambridge: Cambridge University Press, 2011), 100.

13 **"unwholesome atmosphere":** B. Rush, *Medical Inquiries and Observations* (Philadelphia, PA: Prichard & Hall, 1789), 90.

13 **"Philadelphia, from having been":** Rush, *Medical Inquiries and Observations*, 86.

13 **"The first cases":** J. McFarland, *The Epidemic of Yellow Fever in Philadelphia in 1793 and Its Influence upon Dr. Benjamin Rush* (New York: Medical Life Press, 1929), 454.

14 **"the miasmata from our atmosphere":** Benjamin Rush to Thomas Jefferson, August 29, 1804, Founders Online, https://founders.archives.gov/documents/Jefferson/01-44-02-0289.

14 **"A new era":** T. Apel, *Feverish Bodies, Enlightened Minds* (Stanford, CA: Stanford University Press, 2016), 144.

14 **"Like a primal fixed star":** W. Staughton, *An Eulogium in Memory of the Late Dr. Benjamin Rush* (Philadelphia, PA: Published by request of the graduates, 1813), 24.

14 **"He has done more good":** A. Brodsky, *Benjamin Rush: Patriot and Physician* (New York: Truman Talley Books / St. Martin's Press, 2004), 365.

14 **"little short of willful murder":** Brodsky, *Benjamin Rush*, 523.

14 **"This unknown and incomprehensible power":** C. Maclean, *Evils of Quarantine Laws* (London: T. & G. Underwood, 1824), 208.

15 **"Of small things":** M. Grote, "Microbes Before Microbiology: Christian Gottfried Ehrenberg and Berlin's Infusoria," *Endeavour* 46 (2022): 100815.

15 **"How many thousand millions":** C. Ehrenberg, *Passat-Staub und Blut-Regen* (Berlin: Königliche Akademie der Wissenschaften, 1849), 57.

16 **"the falling of impalpably fine dust":** C. Darwin, *The Voyage of the Beagle* (New York: P. F. Collier & Son, 1909), 14.

16 **"I am truly astonished":** Darwin to Ehrenberg, July 4, 1844, Darwin Correspondence Project, https://www.darwinproject.ac.uk/letter/?docId=letters/DCP-LETT-760.xml.

16 **"I will feast":** Ehrenberg to Darwin, July 11, 1844, Darwin Correspondence Project, https://www.darwinproject.ac.uk/letter/?docId=letters/DCP-LETT-762.xml.

16 Darwin **published an account:** C. Darwin, "An Account of the Fine Dust Which Often Falls on Vessels in the Atlantic Ocean," *Quart J Geol Soc London* 2 (1846): 26–30.

17 **"proved the actual existence":** D. D. Cunningham, *Microscopic Examinations of Air* (Calcutta: Superintendent of Government Printing, 1873), 2.

17 **"destruction by absorption of miasm":** *Gardeners' Chronicle*, September 5, 1846, 595.

18 **"The decay is the consequence":** M. J. Berkeley, "Observations, Botanical and Physiological, on the Potato Murrain," *J Hortic Soc* 1 (1846): 9–35.

18 **"are wafted about":** M. J. Berkeley, *Introduction to Cryptogamic Botany* (London: H. Baillière, 1857), 258.

20 **"Their presence is always":** J.-M. Cavaillon and S. Legout, "Louis Pasteur: Between Myth and Reality," *Biomolecules* 12 (2022): 596.

20 **"unfair, arrogant, haughty":** Cavaillon and Legout, "Louis Pasteur," 596.

20 **"The most brazen plagiarist":** Cavaillon and Legout, "Louis Pasteur," 596.

20 **"Where do they come from":** P. Debré, *Louis Pasteur*, trans. Elborg Forster (Baltimore, MD: Johns Hopkins University Press, 1998), 148.

21 **"Spontaneous generation is the production":** Debré, *Louis Pasteur*, 157.

21 **"The world into which":** R. Vallery-Radot, *The Life of Pasteur*, trans. R. L. Devonshire (London: Archibald Constable and Co., Ltd., 1902), 99.

22 **"The air in which":** Debré, *Louis Pasteur*, 163.

23 **"gives us the indubitable proof":** L. Pasteur, *Oeuvres de Pasteur* (Paris: Masson, 1922), 204.

24 **The Sorbonne:** Quotations from Pasteur's 1864 lecture are from Pasteur, *Oeuvres*, 328–46.

CHAPTER 2: THE SANITARIANS

26 **"They sometimes carry sickness":** Pasteur, *Oeuvres*, 338.

27 **"simplicity and perfect conclusiveness":** J. Lister, *The Collected Papers of Joseph, Baron Lister*, Vol. 2 (Oxford: Clarendon Press, 1909), 486.

27 **"some material capable":** J. Lister, "On the Antiseptic Principle in the Practice of Surgery," *Br Med J* 2 (1867): 246–48.

27 **"I do not think":** Debré, *Louis Pasteur*, 278.

28 **"No living epidemiologist":** M. von Pettenkofer, "Von Pettenkofer's Latest Views on Cholera," *Lancet* 130 (1887): 339–40.

28 **"was known throughout":** "Max Josef von Pettenkofer," *Nature* 63 (1901): 399.

28 ***"das Cholera-Miasma":*** M. von Pettenkofer, *Untersuchungen und Beobachtungen über die Verbreitungsart der Cholera* (Munich: J. G. Cotta, 1855), 104.

28 **"the monster":** F. Snowden, *Epidemics and Society* (New Haven, CT: Yale University Press, 2020), 235.

29 **"It can scarcely be necessary":** Maclean, *Evils of Quarantine*, 425.

29 **"The evidence brought forward":** "Final Debate on the Cholera at the Westminster Medical Society," *Lancet* 2 (1832): 146–50.

30 **"inhaled poison":** J. Snow, *On the Mode of Communication of Cholera* (London: J. Churchill, 1849), 4.

31 **"I have disposed":** C.-E. A. Winslow, *The Conquest of Epidemic Disease* (New York: Hafner, 1967), 314.

31 **"It looked as if":** K. Kisskalt, *Max von Pettenkofer* (Stuttgart: Wissenschaftliche Verlagsanstalt, 1948), 44.

32 **"We live in the air":** M. von Pettenkofer, *Cholera: How to Prevent and Resist It* (London: Baillière, Tindall & Cox, 1875), 51.

32 **"Cleanliness acts as a deterrent":** M. von Pettenkofer, "Cholera," *Lancet* 124 (1884): 992–94.

33 **"It is possible by increasing":** Pettenkofer, *Cholera*, 52–53.

33 **"a great laboratory":** J. Duffy, *A History of Public Health in New York City, 1625–1866* (New York: Russell Sage Foundation, 1968), 113.

34 **"Suffice it to say":** F. Nightingale, *Notes on Hospitals* (London: Longman, 1863), 9.

34 **"drinking-water faith":** N. Howard-Jones, "Gelsenkirchen Typhoid Epidemic of 1901, Robert Koch, and the Dead Hand of Max von Pettenkofer," *Br Med J* 1 (1973): 103–5.

36 **"They looked like a pile":** R. Koch, *Essays of Robert Koch*, trans. K. C. Carter (Greenwood, NY: Greenwood Press, 1987), 5.

38 **"The man has made":** Quoted in T. Brock, *Robert Koch: A Life in Medicine and Bacteriology* (Washington, DC: ASM Press, 1999), 46.

39 **"who do not believe":** W. Hunter, "Remarks on the Epidemic of Cholera in Egypt," *Br Med J* 1 (1884): 91–96.

41 **"Koch's discovery of the comma bacillus":** Howard-Jones, "Gelsenkirchen," 104.

41 **"If a weather storm":** Pettenkofer, "Cholera."

42 **"I felt as if":** R. Evans, *Death in Hamburg* (New York: Penguin Books, 2005), 312.

44 **"Herr Doktor Pettenkofer":** I. Sherman, *The Power of Plagues* (Washington, DC: ASM Press, 2020), 165.

44 **"we could guess":** Kisskalt, *Pettenkofer*, 118.

45 **"Thus, it is probable":** Koch, *Essays*, 94.

46 **"The dissemination of germs":** C. Flügge, *"Ueber Luftinfection," Zeitschr Hyg Infectionskr* 25 (1897): 179–224.

48 **"Air was the chief vehicle":** C. L. Scamman, ed., *Papers of Charles V. Chapin, M.D.: A Review of Public Health Realities* (New York: Commonwealth Fund, 1934), 14.

48 **"The most natural explanation":** C. Chapin, "Air Infection of Minor Importance," *JAMA* 51 (1908): 2048–51.

48 **"A bundle comes home":** "Well Men May Carry Diseases," *Boston Post*, March 27, 1911. CC/RI, Box 13, Vol. 30.

48 **"Many a lifelong infection":** "Tells of Contagion in Personal Contact," *Providence Journal*, March 27, 1911. CC/RI, Box 13, Vol. 30.

48 **"The breath of a patient":** "Stop Kissing, Keep Healthy," *Boston Evening News*, March 27, 1911. CC/RI, Box 13, Vol. 30.

49 **"They have not wings":** "Well Men May Carry Diseases."

49 **"The death of Professor Pettenkofer":** "Max von Pettenkofer," *Lancet* 157 (1901): 490–91.

49 **"Pettenkofer and his teaching":** M. Greenwood, *Epidemics and Crowd-diseases: An Introduction to the Study of Epidemiology* (London: Williams & Norgate, 1935), 60.

49 **"To men of this generation":** "Pettenkofer and His Theory," *Br Med J* 2 (1935): 518.

49 **"It will be a great relief":** C. Chapin, *The Sources and Modes of Infection* (New York: John Wiley & Sons, 1910), 264.

50 **"I believe that perhaps":** "Dr. Chapin Paid Greatest Honor by Contemporaries," *Providence Journal*, April 16, 1930.

CHAPTER 3: A WATERMELON DOCTOR

51 **"They have flown so much":** R. Owen, "The Lindberghs: Bold Flying Partners," *New York Times*, July 7, 1933.

52 **"We circle":** A. Lindbergh, *Locked Rooms, Open Doors* (New York: Harcourt Brace), 46.

52 **"the government's ace-spore hunter":** "Flying Scientist Chases Spores," *El Paso Herald-Post*, May 12, 1933.

54 **"There are too few scientists":** P. Miquel, *Les organismes vivants de l'atmosphère* (Paris: Gauthier-Villars, 1883), vii.

54 **"an inference made yesterday":** Miquel, *Les organismes*, vi.

54 **"The micrographer who has leisure":** P. H. Gregory, *The Microbiology of the Atmosphere* (London: Leonard Hill, 1961), 10.

54 **"I have succeeded":** Miquel, *Les organismes*, vii.

55 **"When I had got":** C. H. Blackley, *Experimental Researches on the Causes and Nature of Catarrhus Æstivus* (London: Baillière, Tindall & Cox, 1873), 43.

55 **"A small cloud of pollen":** Blackley, *Experimental Researches*, 44.

56 **"The power of pollen":** Darwin to Blackley, July 5, 1873, Darwin Correspondence Project, https://www.darwinproject.ac.uk/letter/?docId=letters/DCP-LETT-8965.xml.

56 **"Hay-fever is essentially a neurosis":** K. J. Waite, "Blackley and the Development of Hay Fever as a Disease of Civilization in the Nineteenth Century," *Med Hist* 39 (1995): 186–96.

57 **"Absolute *silence*":** J. Glaisher, *Travels in the Air* (London: Richard Bentley, 1871), 147.

58 **"What goes on up in the air?":** OHECS, Vol. 7, 1350.

58 **"The terrible devastation":** E. Stakman, "The Promise of Modern Botany for Man's Welfare Through Plant Protection," *Sci Monthly* 44 (1937): 117–30.

59 **"This epidemic of late blight":** J. Garnett, "'Wheat Man's Burden'" (master's diss., George Mason University, 2021), 113.

60 **"We don't know":** OHECS, Vol. 2, 270.

60 **"This barberry bush":** P. D. Peterson, "'The Barberry or Bread': The Public Campaign to Eradicate Common Barberry in the United States in the Early 20th Century," APS Features (2013).

60 **"Thousands of bushes":** E. Stakman, *The Black Stem Rust*, USDA 100.

60 **"If there was a virulent race":** OHECS, Vol. 7, 1345.

62 **"The 8 checks":** F. Meier, "Watermelon Stem-End Rot," *J Agric Res* 6 (1916): 49–52.

62 **"This piece of research":** *Harvard College Class of 1916: Secretary's Third Report* (Cambridge, MA: Harvard University Press, 1922), 293.

62 **"to be of as much service":** *Harvard College Class of 1916*, 294.

62 **"In March 1918":** *Harvard College Class of 1916*, 293.

63 **"weakling parasite":** W. Orton and F. Meier, "Diseases of Watermelons," *USDA Farmers' Bulletin* 1277 (1922): 21.

64 **"Excellent progress has been made":** F. Meier, "The Stem Rust Control Program," *J Econ Entomol* 26 (1933): 653–59.

64 **"To an increasing degree":** F. Meier and W. Popham, "Progress in Barberry Eradication in 1932 and Summarized Results Covering the Period 1918–1932" (Washington, DC: USDA Bureau of Plant Industry, 1932).

64 **"Too often have farmers":** F. Meier, "Seeking Plant Diseases Among the Clouds," *Progress* 1 (1933): 11.

65 **"The crew of the *Los Angeles*":** Meier, "Seeking Plant Diseases," 11.

65 **"Judging from our success":** Meier, "Seeking Plant Diseases," 11.

66 **"ridiculed the idea":** FM to Anne M. Lindbergh, February 23, 1935. CAL/FM.

66 **"girl-meeting project":** D. Friedman, *The Immortalists: Charles Lindbergh, Dr. Alexis Carrel, and Their Daring Quest to Live Forever* (New York: Ecco, 2008), 46.

66 **"baseball-player type":** S. Hertog, *Anne Morrow Lindbergh: Her Life* (New York: Doubleday, 1999), 32.
67 **"I was spellbound":** C. Lindbergh, *Autobiography of Values* (New York: Harcourt Brace Jovanovich, 1978), 129.
67 **"In Carrel":** C. Lindbergh, *Autobiography of Values*, 17.
68 **"attempting to build":** C. Lindbergh, *Autobiography of Values*, 17.
68 **"You could evade":** C. Lindbergh, *Autobiography of Values*, 108–9.
69 **"an extremely simple device":** CL to Arthur Train Jr., May 12, 1938. CAL, Box 36, Folder 1938, May 10–14.
69 **"While it is realized":** Henry Wallace to CL, July 5, 1933. USDA/RSA, RG 16, Entry PI- 191 17, General Correspondence, 1933, Location 170/6/7/5.
70 **"That dreadful ache":** A. Lindbergh, *Locked Rooms*, 47.
70 **"Very bad for you":** D. Herrmann, *Anne Morrow Lindbergh: A Gift for Life* (New York: Ticknor & Fields, 1992), 130.
70 **"The mountains towered":** A. Lindbergh, "Flying Around the North Atlantic," *National Geographic* 66 (1934): 259–337.
71 **"This may be sufficient":** F. Meier, "Microorganisms in the Atmosphere of Arctic Regions," *Phytopathology* 25 (1935): 27.
71 **"He had a brief":** "Lindberghs Safe, No Plane Crash," *Corvallis Gazette-Times*, August 11, 1933.
71 **"Lindy & Ann":** "'Lindy & Ann' Home Again!" British Pathé, aired January 1, 1934, britishpathe.com/asset/40553/.

CHAPTER 4: ETHEREAL SPACE

73 **"Here our ascent":** A. Piccard, "Piccard Tells His Story of Trip to Stratosphere; Escaped Death Narrowly," *New York Times*, May 30, 1931.
74 **"The problem was settled":** FM to CL, December 20, 1933. CAL/FM.
74 **"It was the first time":** "Fungus Spores Didn't Die in the Stratosphere," *Boston Globe*, November 28, 1933.
75 **"Your slide exposures":** FM to CL, December 20, 1933.
75 **"Definitive evidence has been obtained":** F. Meier and C. Lindbergh, "Collecting Microorganisms from the Arctic Atmosphere: With Field Notes and Material," *Sci Mon* 40 (1935): 5–20.
75 **"Could you find time":** FM to CL, December 20, 1933.
75 **"We have made photographs":** FM to Anne M. Lindbergh, February 6, 1934. CAL/FM.
76 **"A tremendous amount":** FM to M. A. McCall, May 3, 1935. USDA/ES, RG 33, Entry PI-83 3, Gen. Corres., Box 293, Folder-Plant Industry, 7-1-34 to 6-30-35.
76 **"that Mr. F. C. Meier shall serve":** Memo to Willis Gregg, August 10, 1934. USDA/WB, RG 27, Entry NC3 50, Gen. Corres., 1912–42, Box 3001, Decimal 070.1.
77 **"The weather was ideal":** Quotations from A. Stevens, "Exploring the Stratosphere," *National Geographic* 66 (1934): 397–434.
80 **"Results were somewhat inconclusive":** F. Meier, "Effects of Conditions in the Stratosphere on Spores of Fungi," *Natl Geogr Soc Stratosph Ser* 2 (1936): 152–53.
80 **"All came back":** Stevens, "Exploring the Stratosphere," 397–434.
80 **"an unusual botanical collecting trip":** Meier and Lindbergh, "Collecting Microorganisms," 5.
80 **"A vast unseen world":** "Lindbergh's Arctic Flight Proved Bacteria Travel in Air Across Sea," *New York Times*, December 24, 1934.
81 **"In the brave days":** "Germs and Winds," *New York Times*, December 25, 1934.
82 **"Perhaps this work":** FM to Anne Lindbergh, February 23, 1935. CAL/FM.
82 **"Have told him":** CL to FM, April 22, 1935. CAL, Box 34, Folder 1935 Jan–Aug.

83 **"As the Bureau considers":** F. D. Richey to C. W. Warburton, May 20, 1935. USDA/ES, RG 33, Entry PI-83 3, Gen. Corres., Box 293, Folder-Plant Industry, 7-1-34 to 6-30-35.

83 **"at such time":** C. W. Warburton to F. D. Richey, May 29, 1935. USDA/ES, RG 33, Entry PI-83 3, Gen. Corres., Box 293, Folder-Plant Industry, 7-1-34 to 6 30-35.

84 **"first attempt to explore":** L. Rogers and F. Meier, "The Collection of Micro-Organisms Above 36,000 Feet," *Natl Geogr Soc Stratosph Ser* 2 (1936): 146–51.

84 **"Camp life, even in heated tents":** A. Stevens, "The Scientific Results of the World-Record Stratosphere Flight," *National Geographic* 69 (1936): 693–714.

85 **"Plant Diseases Live":** "Plant Diseases Live 13 Miles Up," *New York Times*, November 26, 1935.

85 **"Tubes containing the spores":** "Plant Diseases Live 13 Miles Up."

86 **"Perhaps we see from the results":** Rogers and Meier, "The Collection of Micro-Organisms."

86 **"Scientists have debated":** "New Evidence of Life upon Planets Bared," *Miami Tribune*, November 26, 1935.

87 **Amelia Earhart:** "Earhart Tragedy Research Loss," *Victoria Daily Times*, August 7, 1937.

87 **"the extent to which":** FM to Secretary of Agriculture Henry Wallace, October 25, 1937. USDA/ES RG 16, Entry PI-191 17, Gen. Corres., 1937, Location 170/6/23/3, Box 2528, Committees.

87 **"Miss Earhart, in this phase":** "Earhart Plane Loss Stays Collection of Micro-Organisms," *Sci News*, August 9, 1937.

88 **the committee gathered:** Quotations from Committee on Aerial Dissemination of Pathogens, minutes of meeting, November 12, 1937. NAS, Meetings: 1937-1938.

89 **"The Doctors Wells":** Only Mildred was a doctor, it's worth noting.

89 **an ambitious plan:** Outline for Special Research Project, Six Months' Duration—July, August 1938—March, April, May, June 1939, Department of Agriculture, Project No. SRF-2-6.

CHAPTER 5: A PERFECT CYCLE

92 **"scare us to death":** G. Whipple, "The Public Health Work of Professor Sedgwick," *Science* 53 (1921): 171–78.

92 **"He strove to inspire":** WW to Geddes Smith, March 28, 1941. CF, Series 13.9, Box 68, Folder 813.

93 **"the most thoughtful":** William Sedgwick to CC, October 28, 1902. CC/BU, Box 1, 1901–1913, Folder 3.

93 **"I consider it":** William Sedgwick to CC, April 3, 1913. CC/RI, Box 10, Vol. 10.

93 **"Any thoughtful student":** Sedgwick to CC, October 28, 1902.

93 **"This is as it should be":** G. Tucker, "The Number and Distribution of Micro-Organisms in the Air of the Boston City Hospital," *Public Health* 3 (1890): 78–85.

94 **"Grave danger":** "Oysters Under Fire from U.S. Attorney," *Washington Times*, October 8, 1910, 1.

94 **"He did excellent work":** Wade H. Frost to Barbara S. Quin, September 11, 1937. CF, Series 18, Box 269, Folder 2566.

94 **"paranoiac":** Lester G. Evans, September 17, 1937. Interview with Earle B. Phelps. CF, Series 18, Box 269, Folder 2566.

95 **"special skills in sanitation":** R. Ginn, *The History of the U.S. Army Medical Service Corps* (Washington, DC: US Office of the Surgeon General, 1997), 57.

96 **"massed singing":** US Department of War, *Annual Report of the Surgeon General, U.S. Army, to the Secretary of War*, Vol. 1, Part 2 (USGPO, 1920), 2152.

97 **"some generally-diffused poison":** T. B. Peacock, *On the Influenza, or Epidemic Catarrhal Fever of 1847–8* (London: John Churchill, 1848), 106.

97 **"I consider myself justified":** R. Pfeiffer, "The Influenza Baccilus. I.—Preliminary Communication on the Exciting Causes of Influenza," *Br Med J* 1 (1892): 128.
97 **"the bacillus influenza of Pfeiffer":** "Government Seeks to Check Influenza," *Providence Journal*, September 14, 1918.
97 **"The air has no part":** C. Chapin, "The Air as a Vehicle of Infection," *JAMA* 62 (1914): 423–30.
97 **"Influenza is chiefly sprayed":** "How to Keep from Getting Influenza," *Providence Sunday Journal*, October 6, 1918.
98 **"One may fool the people":** C. Chapin, "A Bad General," *Am J Public Health* 9 (1919): 784–85.
98 **"Otherwise, the germs present":** Wu Lien-Teh, *Plague Fighter* (Cambridge: W. Heffer & Sons, 1959), 21.
99 **"Their use during":** M. Barber and O. Teague, "Studies on Pneumonic Plague and Plague Immunization, XII," in *Some Experiments to Determine the Efficacy of Various Masks for Protection Against Pneumonic Plague* (Manila: Philippine Bureau of Printing, 1912), 268.
99 **"The mask not only protects":** G. Weaver, "The Value of the Face Mask and Other Measures: In Prevention of Diphtheria, Meningitis, Pneumonia, Etc.," *JAMA* 70 (1918): 76–78.
99 **"The evidence before the committee":** "A Working Program Against Influenza Prepared by an Editorial Committee of the American Public Health Association," *Am J Public Health* 9 (1919): 1–13.
99 **"During the compulsory":** W. Kellogg, *Influenza: A Study of Measures Adopted for the Control of the Epidemic* (Sacramento: California State Printing Office, 1919), 12.
100 **"the failure of public health":** WW, n.d. Author's questionnaire. CF, Series 13.9, Box 68, Folder 821.
101 **"Oyster culture, the most valuable fishery":** W. F. Wells, *Studies in Oyster Culture* (New York: J. B. Lyon Co., 1922), 42.
101 **"I was completely stumped":** H. G. Borland, "The Rescue of Oysters," *New York Tribune*, July 9, 1922.
101 **"I have had to be":** "How to Raise a Large Family," *Suffolk County News*, August 20, 1920.
102 **"eating with obvious gusto":** "The Sad Story of a Social Downfall," *St. Louis Post-Dispatch*, February 15, 1925.
102 **"Artificial oysters!":** "Artificial Oysters," *Pittsburgh Gazette Times*, December 25, 1924.
102 **"The super oyster":** "Super-Oyster Is on Its Way to Dinner Table: Bigger and Better Bivalve Sports Pedigree," *Hartford Courant*, March 27, 1927.
103 **"Mildred has a widespread reputation":** *Cactus Yearbook 1915* (Austin: University of Texas, 1915), 349.
104 **"that he believed the case":** "W. F. Weeks: Former Steward at the Insane Asylum, Arrested by Sheriff White," *Austin American-Statesman*, December 19, 1889.
104 **"After the business meeting":** "Town Talk," *Suffolk County News*, October 26, 1928.
105 **"confined at home":** Lester G. Evans, September 13, 1937. 3101 University of Pennsylvania—Study of the Prevention and Control of Air-Borne Infections. CF, Series 18, Box 269, Folder 2566.
105 **"The medical profession":** "Studying Child Paralysis," *New York Times*, August 5, 1899.
105 **"The lack of obvious connection":** J. Milbank, ed., *Poliomyelitis* (Baltimore, MD: Williams & Wilkins Co., 1932), 370.
105 **"does not behave":** Milbank, *Poliomyelitis*, 444.
106 **"Absence of infection":** Chapin, "The Air as a Vehicle," 428.
106 **"the various air-borne":** Milbank, *Poliomyelitis*, 447.
106 **"It is conveyed":** Milbank, *Poliomyelitis*, 442.
106 **"It is interesting to note":** "Review of *Poliomyelitis*," *N Engl J Med* 208 (1933): 616–17.

106 **"No new conclusions":** "Notices of Recent Publications," *Brain* 55 (1933): 604–8.
107 **"This slender, middle-aged man":** E. Riley, "A Tribute to William Firth Wells," *Ann NY Acad Sci* 353 (1980): 81–82.
107 **"He has given":** Wilson Smillie to John A. Ferrell, May 28, 1931. CF, Series 200, Box 20, Folder 233.
107 **peculiar:** Lester G. Evans, September 7, 1937. Memo to Miss Quin. CF, Series 18, Box 269, Folder 2566.
109 **"He won . . . not by accident":** EW to RH, May 6, 1947. CF, Series 18, Box 270, Folder 2577.
110 **"a wizard whom":** R. Riley, "Aerial Dissemination of Pulmonary Tuberculosis," *Am Rev Tuberc* 76 (1957): 931–41.
111 **"The contrast between":** W. Wells, *Airborne Contagion and Air Hygiene: An Ecological Study of Droplet Infections* (Cambridge, MA: Harvard University Press, 1955), vii.
111 **"It appears . . . that transmission of infection":** W. Wells, "On Air-Borne Infection: Study II. Droplets and Droplet Nuclei," *Am J Epidemiol* 20 (1934): 611–18.
112 **"This is the type of work":** E. R. Long, JS, and SM, August 4, 1937. A Project for Further Study of the Prevention and Control of Air-Borne Infections. CF, Series 200, Box 20, Folder 233.
112 **"A practical joke":** W. Wells and M. Wells, "Air-Borne Infection," *JAMA* 107 (1936): 1698–1703.
113 **"'Air-borne infection' might revive":** Wells and Wells, "Air-Borne Infection."
115 **"the most prevalent":** M. Rosenau, *Preventive Medicine and Hygiene* (New York: D. Appleton, 1935), 1.
115 **"The hazard is much less":** Rosenau, *Preventive Medicine*, 917.
115 **"If, on the contrary":** Wells and Wells, "Air-Borne Infection."
116 **"The complacency of those":** WW to RC, February 10, 1958. CF, Series 13.9, Box 68, Folder 818.
116 **"shatter this comforting doctrine":** W. Kaempffert, "The Week in Science: Star of Bethlehem a Nova?" *New York Times*, December 23, 1934.
116 **"Germs and Winds":** "Germs and Winds."
117 **"The indoor air":** WW to BC, July 20, 1954. CF, Series 13.9, Box 68, Folder 821.
118 **"Ultra Violet Air Dooms Germs":** *Brooklyn Times Union*, September 20, 1935.
118 **"Scientists Fight Flu Germs with Violet Ray":** *Decatur Daily Review*, July 30, 1936. See also "Light on Influenza," *New York Times*, July 23, 1936.
118 **"In time the lamps":** "Bacteriology: Researchers Measure Light's Antiseptic Power," *Newsweek*, September 28, 1935.
118 **"Chief among his aids":** "Scientists Turn to Schools in Fight on Infections in Air," *South Bend Tribune*, June 25, 1937.
118 **"In his fast scientific stride":** "Medicine: Light on Disease," *Time*, August 3, 1936.
118 **"the intellectual center":** "World Scholars to Honor Harvard," *New York Times*, August 30, 1936.
118 **"The theater of operation":** WW and MW, "Air-Borne Infection," *JAMA* 107 (1936): 1698–1703.
119 **"The growth of these ideas":** WW to CW, September 30, 1936. CWP.
120 **"has his feet on the ground":** Wade H. Frost to Barbara S. Quin, September 11, 1937.
120 **"I had an extraordinary amount":** JS to Cretyl Mills, December 14, 1964. APS/JS.
121 **"have all had a good deal":** SM to Barry C. Smith, August 4, 1937. CF, Series 200, Box 20, Folder 233.
121 **"there is a brilliant":** RH, August 16, 1941. Visit to project. CF, Series 18, Box 269, Folder 2573.
121 **"So far as":** MW to SM, September 10, 1937. CF, Series 18, Box 269, Folder 2566.
122 **"I am afraid":** MW to CW, May 19, 1938. CWP.
122 **the Infection Machine:** "A Robot That Sneezes Microbes," *Tennessean*, March 10,

1940; J. Shaltz, "'Sneezer' Unmasks Disease," *Philadelphia Record*, (date unknown) 1941.

123 **"Pupils come to this school":** T. Wilder, "A School Physician Studies Contagious Diseases," *Child Educ* 11 (1935): 199–204.

123 **"Black Light Kills":** S. Spencer, "Black Light Kills 'Flying' Germs," *Philadelphia Bulletin*, March 7, 1938.

CHAPTER 6: THE SCATTERED WORKERS

125 **"The *Hawaii Clipper*":** "'Mercy' Flight for PAA Plane," *Honolulu Star-Bulletin*, July 23, 1938.

125 **"Fred had left Washington":** R. Haskell and H. Barss, "Fred Campbell Meier, 1893–1938," *Phytopathology* 29 (1939): 293–302.

126 **"ARRIVED HONOLULU 7MORNING":** US Department of Agriculture, "Memorandum Regarding Mr. Fred C. Meier, Principal Agriculturalist, Bureau of Plant Industry, U.S. Department of Agriculture, Who While on Official Duty, Was a Passenger on the Hawaii Clipper When It Was Lost at Sea July 28, 1938," September 10, 1938, NAL Special Collections, USDA Historical Collection.

126 **"Stand by for one minute":** Civil Aeronautics Authority, "Preliminary Report of Investigation of the Disappearance of an Aircraft of Pan American Airways, Incorporated, in the Vicinity of Latitude 12 27, North, Longitude 130 40, East, on July 29, 1938."

127 **"The present problem":** "Distinguished Men on Board Clipper," *New York Times*, July 30, 1938.

127 **"Strangely enough, Dr. Meier":** "Cling to Hope: D.C. Men Alive," *Washington Evening Star*, July 30, 1938.

127 **"The loss to our Committee":** R. Coker to Committee on Aerobiology, August 8, 1938. NAS, Death of chairman, 1938.

128 **"Fantastic tales are being passed":** P. Mallon, "Experts Connect Crash of Clipper to Scientists' Study of Microbes," *Philadelphia Inquirer*, September 7, 1938.

128 **"He has of course":** Albert L. Barrows to Esmond R. Long, September 13, 1938. NAS, Death of chairman, 1938.

128 **"I think probably":** National Research Council, Division of Biology and Agriculture, Committee on Aerobiology, minutes of meeting, October 1, 1938. NAS, Meetings: 1937–1938.

129 **"Pending the discovery":** Civil Aeronautics Authority, "Preliminary Report."

129 **"If it had to be":** N. Smith, "Fred Campbell Meier," *Science* 88 (1938): 233.

130 **"if anything of value":** E. C. Auchter to R. E. Coker, October 5, 1938. USDA/BPI, Record Group 54, Entry PI-66 167, Gen. Corres. of the Chief, 1908–39, Box 822, File 31754.

130 **"We are existing":** National Research Council, Division of Biology, minutes of meeting, October 1, 1938.

131 **"not quite the book":** R. McLean, "Microbiology of the Air," *Nature* 152 (1943): 258–59.

CHAPTER 7: WAR AT HOME

135 **"Misunderstanding, or even misrepresentation":** WW and MW, August 1940, Annual Report, August 1, 1939—July 31, 1940. CF, Series 18, Box 271, Folder 2584.

137 **"It is often forgotten":** F. Weyrauch and J. Rzymkowski, "*Photographien zur Tröpfcheninfektion*," *Z Hyg* 120 (1938): 444–49.

137 **"A large part of my success":** "Nobel Prize Laureates," Zeiss, https://www.zeiss.com/microscopy/en/about-us/nobel-prizes.html.

137 **"In general, the droplets sink":** Weyrauch and Rzymkowski, *"Photographien."*
138 **"We immediately recognized":** W. Wells, *Airborne Contagion*, 4.
138 **"Infection of Air":** W. Wells, M. Wells, and S. Mudd, "Infection of Air: Bacteriologic and Epidemiologic Factors," *Am J Public Health Nation's Health* 29 (1939): 863–80.
138 **"With our disclosure":** WW to BC, June 24, 1953. CF, Series 13.9, Box 68, Folder 816.
139 **"of a fly rod":** M. W. Jennison and J. W. M. Bunker, "Analysis of the Movement of Cilia from the Clam (Mya) by High-Speed Photography with Stroboscopic Light," *J Cell Comp Physiol* 5 (1934): 189–97.
139 **"We have been able":** M. W. Jennison and H. E. Edgerton, "Droplet Infection of Air: High-Speed Photography of Droplet Production by Sneezing," *Proc Soc Exp Biol Med* 43 (1940), 455–58.
140 **"The muzzle velocity":** Marshall Jennison, "Infection of Air by Sneezing," June 14, 1940. CF, Series 13.9, Box 68, Folder 819.
140 **"proof of air-borne":** F. Moulton, ed., *Aerobiology* (Washington, DC: American Association for the Advancement of Science, 1942), 125.
140 **"This Is Nothing":** "This Is Nothing to Sneeze At," *Plain Speaker*, April 19, 1940.
141 **"Here for the first time":** RH, May 4–5, 1939. Visit of Mr. Geddes Smith and Dr. Heffron to Dr. Mudd and the Wellses. CF, Series 18, Box 269, Folder 2569.
141 **"Now we have a natural way":** Shaltz, "'Sneezer' Unmasks Disease."
142 **"The Wells themselves":** RH, August 14–16, 1941. Visit to project. CF, Series 18, Box 269, Folder 2571.
142 **"Among Mr. Wells' gifts":** SM to RH, August 23, 1939. CF, Series 18, Box 269, Folder 2569.
142 **"Wells is a queer dick":** Alfred N. Richards to Barbara S. Quin, May 25, 1940. CF, Series 18, Box 269, Folder 2571.
142 **"I cannot overlook this":** WW to SM, January 17, 1940. CF, Series 18, Box 270, Folder 2580.
142 **"Creative workers in any field":** SM, August 3, 1938. Report of the Laboratories for the Study of Air-borne Infection, 1937–1938. CF, Series 18, Box 269, Folder 2567.
143 **"Probably no contagious disease":** W. Wells, M. Wells, and T. Wilder, "The Environmental Control of Epidemic Contagion: I. An Epidemiologic Study of Radiant Disinfection of Air in Day Schools," *Am J Epidemiol* 35 (1942): 97–121.
143 **"No more severe test":** WW to RH, April 12, 1940. CF, Series 18, Box 269, Folder 2571.
144 **"Many parents have gotten":** A. Dafoe, "Lax Measles Quarantines Imperil Many Children," *Philadelphia Inquirer*, April 7, 1941.
145 **"perfectly extraordinary":** RH, August 16, 1941. Visit to project.
145 **"proved as conclusively":** EW to Alfred N. Richards, August 19, 1941. CF, Series 18, Box 269, Folder 2573.
145 **"I have been under":** WW to CW, October 28, 1941. CWP.
145 **"This was more":** R. Riley, "Airborne Contagion: Historical Background," *Ann NY Acad Sci* 353 (1980): 3–9.
146 **"There is reason to believe":** R. Hare, "The Expulsion of Haemolytic Streptococci by Nasopharyngeal Carriers," *Can J Public Health* 31 (1940): 539–55.
146 **"reign of terror":** A. Glass, "FDR Calls for 'Quarantine' of Aggressor Nations, Oct. 5, 1937," *Politico*, October 5, 2018.
146 **"an honor which I shall":** CL to Hermann Goering, October 25, 1938. CAL, Box 37, Folder 1938 Oct. 8–31.
146 **"I feel that I understand":** CL to General Erhard Milch, February 14, 1939. CAL, Box 37, Folder 1939 Feb. 11–28.
147 **"A general European war":** CL to Joseph P. Kennedy, September 22, 1938. CAL, Box 37, Folder 1938 Sep.
147 **"It is time to turn":** C. Lindbergh, "Aviation, Geography, and Race," *Reader's Digest* 35 (1939): 64–67.

147 **"Casualties from respiratory disease":** WW and MW, Annual Report, 1940.
147 **"This, of course":** SM to Barbara S. Quin, July 30, 1940. CF, Series 18, Box 269, Folder 2571.
148 **"Dr. Wells feels":** Science Service, "Germ Killing Rays Urged for Barracks Rooms," *Cincinnati Post*, November 2, 1940.
148 **"We may expect":** "Present War May Bring Conquest of Influenza," *York Daily Record*, October 27, 1939.
148 **"The larger the experiment":** WW, August 14, 1941. Memorandum on Environmental Control of Epidemic Disease in the Army. CF, Series 18, Box 269, Folder 2573.
148 **"It might be that":** WW and MW, August 5, 1941. Fourth Annual Report, October 1, 1940–July 31, 1941. CF, Series 18, Box 269, Folder 2573.
149 **"We wasted much time":** WW and MW, April 22, 1943. Progress Report, October 1, 1940–April 15, 1943. CF, Series 18, Box 270, Folder 2575.
149 **"An epidemic of respiratory contagion":** WW to RH, June 10, 1942. CF, Series 18, Box 270, Folder 2574.
149 **"Frankly I'm afraid":** Geddes Smith to Barry C. Smith, October 1, 1942. CF, Series 13.9, Box 68, Folder 813.
150 **"My own feeling":** MW to EW, September 2, 1943. EWP.
150 **"I am, of course, fearful":** MW to EW, August 16, 1943. EWP.
151 **"It is a rather sad commentary":** RH, April 14, 1942. Telephone conversation with Dr. Long. CF, Series 18, Box 270, Folder 2574.
151 **"It may take quite a while":** EW to RH, May 13, 1943. CF, Series 18, Box 270, Folder 2575.
152 **"who enjoys the capable assistance":** S. M. Spencer, "War on the Flying Microbes," *Saturday Evening Post*, February 16, 1946.
152 **"domestic crisis":** RH, May 18, 1945. Summary memorandum. CF, Series 18, Box 270, Folder 2581.
152 **"Rightly or wrongly":** MW to EW, October 6, 1944. EWP.
152 **"The studies have":** MW to EW, October 6, 1944.
153 **"in order to take care":** RH, July 12, 1944. Memorandum of visit. CF, Series 18, Box 270, Folder 2576.
153 **"We have a better set":** EW to RH, May 12, 1944. CF, Series 18, Box 270, Folder 2576.
153 **"I think it is one":** CW to MW, January 26, 1945. CWP.
153 **"Will and I are working":** MW to EW, October 6, 1944.
154 **"He found that ultra-violet lamps":** W. Kaempffert, "Tests of Ultra-Violet Lamps in Checking the Spread of Transmissible Diseases," *New York Times*, October 8, 1944.
154 **"The question is more than":** MW to Theodore Wilder, October 11, 1944. 3558 University of Pennsylvania. CF, Series 18, Box 270, Folder 2576.
155 **"Many of the points":** WW to Geddes Smith, October 9, 1944. 3558 University of Pennsylvania. CF, Series 18, Box 270, Folder 2576.
155 **"The mingling of sailors":** WW, August 18, 1944. Report of the Laboratories for the Study of Air-borne Infection, 1943–1944. CF, Series 18, Box 271, Folder 2587.
156 **"Germ-killing ultraviolet light":** Science Service, "Ultraviolet Irradiation of Barracks Lessens Sickness," *Gazette and Daily* (York, PA), October 11, 1944.
156 **"In view of the fact":** S. Wheeler, H. S. Ingraham, A. Hollaender, et al., "Ultra-Violet Light Control of Air-Borne Infections in a Naval Training Center: Preliminary Report," *Am J Public Health* 35 (1945): 457–68.
156 **"Such a cloud":** WW, April 2, 1947. A Report of the Laboratories for the Study of Air-borne Infection, April 15, 1945–April 7, 1947. CF, Series 18, Box 270, Folder 2577.
156 **"Oiling of floors":** US Army Medical Department, *Preventive Medicine in World War II*, Vol. 2: *Environmental Hygiene* (Washington, DC: Office of the Surgeon General, 1955), 72.

157 **"I envy your great opportunity":** WW to OR, December 19, 1944. APS/OR.
157 **"Only the Wells":** James Perkins to William Holla, March 12, 1946. CF, Series 18, Box 270, Folder 2576.
158 **"I think it would go":** WW to CW, April 10, 1945. CWP.
158 **"It is not to be construed":** J. Perkins, A. Bahlke, and H. Silverman, "Effect of Ultra-Violet Irradiation of Classrooms on Spread of Measles in Large Rural Central Schools: Preliminary Report," *Am J Public Health Nation's Health* 37 (1947): 529–37.
158 **"The striking success":** "The Control of Air-Borne Infection," *Br Med J* 2 (1946): 820–21.
159 **"Unless extremely violent":** R. Hare and D. Mackenzie, "The Source and Transmission of Nasopharyngeal Infections Due to Certain Bacteria and Viruses," *Br Med J* 1 (1946): 865–70.
159 **"ravages of misinterpretation":** WW to RH, October 25, 1945. CF, Series 18, Box 270, Folder 2576.
159 **"He has now fallen out":** RH, June 12, 1947. 3765 University of Pennsylvania—Study of the Prevention and Control of Air-borne Infections. CF, Series 18, Box 270, Folder 2577.
159 **"I don't understand":** EW to RH, May 11, 1945. CF, Series 18, Box 270, Folder 2581.
159 **"The split-up":** RH, May 9, 1947. Interview with Doctors Kenneth Maxie and O. H. Robertson. CF, Series 18, Box 270, Folder 2577.
160 **"Most dust-born bacteria":** WW et al., August 1, 1946. Progress Report, April 1, 1945—August 1, 1946. CF, Series 18, Box 271, Folder 2588.
161 **"The relative importance":** Alexander Langmuir to WW, May 28, 1949. CWP.
161 **"Thus, the challenge":** A. Langmuir, "Epidemiology of Airborne Infection," *Bacteriol Rev* 25 (1961): 173–81.
162 **"I cannot say":** WW to RH, June 17, 1947. CF, Series 18, Box 270, Folder 2577.

CHAPTER 8: WINGS FOR DEATH

163 **"is almost literally":** E. Stakman, "The Field of Extramural Aerobiology," in *Aerobiology*, ed. F. R. Moulton (Washington, DC: American Association for the Advancement of Science, 1942), 1.
163 **"They ordered corpses":** M. Wheelis, "Biological Warfare at the 1346 Siege of Caffa," *Emerg Infect Dis* 8 (2002): 971–75.
164 **"All those people":** H. G. Wells, *The Stolen Bacillus and Other Incidents* (London: Methuen, 1895), 11.
164 **"Distribution from airships":** C. Sims-Dawson, "How Deadly New Bacilli May Be Bred to Take the Place of Bullets," *San Francisco Examiner*, September 5, 1915.
165 **"The extraordinarily high proportion":** "Germ-Laden Air-Bombs," *Aeronautics* 9 (1915): 415.
165 **"Great masses of men":** League of Nations, *Proceedings of the Conference for the Supervision of the International Trade in Arms and Ammunition and in Implements of War. Held in Geneva, May 4th to June 17th, 1925* (Lausanne: Imp. Réunies, 1925), 340.
165 **"bacteriological methods of warfare":** "1925 Geneva Protocol," United Nations Office for Disarmament Affairs, https://www.un.org/disarmament/wmd/bio/1925-geneva-protocol/.
166 **"Indeed, to fail":** Elvin Stakman, February 17, 1942. "Plant Pathology in War." Exhibit F, Report of the W.B.C. Committee. National Academy of Sciences Archives, Committees on Biological Warfare, Series 1, Box 2.
167 **"A SHOT/GUN ATTACK":** Stakman, "Plant Pathology."
167 **"distinctly feasible":** E. B. Fred. Report of the W.B.C. Committee, July 18, 1942. National Academy of Sciences Archives, Committees on Biological Warfare, Series 1, Box 2.

168 **"Biological warfare is":** S. Whitby, *Biological Warfare Against Crops* (Basingstoke, UK: Palgrave Macmillan, 2002), 72.
168 **"The world was like":** S. Harrison, "C&B Tactics Denounced," *Morning Herald* (Hagerstown, MD), November 20, 1968.
168 **"Mr. Biological Warfare":** M. Shepherd, "Germ Warfare Has Terrible Potential," *St. Louis Globe-Democrat*, December 13, 1959.
169 **"looked to my dental classmates":** TR, n.d. "Autobiographical Sketch of Theodor Rosebury, Author of 'Peace or Pestilence.'" TRP, Box 1, Folder 1.
170 **"with a fallen log":** TR, 1972. "Experiences at the Columbia Medical Center in Its Early Years," SDOS lecture. TRP, Box 7, Folder 6.
170 **"White man's food":** "Footnotes," *Lincoln Nebraska State Journal*, December 12, 1938.
170 **"I began to read":** TR, "Experiences at the Columbia Medical Center."
171 **"neither recognize nor stop":** "Influenza Sure to Follow War," *Union County Journal* (Marysville, OH), January 8, 1940.
171 **"At most, this was a subject":** T. Rosebury, *Peace or Pestilence: Biological Warfare and How to Avoid It* (New York: Whittlesey House / McGraw-Hill, 1949), 5.
171 **"I shall never forget":** TR, 1951. "Fear, War, and Science." TRP, Box 9, Folder 2.
172 **"The air-borne route":** T. Rosebury and E. Kabat, "Bacterial Warfare: A Critical Analysis of the Available Agents, Their Possible Military Applications, and the Means for Protection Against Them," *J Immunol* 56 (1947): 7–96.
172 **"By the air-borne route":** TR, *Peace or Pestilence*, 121.
174 **"Scarcely a month passes":** J. Bell, "On 'Woolsorter's Disease,'" *Lancet* 1 (1880): 871–73.
174 **"The fluids from this animal":** J. Bell, "Anthrax: Its Relation to the Wool Industry," in *Dangerous Trades: The Historical, Social, and Legal Aspects of Industrial Occupations as Affecting Health, by a Number of Experts*, ed. T. Oliver (New York: E. P. Dutton, 1902), 635.
174 **"Its properties make it":** Rosebury and Kabat, "Bacterial Warfare."
175 **"The immense amount":** E. H. Cushing to TR, June 11, 1942. TRP, Box 1, Folder 20.
175 **"I have no doubt":** TR, May 12, 1942. Note to be attached to Form 38746 (C.S.C.). TRP, Box 1, Folder 21.
175 **"This is so ridiculous":** TR to "George," January 29, 1943. TRP, Box 1, Folder 21.
175 **"It was a cause":** TR, "Five Morbid Pieces."
176 **"These germs may have constituted":** Rosebury, *Peace or Pestilence*, 115.
177 **Camp Detrick's officers' club:** T. Rosebury, H. Ellingson, and G. Meiklejohn, "A Laboratory Infection with Psittacosis Virus Treated with Penicillin and Sulfadiazine, and Experimental Data Bearing on the Mode of Infection," *J Infect Dis* 80 (1947): 64–77. Rosebury also recounts his illness in "Five Morbid Pieces."

CHAPTER 9: A FINE FROZEN DAIQUIRI

181 "for use if this mode": E. Willis, "Landscape with Dead Sheep: What They Did to Gruinard Island," *Med Confl Surviv* 18 (2002): 199–210.
183 **"significant contributions":** G. Merck, "Report to the Secretary of War: Biological Warfare," *Mil Surg* 98 (1946): 237–42.
184 **"I hope that I have an opportunity":** TR to Willard C. Rappleye, October 3, 1945. TRP, Box 1, Folder 21.
184 **"throw light":** Rosebury, Ellingson, and Meiklejohn, "A Laboratory Infection with Psittacosis Virus."
184 **"It still smacks":** Oram C. Woolpert to TR, June 11, 1946. TRP, Box 1, Folder 21.
184 **"I think they have":** TR to Oram C. Woolpert, June 15, 1946. TRP, Box 1, Folder 21.
185 **"We were in a crisis":** T. Rosebury, "Medical Ethics and Biological Warfare," *Perspect Biol Med* 6 (1963): 512–23.

185 **"to bring the subject":** TR to Waldemar Kaempffert, May 19, 1947. TRP, Box 3, Folder 14.
185 **"Frankly, I believe":** Oram C. Woolpert to TR, March 30, 1946. TRP, Box 1, Folder 20.
186 **"Publication of this paper":** Willard Rappleye to TR, April 12, 1946. TRP, Box 1, Folder 20.
186 **"It is now presented":** W. Kaempffert, "Germ Warfare Brought to the Fore Again by Publication of Rosebury-Kabat Paper," *New York Times*, May 25, 1947.
186 **"Set down with":** "Medicine: Death in Convenient Bottles," *Time*, May 26, 1947.
187 **"It was clear":** TR, "Five Morbid Pieces."
187 **"I had returned":** TR, "Five Morbid Pieces."
187 **"Leaving my lab":** TR to Elvin Kabat, July 19, 1949. TRP, Box 3, Folder 14.
187 **"The integrity of science":** T. Rosebury and M. Phillips, "Two Aspects of the Loyalty Problem," *Science* 110 (1949): 123–24.
187 **"We need not doubt":** Rosebury, *Peace or Pestilence*, 116.
188 **"It seems to me":** Rosebury, *Peace or Pestilence*, 167.
188 **"atomic Pearl Harbor":** L. McEnaney, *Civil Defense Begins at Home* (Princeton, NJ: Princeton University Press, 2020), 41.
188 **"There are indications":** US Department of Defense, "Report of the Secretary of Defense's Ad Hoc Committee on Chemical, Biological and Radiological Warfare" (USGPO, 1950), 15.
188 **"The big danger":** US Federal Civil Defense Administration, *What You Should Know About Biological Warfare: The Official U.S. Government Booklet* (USGPO, 1951), 8.
189 **"Specially designed bombs":** A. Langmuir, "The Potentialities of Biological Warfare Against Man—An Epidemiological Appraisal," *Public Health Rep* 66 (1951): 387–99.
189 **"Epidemiology means":** A. Fairchild, R. Bayer, and J. Colgrove, *Searching Eyes: Privacy, the State, and Disease Surveillance in America* (Berkeley: University of California Press, 2007), 17.
189 **"When we see that smoking":** E. Etheridge, *Sentinel for Health: A History of the Centers for Disease Control* (Berkeley: University of California Press, 1992), 127.
190 **"What You Should Know":** *What You Should Know About Biological Warfare* (Ray Film Industries, 1952), https://digital.library.jhu.edu/node/11773.
192 **"The ranks of American scientists":** J. McCarthy, "Communist Infiltration of the Scientific Profession," in *Congressional Record: Proceedings and Debates of the 81st Congress, Second Session*, Vol. 96, Part 18 (USGPO, 1950), A7255.
192 **"It may be well":** TR, "Fear, War, and Science."
193 **"I came to feel":** TR, "Five Morbid Pieces."
193 **"If we could tolerate":** TR, "Five Morbid Pieces."
193 **"believed to be Communists":** "Red Quiz Names 38 U.S. Employees in U.N.," *Los Angeles Times*, January 2, 1953.
193 **"Some of my colleagues":** TR, "Five Morbid Pieces."
194 **"It is ironic":** TR, "Five Morbid Pieces."
194 **"looks like a spaceship":** M. Kortepeter, *Inside the Hot Zone: A Soldier on the Front Lines of Biological Warfare* (Lincoln, NE: Potomac Books, 2020), 144.
194 **"We were pioneers":** C. Mabeus, "Fort Detrick's Eight-Ball—A Relic of Cold War Bio-Warfare," *Frederick (MD) News-Post*, October 11, 2013.
195 **"a curious clinical observation":** R. Wheat et al., "Infection Due to Chromobacteria," *AMA Arch Intern Med* 88 (1951): 4610466.
197 **"Spare me your sermon":** T. Rosebury, "Five Morbid Pieces."
198 **"I was back":** TR, "Five Morbid Pieces."
198 **"Science has brought forth":** T. Rosebury, "Some Historical Considerations," *Bull At Sci* 16 (1960): 227–36.
198 **"to dispel the miasma":** B. Chisholm et al., "Pugwash International Conference of Scientists: Statement on Biological and Chemical Warfare," *Nature* 184 (1959): 1018–20.

199 **"Is it true":** J. Ostergren, "Man Mobbed by Multitudes of Microbes," *The Oregonian*, April 3, 1964.
200 **"Although I had never":** TR, "Five Morbid Pieces."
200 **"The control":** US Senate Subcommittee to Investigate the Administration of the Internal Security Act and Other Internal Security Laws, *The Anti-Vietnam Agitation and the Teach-in-Movement: The Problem of Communist Infiltration and Exploitation; A Staff Study* (USGPO, 1965).
201 **"If Rosebury is not a Communist":** "Disgrace to St. Louis," *St. Louis Globe-Democrat*, October 17, 1965.
201 **"My position at the school":** TR, "Five Easy Pieces," 204.
201 **"the intermittent attack":** "A Distinguished Scientist," *St. Louis Post-Dispatch*, April 17, 1966.
202 **"sane, elegant, and informative":** J. Leonard, "Somebody in There Likes Us," *New York Times*, June 26, 1969.
202 **"Since the United States":** T. Rosebury, "U.S. May Use Stronger Chemical-Biological Weapons in Vietnam," *Gazette and Daily* (York, PA), July 13, 1967.
203 **"Biological weapons have massive":** R. Nixon, "Statement on Chemical and Biological Defense Policies and Programs," Public Papers of the Presidents, 461 (1969), 968–69.
203 **"Our entire stockpile":** G. Ford, "Statement on the Geneva Protocol of 1925 and the Biological Weapons Convention," American Presidency Project, January 22, 1975.
204 **"He knew the stuff":** TR, "Five Morbid Pieces."
204 **"But if our enemies":** M. Leitenberg and R. Zilinskas, *The Soviet Biological Weapons Program: A History* (Cambridge, MA: Harvard University Press, 2012), 25.
205 *Biohazard*: All quotations from Alibek in this section are from K. Alibek with S. Handelman, *Biohazard* (New York: Random House, 1999).
206 **"inadvertent exposure":** B. Gwertzman, "Soviet Mishap Tied to Germ-War Plant," *New York Times*, March 19, 1980.
210 **"Luckily, I was not drafted":** WW to RC, October 25, 1951. CF, Series 13.9, Box 68, Folder 814.
210 **"the suicide of bacteriology":** WW to R. M. Ferry, January 18, 1949. CF, Series 13.9, Box 68, Folder 813.
210 **"I only wish":** WW to RC, October 25, 1951.

CHAPTER 10: LOCH RAVEN

213 **"This is a rather quiet corner":** WW to Cyril Tasker, March 23, 1948. CWP.
214 **"I submit":** WW to RH, September 17, 1948. CF, Series 13.9, Box 68, Folder 813.
214 **"The sheer multitude":** RH to RC, September 21, 1948. CF, Series 13.9, Box 68, Folder 813.
214 **"I am not too sure":** RH, July, 29, 1949. William F. Wells's Manuscript on Dynamics of Air-borne Contagion. CF, Series 13.9, Box 68, Folder 813.
215 **"We must feel our way":** WW to JS, August 19, 1948. APS/JS.
215 **"He is a true pioneer":** JS to John Barnwell, July 6, 1948. APS/JS.
216 **"When I get the outline":** MW to EW, October 6, 1944. EWP.
216 **"There is one opportunity":** EW to MW, October 9, 1944. EWP.
216 **"One can well cite":** W. Holla, *Annual Report* (White Plains, NY: Westchester County Department of Health, 1949), 3. WCA.
217 **"I feel wholly inadequate":** MW to EW, February 14, 1945. EWP.
217 **"in the control":** "Experiment with Health Rays Pleases County Commissioner," *Daily Times* (Mamaroneck, NY), June 27, 1946.
217 **"difficulties encountered":** Informal Discussion of Westchester Board of Health Commissioner's Office, May 23, 1946. Board of Health Minutes, 1948. Series 390 A-0540 (2) F. WCA.

218 **"the most patronized"**: M. Wells and W. A. Holla, "Ventilation in the Flow of Measles and Chickenpox Through a Community," *JAMA* 142 (1950): 1337–44.
218 **"In the next three years"**: W. A. Holla, "Invisible Lifeguard," *American Weekly*, March 10, 1946.
218 **"These germs"**: "Special Lights Cut School Illnesses," *New York Times*, June 25, 1946.
219 **"We are dubious"**: MW to Peter Huntington, February 10, 1947. APS/JS.
219 **"I think you have a big story"**: Geddes Smith to WW, September 21, 1948. CF, Series 13.9, Box 68, Folder 813.
220 **"Until now she has been"**: WW to JS, November 10, 1949. APS/JS.
220 **"He was an aristocrat"**: S. Permutt, "Richard Lord Riley, 1911–2001: An Appreciation," *Am J Respir Crit Care Med* 166 (2002): 257.
221 **"Do you suppose the Fund might"**: WW to RH, January 3, 1950. CF, Series 13.9, Box 68, Folder 813.
221 **"Ultra-violet radiation disinfects"**: "Violet Rays Head Off Measles and Chicken Pox in Trial Run," *New York Herald Tribune*, May 24, 1950, 23.
222 **"It is apparent"**: J. Downes, "Control of Acute Respiratory Illness by Ultra-Violet Lights," *Am J Public Health Nation's Health* 40 (1950): 1512–20.
222 **"It proved inconclusive"**: Westchester County Health Department, 1955. *To Westchester's Good Health: A Silver Anniversary Report on One County Department's War Against Disease and Disability*. WCA.
222 **"it contributed to the blackballing"**: RR to Don Smith, July 9, 1985. RRP, Box 6.
223 **"We can't make a really good book"**: RC, April 25, 1950. Memorandum to Geddes Smith and Doctor Heffron. CF, Series 13.9, Box 68, Folder 813.
223 **"improve the form"**: WW to RC, November 27, 1950. CF, Series 13.9, Box 68, Folder 814.
223 **"I hope that I can find"**: WW to RC, January 8, 1951. CF, Series 13.9, Box 68, Folder 814.
223 **"Will seems to have"**: MW to RH, March 30, 1951. CF, Series 13.9, Box 68, Folder 814.
224 **"as I had to get in a crop"**: WW to RH, May 17, 1951. CF, Series, 13.9, Box 68, Folder 814.
224 **"The sad truth"**: WW to RC, February 27, 1952. CF, Series 13.9, Box 68, Folder 814.
224 **"a goal beyond anything"**: WW to RC, June 29, 1951. CF, Series 13.9, Box 68, Folder 814.
224 **"I hope"**: WW to RC, July 9, 1951. CF, Series 13.9, Box 68, Folder 814.
224 **"at last it has really"**: WW to RC, December 7, 1951. CF, Series 13.9, Box 68, Folder 814.
224 **"I have read it"**: RC to RH, December 7, 1951. CF, Series 13.9, Box 68, Folder 814.
225 **"a bilious attack"**: WW to RC, March 24, 1952. CF, Series 13.9, Box 68, Folder 814.
226 **"I am sorry"**: RH to JS, May 29, 1952. ASP/JS.
226 **"that a man"**: JS to RH, June 16, 1952. ASP/JS.
226 **"Let's hope"**: RC. Memo, June 18, 1952. CF, Series 13.9, Box 68, Folder 815.
226 **"Your patience"**: WW to BC, January 28, 1953. CF, Series 13.9, Box 68, Folder 815.
226 **"As I recede"**: WW to RC, May 7, 1952. CF, Series 13.9, Box 68, Folder 814.
226 **"The whole of 'Book One'"**: WW to RC, October 6, 1952. CF, Series 13.9, Box 68, Folder 815.
226 **"This about exhausts"**: WW to RC, December 10, 1952. CF, Series 13.9, Box 68, Folder 815. (Crane added his note to this letter.)
227 **"Irradiation of a village"**: MW, "969. Evaluation of Air Disinfection," VI Congresso Internazionale di Microbiologia, Roma, 6–12 Septembre 1953, *Riassunti delle Comunicazioni* 3 (1953): 265–66.
227 **"We added one more"**: WW to BC, August 25, 1953. CF, Series 13.9, Box 68, Folder 819.
227 **"it would be nice"**: WW to RC, September 11, 1950. CF, Series 13.9, Box 68, Folder 814.
227 **"I cannot believe"**: WW to BC, March 14, 1954. CF, Series 13.9, Box 68, Folder 816.

228 **"Sanitary principles":** WW to BC, July 20, 1954.
228 **"Prejudice dies":** WW to BC, March 14, 1954.
228 **"I was shocked":** MW to RC, April 17, 1954. CF, Series 13.9, Box 68, Folder 816.
228 **"It's hard":** BC to RC, April 19, 1954. CF, Series 13.9, Box 68, Folder 816.
228 **"As you know":** BC to MW, April 19, 1954. CF, Series 13.9, Box 68, Folder 816.
228 **"Thank you for your nice letter":** MW to BC, April 1954. CF, Series 13.9, Box 68, Folder 816.
229 **"Mr. Wells told me":** RC to BC, April 26, 1954. CF, Series 13.9, Box 68, Folder 816.
229 **"I think we broke through":** WW to BC, August 17, 1954. CF, Series 13.9, Box 68, Folder 816.
229 **"has been somewhat":** JS to RR, February 12, 1954. APS/JS.
230 **"This arrangement":** R. Riley, "Air-Borne Infections," *Am J Nurs* 60 (1960): 1246–68.
230 **"In a word":** R. Riley, "What Nobody Needs to Know About Airborne Infection," *Am J Respir Crit Care Med* 163 (2001): 7–8.
231 **"Even though page proofs":** WW to BC, January 23, 1955. CF, Series 13.9, Box 68, Folder 817.
231 **"After I looked into it":** WW to RC, April 19, 1955. CF, Series 13.9, Box 68, Folder 817.
231 **"The book is not for here":** WW to BC, November 9, 1954. CF, Series 13.9, Box 68, Folder 817.
231 **John J. Phair:** J. Phair, "Airborne Contagion and Air Hygiene," *Am J Public Health Nation's Health* 45 (1955): 1495–96.
231 **Theodor Rosebury:** T. Rosebury, "Review of 'Airborne Contagion and Air Hygiene: An Ecological Study of Droplet Infections.' By William Firth Wells," *Q Rev Biol* 31 (1956): 161–62.
232 **"The only fly":** WW to BC, June 14, 1955. CF, Series 13.9, Box 68, Folder 817.
232 **"the deepest satisfaction":** WW to RC, October 9, 1956. CF, Series 13.9, Box 68, Folder 817.
232 **"How long, oh Lord":** WW to JS, December 27, 1955. ASP/JS.
232 **"You will be pleased":** RR to JS, December 11, 1956. ASP/JS.
233 **"These ideas":** R. Riley, "Aerial Dissemination."
233 **"It always does me good":** JS to RR, January 14, 1958. ASP/JS.
233 **"If and when":** WW to RC, October 9, 1956.
233 **"I have come back":** WW to RC, July 1, 1957. CF, Series 13.9, Box 68, Folder 817.
234 **"Wells kept in touch":** R. Riley, "What Nobody Needs to Know."
235 **"Maybe the guinea pigs":** R. Riley, "What Nobody Needs to Know."
235 **"This may profoundly modify":** WW to JS, May 6, 1957. ASP/JS.
235 **"Throughout each day":** "Hong Kong Battling Influenza Epidemic," *New York Times*, April 17, 1957.
236 **"The Livermore study":** RR to JS, September 25, 1957. ASP/JS.
236 **"will not be ended":** WW to RC, July 19, 1957. CF, Series 13.9, Box 68, Folder 817.
236 **"The mission of the sanitarian":** WW to RC, February 10, 1958.
236 **"I hope these running notes":** WW to RC, December 7, 1957. CF, Series 13.9, Box 68, Folder 817.
237 **"I have been wanting":** JS to RR, January 14, 1958.
237 **"having weathered":** RR to JS, March 19, 1958. ASP/JS.
237 **"We continue to discuss":** RR to JS, September 22, 1958. ASP/JS.
238 **"It was impossible":** M. A. Tarumianz to N. Maxon Terry, February 23, 1959. ASP/JS.
238 **"This came as no surprise":** RR to JS, March 2, 1959. ASP/JS.
239 **"To achieve the prerequisites":** R. Riley, "Protective Measures; Reasonable or Ritualistic," *Nurs Outlook* 7 (1959): 38–39.
239 **"The individual":** R. Riley and F. O'Grady, *Airborne Infection: Transmission and Control* (New York: Macmillan, 1961), 168.
240 **"I make note of these changes":** WW to RC, November 19, 1959. CF, Series 13.9, Box 68, Folder 818.

240 **"It seems therefore":** RC to WW, November 20, 1959. CF, Series 13.9, Box 68, Folder 818.

240 **"This study was designed":** W. Jordan, "The Mechanism of Spread of Asian Influenza," *Am Rev Respir Dis* 83 Pt. 2 (1961): 29–40. See also A. Wehrwein, "Ultraviolet Irradiation of Air Is Said to Curb the Spread of Flu," *New York Times*, May 27, 1959.

241 **"Their epidemic control":** Ross McLean to William Tucker, February 23, 1960. APS/JS.

241 **"The question is":** Riley and O'Grady, *Airborne Infection*, 163.

241 **"I was hospitalized myself":** Cretyl Mills to RC, April 8, 1962. CF, Series 13.9, Box 68, Folder 818.

242 **"The end came":** Clara W. Masland to JS, 1963. APS/JS.

242 **"the final proof":** APS/JS to Cretyl Mills, December 14, 1964.

243 **"Despite the contentions":** R. Hare, "The Transmission of Respiratory Infections," *Proc R Soc Med* 57 (1964): 221–30.

245 **"The future appears bright":** S. Waksman, *The Conquest of Tuberculosis* (Berkeley: University of California Press, 1964), 214.

CHAPTER 11: THE BROTHERS RILEY

247 **"I shared Wells' conviction":** E. Riley, "The Role of Ventilation in the Spread of Measles in an Elementary School," *Ann NY Acad Sci* 353 (1980): 25–34.

248 **"I have rarely":** A. Langmuir, "The Territory of Epidemiology: Pentimento," *J Infect Dis* 155 (1987): 349–58.

249 **"In my opinion":** R. Riley, "Rising Tuberculosis Rate in Baltimore City," *Am Rev Respir Dis* 113 (1976): 577–78.

250 **"This is an original contribution":** RR to John P. Fox, September 4, 1977. RRP, Box 1.

250 **"Airborne Spread of Measles":** E. Riley, G. Murphy, and R. Riley, "Airborne Spread of Measles in a Suburban Elementary School," *Am J Epidemiol* 107 (1978): 421–32.

250 **"This is something":** RR to ER, June 3, 1977. RRP, Box 1.

251 **"rather egregious blunder":** A. Langmuir, "Changing Concepts of Airborne Infection of Acute Contagious Diseases," *Ann NY Acad Sci* 353 (1980): 35–44.

251 **"sinned grievously":** Langmuir, "Territory of Epidemiology."

251 **"I am happy":** ER to RR, July 26, 1993. RRP, Box 6.

251 **"carry on the gospel":** ER to RR, June 20, 1993. RRP, Box 6.

252 **"These are the grubby kinds":** RR to Don Smith, May 22, 1992. RRP, Box 6.

252 **"The construction of tighter buildings":** R. Riley, "Ultraviolet Air Disinfection for Control of Respiratory Contagion," in *Architectural Design and Indoor Microbial Pollution*, ed. R. Kundsin (Oxford: Oxford University Press, 1988), 193.

252 **"The medical profession":** R. Riley, "Airborne Contagion: Historical Background," *Ann NY Acad Sci* 353 (1980): 3–9.

254 **"the gay plague":** J. Eilperin, "How Attitudes Toward AIDS Have Changed, in the White House and Beyond," *Washington Post*, December 4, 2013.

255 **"If just one case":** R. Riley and E. Nardell, "Clearing the Air," *Am Rev Respir Dis* 139 (1989): 1286–94.

255 **"He is courageous":** RR to L. Jack Failing, December 2, 1993. RRP, Box 6.

257 **"Unfortunately, I've witnessed":** Mark Nicas to RR, June 29, 1993. RRP, Box 6.

259 **"Thus ended the career":** R. Riley, "What Nobody Needs to Know."

259 **"conditional recommendation":** World Health Organization, "Guidelines on Tuberculosis Infection Prevention and Control: 2019 Update" (Geneva: World Health Organization, 2019), xvi.

260 **"airborne precautions":** J. S. Garner, "Guideline for Isolation Precautions in Hospitals. Part I. Evolution of Isolation Practices, Hospital Infection Control Practices Advisory Committee," *Am J Infect Control* 24 (1996): 24–31.

260 **"may remain suspended":** J. S. Garner, "Guideline for Isolation Precautions in Hospitals. The Hospital Infection Control Practices Advisory Committee," *Infect Control Hosp Epidemiol* 17 (1996): 53–80.
260 **"generally travel only short distances":** Garner, "Guideline for Isolation Precautions."

CHAPTER 12: SACKS OF SUGAR, VIALS OF POWDER

261 **"It is hard to imagine":** J. Lederberg, "Pandemic as a Natural Evolutionary Phenomenon," *Soc Res* 55 (1988): 343–59.
261 **"suicidal policies of biological warfare":** J. Lederberg, "Biological Goal: Human Welfare," *New York Times*, January 12, 1970.
262 **"The idea of a totally new infectious agent":** A. Fauci, *On Call* (New York: Viking, 2024), 36.
262 **"There is no medical authority":** E. Thomas, "The New Untouchables," *Time*, September 23, 1985.
262 **"lead the way":** The New School Center for Public Scholarship, "In Time of Plague: The History and Social Consequences of Lethal Epidemic Disease," 2016.
263 **"Our principal competitors":** Lederberg, "Pandemic."
264 **"Interesting, he thought":** M. Crichton, *The Andromeda Strain* (New York: Knopf, 1969), 127.
264 **"you might find ways":** Letter from Joshua Lederberg to D. D. Coradimi (?), 1973, https://mhm.nlm.nih.gov/spotlight/bb/catalog/nlm:nlmuid-101584906X18804-doc.
264 **"I think that we can":** R. Weiss, "The Viral Advantage," *Science News*, September 23, 1989.
265 **"certainly induced a lot":** Institute of Medicine, *Microbial Evolution and Co-Adaptation* (Washington, DC: National Academies Press, 2009), 85.
266 **"I can tell you":** Weiss, "Viral Advantage."
268 **"In principle, biological weapons":** US Congress Office of Technology Assessment, *Proliferation of Weapons of Mass Destruction: Assessing the Risks*, OTA-ISC-559 (USGPO, 1993), 52.
269 **"Crisis in the Hot Zone":** R. Preston, "Crisis in the Hot Zone," *New Yorker*, October 18, 1992.
269 **"a kind of airborne Ebola flu":** R. Preston, *The Hot Zone: A Terrifying True Story* (New York: Anchor, 1994).
269 **"The single biggest threat":** W. Petersen, dir., *Outbreak* (Los Angeles: Arnold Kopelson Productions, 1995).
270 **"the dangerous link":** S. Wright, "Terrorists and Biological Weapons: Forging the Linkage in the Clinton Administration," *Politics Life Sci* 25 (2007): 57–115.
271 **"can do a lot of damage":** Quoted in S. Coll, *The Achilles Trap* (New York: Penguin Press, 2024), 394.
271 **"The likelihood of there being":** Wright, "Terrorists and Biological Weapons."
272 **"I don't know how many times":** Wright, "Terrorists and Biological Weapons."
272 **"Everything I heard":** B. Clinton, *My Life* (New York: Knopf, 2004), 789.
272 **"to aid our preparedness":** W. Clinton, *Public Papers of the Presidents of the United States: William J. Clinton, 1999, Book 1* (USGPO, 1999), 828.
272 **"the first-ever civilian stockpile":** Clinton, *Public Papers of Presidents*, 85.
273 **"They just threw":** K. Dilanian et al., "From Clinton to Trump, 20 Years of Boom and Mostly Bust in Prepping for Pandemics," NBC News, April 13, 2020.
274 **"Smallpox could bring":** G. Gordon and S. Cruz, "Biological Warfare May Be Next, Experts Say," *Star Tribune* (Minneapolis, MN), September 18, 2001.
275 **"You can not stop us":** J. Guillemin, *American Anthrax* (New York: Times Books, 2011), 79.
275 **"I was struck":** G. Bush, *Decision Points* (New York: Crown, 2010), 157.

275 **"It appears to have foreshadowed":** US Senate Committee on Armed Services, *The Dark Winter Scenario and Bioterrorism* (USGPO, 2002).
275 **"We have copies":** R. Graysmith, *Amerithrax: The Hunt for the Anthrax Killer* (New York: Berkley Books, 2003), 79. It would later come to light that al Qaeda leader Ayman al-Zawahiri had read Theodor Rosebury's *Peace or Pestilence.* See A. Cullison, "Inside Al-Qaeda's Hard Drive," *Atlantic*, September 2004.
276 **"America's war on terror":** Graysmith, *Amerithrax.*
276 **"Less than a teaspoon":** C. Powell, "A Policy of Evasion and Deception," *Washington Post*, February 5, 2003.
277 **"We believed it":** K. DeYoung, *Soldier: The Life of Colin Powell* (New York: Knopf, 2006), 493.
278 **"Put simply":** Bipartisan Commission on Biodefense, *The National Blueprint for Biodefense* (Washington, DC: Bipartisan Commission on Biodefense, 2024), 10.
278 **"The American people":** D. Koplow, "Losing the War on Bioterrorism," *Georgetown Security Law Commentary*, October 6, 2008.

CHAPTER 13: SEA, LAND, FIRE, CLOUDS

281 **Comtois and Isard:** P. Comtois, "A Fifteen Years History at the Dawn of a New Millennium," *Aerobiologia* 15 (1999): iii; P. Comtois and S. Isard, "Aerobiology: Coming of Age in a New Millennium," *Aerobiologia* 15 (1999): 259–66.
289 **"There were so many":** Athenaeus, *The Learned Banqueters*, ed. and trans. S. D. Olson (Cambridge, MA: Harvard University Press, 2010).
293 **"Clouds are environments":** P. Amato, M. Joly, L. Besaury, et al., "Active Microorganisms Thrive Among Extremely Diverse Communities in Cloud Water," *PLoS One* 12 (2017): e0182869.

CHAPTER 14: WE'RE ALL GOING TO GET IT

301 **"It is beginning":** K. Bradsher, "A Deadly Virus on Its Mind, Hong Kong Covers Its Face," *New York Times*, March 31, 2003.
301 **"Using it to cover":** M. Zhang, "Mass Masking Without Mandates—The Role of Gender in Mask Use in China," *N Engl J Med* 389 (2023): 874–75.
304 **"We're not talking":** "SARS: How One Man Infected 324 Others," *Sydney Morning Herald*, April 18, 2003.
306 **"If we don't have a vaccine":** K. Crowe, "The Mysteries of Testing, Treatment and Transmission," CBC News, April 25, 2003.
306 **"There is compelling evidence":** J. Conly, "Personal Protective Equipment for Preventing Respiratory Infections: What Have We Really Learned?" *CMAJ* 175 (2006): 263–64.
306 **"Part of the heated debate":** A. Campbell, "Spring of Fear: The SARS Commission Final Report" (Toronto: SARS Commission, 2006), 11.
307 **"It was eerie":** L. Altman, "The Doctor's World; Behind the Mask, the Fear of SARS," *New York Times*, June 24, 2003.
307 **"No . . . Canadians":** "Bill Schorr Cartoon for 04/24/2003," Cagle Cartoons.
307 **"As April turned":** K. Greenfeld, "When SARS Ended," *New Yorker*, April 17, 2020.
308 **"We were lucky":** World Health Organization, "SARS: How a Global Epidemic Was Stopped," *Bull World Health Organ* 85 (2007): 324.
309 **"The main mode":** World Health Organization, "SARS."
311 **"less than basic":** L. Morawska, "Droplet Fate in Indoor Environments, or Can We Prevent the Spread of Infection?" *Indoor Air* 16 (2006): 335–47.
313 **"If nature were not beautiful":** L. Bourouiba, "Numerical and Theoretical Study of Homogeneous Rotating Turbulence" (PhD diss., McGill University, 2008).
314 **"I am definitely":** "Nothing to Sneeze At," *Science Friday*, May 2, 2014.

314 **"seems overly simplified":** L. Bourouiba, "Turbulent Gas Clouds and Respiratory Pathogen Emissions: Potential Implications for Reducing Transmission of COVID-19," *JAMA* 323 (2020): 1837–38.

CHAPTER 15: A STATE OF PREPAREDNESS

315 **"Three years ago":** G. Bush, "President Outlines Pandemic Influenza Preparations and Response," White House, November 1, 2005.

316 **"You've got to read this":** M. Mosk, "George W. Bush in 2005: 'If We Wait for a Pandemic to Appear, It Will Be Too Late to Prepare,'" ABC News, April 5, 2020.

317 **"The amount of direct scientific":** "National Strategy for Pandemic Influenza" (S4-3), Homeland Security Council, November 2005.

318 **"Recommending the use":** R. Tellier, "Review of Aerosol Transmission of Influenza A Virus," *Emerg Infect Dis* 12 (2006): 1657–62.

318 **"Effective aerosol transmission":** C. Lemieux et al., "Questioning Aerosol Transmission of Influenza," *Emerg Infect Dis* 13 (2007): 173–75.

324 **"What I knew":** B. Obama, *A Promised Land* (New York: Crown, 2020), 385.

324 **"a gaffe machine":** A. Saenz, "Joe Biden Believes He Is the 'Most Qualified Person in the Country to Be President,'" CNN, December 4, 2018.

325 **"I wouldn't go anywhere":** C. Lee and A. Parnes, "Biden Says Avoid Planes, Subways; Puts Out Clarifying Statement," *Politico*, April 30, 2009.

325 **"Debate continues":** Institute of Medicine, *Preparing for an Influenza Pandemic* (Washington, DC: National Academies Press, 2007), 46–47.

325 **"The vice president's advice":** D. Seifman, "Man Up, Biden!" *New York Post*, May 1, 2009.

326 **"We can't hasten":** H. Branswell, "Bringing Pandemic Vaccine to Flu Clinics Requires Testing," *Fort McMurray Today* (Alberta: Canadian Press), June 26, 2009.

330 **"It is surprising":** D. Snider et al., "Meeting Summary of the Workshop 'Approaches to Better Understand Human Influenza Transmission,'" Centers for Disease Control, https://web.archive.org/web/20111021150803/http://www.cdc.gov/influenzatransmissionworkshop2010/pdf/Influenza_Transmission_Workshop_Summary_508.pdf.

332 **"Research over the years":** W. Seto et al., "Infection Prevention and Control Measures for Acute Respiratory Infections in Healthcare Settings: An Update," *East Mediterr Health J* 19 (2013): S39–47.

333 **"We talked about":** M. Dulaney, "West Africa's Ebola Outbreak Was 'Catastrophic'— Here's the Story of How It Was Contained," ABC News (Australia), April 6, 2020.

333 **"We must consider":** M. Osterholm, "What We're Afraid to Say About Ebola," *New York Times*, September 11, 2014.

334 **"Tony, explain to me":** R. Wilson, *Epidemic: Ebola and the Global Scramble to Prevent the Next Killer Outbreak* (Washington, DC: Brookings Institution Press, 2018), 124.

334 **"rapid airborne transmission":** National Security Council, "Playbook for Early Response to High-Consequence Emerging Infectious Disease Threats and Biological Incidents" (Washington, DC: Executive Office of the President of the United States, 2016), 30.

335 **"really stupid":** N. Toosi et al., "Before Trump's Inauguration, a Warning: 'The Worst Influenza Pandemic Since 1918,'" *Politico*, March 16, 2020.

336 **"How does this community":** Linsey Marr, August 26, 2022. Personal communication with the author.

336 **"We may even be able to prevent":** L. Bourouiba, "How Diseases and Epidemics Move Through a Breath of Air," TEDMED, 2018.

336 **"What a machine like the Gesundheit":** "Dr. Fauci: A Universal Flu Vaccine Is Possible," *Matter of Fact with Soledad O'Brien*, February 10, 2018.

337 **Disease X:** World Health Organization, "2018 Annual Review of Diseases Prioritized Under the Research and Development Blueprint."

CHAPTER 16: DISEASE X

344 **"Something is definitely wrong":** D. Yang, *Wuhan: How the COVID-19 Outbreak in China Spiraled Out of Control* (Oxford: Oxford University Press, 2024), 53.

344 **"At the end of December":** Yang, *Wuhan*, 68.

344 **"So far":** "Wuhan Municipal Health Commission's Briefing on the Current Pneumonia Epidemic Situation in Our City," December 31, 2019.

345 **"that the virus":** US Senate Committee on Homeland Security & Governmental Affairs, "Historically Unprepared: Examination of the Federal Government's Pandemic Preparedness and Initial Covid-19 Response," 2022, 80.

347 **"I think it's barbaric":** K. Lawrence, *The World According to Trump* (Kansas City, MO: Andrews McMeel, 2005), 71.

347 **"We have it":** M. Belvedere, "Trump Says He Trusts China's Xi on Coronavirus and the US Has It 'Totally Under Control,'" CNBC, January 22, 2020.

348 **"Forget SARS":** Covid Crisis Group, *Lessons from the Covid War: An Investigative Report* (New York: PublicAffairs, 2023), 88.

348 **"We pretty much":** P. Bump, "What Trump Did About Coronavirus in February," *Washington Post*, April 20, 2020.

348 **"very tricky":** B. Woodward, *Rage* (New York: Simon & Schuster, 2020), xix.

350 **"First of all":** World Health Organization, Coronavirus Press Conference, February 11, 2020.

351 **"Covid-19 is transmitted":** World Health Organization, "Report of the WHO-China Joint Mission on Coronavirus Disease 2019 (Covid-19)," 2020.

352 **"Maintain at least":** World Health Organization, "Coronavirus Disease (Covid-19) Advice for the Public," 2020.

352 **"I don't think they know":** N. Greenfieldboyce, "WHO Reviews 'Current' Evidence on Coronavirus Transmission Through Air," National Public Radio, March 28, 2020.

354 **"I can't believe":** R. Bright, "Complaint of Prohibited Personnel Practice or Other Prohibited Activity (OMB No. 3255-0005)," US Office of Special Counsel, 2020.

355 **"Should you wear a mask":** M. Cramer and K. Sheikh, "Surgeon General Urges the Public to Stop Buying Face Masks," *New York Times*, February 29, 2020. As Cramer and Sheikh write, Mike Ryan echoed Redfield's comment at a WHO briefing. "The most important thing everyone can do is wash your hands, keep your hands away from your face and observe very precise hygiene," he said.

355 ***60 Minutes*:** B. McCandless Farmer, "Dr. Anthony Fauci Talks with Dr. Jon LaPook About COVID-19," CBS News, March 8, 2020.

356 **"Avoid the 'Three Cs'!":** Government of Japan, "Avoiding the Three Cs: A Key to Preventing the Spread of COVID-19," 2020.

359 **"consistent with aerosolization":** National Academies of Sciences, Engineering, and Medicine, *Rapid Expert Consultations on the COVID-19 Pandemic: March 14, 2020–April 8, 2020* (Washington, DC: National Academies Press, 2020), 38.

359 **"I wouldn't go":** R. Xia, "Beachgoers Beware: Virus May Be Lurking," *Los Angeles Times*, April 3, 2020.

360 **"There is a professor":** US Department of State, "March 31, 2020: Members of the Coronavirus Task Force Hold a Press Briefing."

361 **"*FACT: COVID19*":** https://twitter.com/WHO/status/1243972193169616898.

362 **"Dear Dr. Tedros":** L. Morawska, W. Bahnfleth, P. M. Bluyssen, et al., "Coronavirus Disease 2019 and Airborne Transmission: Science Rejected, Lives Lost. Can Society Do Better?" *Clin Infect Dis* 76 (2023): 1854–59.

363 **"A Choir Decided":** R. Read, "A Choir Decided to Go Ahead with Rehearsal. Now Dozens of Members Have COVID-19 and Two Are Dead," *Los Angeles Times*, March 29, 2020.

364 "Skagit County Public Health": Skagit County, "Declaration of Emergency."
364 "Nicki Hamilton": K. Stern, "Taking Flattening the Curve Seriously," *La Conner Weekly News* (Washington), March 25, 2020.
365 "The hardest thing": M. Valdes, "Siblings Find Closure a Year After COVID-19 Thrashed Choir," Associated Press, April 9, 2021.
365 "One can only imagine": K. Tingley, "How the Skagit Valley Chorale Learned to Sing Again Amid Covid," *New York Times Magazine*, April 8, 2021.

CHAPTER 17: HISTORY SET US UP

367 "UR WEARING": M. Richtel, "Frightened Doctors Face Off with Hospitals over Rules on Protective Gear," *New York Times*, March 31, 2020.
367 "I feel like": M. Schwirtz, "Nurses Die, Doctors Fall Sick and Panic Rises on Virus Front Lines," *New York Times*, March 30, 2020.
368 "The CDC is recommending": Trump White House Archives, "Remarks by President Trump, Vice President Pence, and Members of the Coronavirus Task Force in Press Briefing," April 3, 2020.
368 "The thinking is": "What Dr. Fauci Wants You to Know About Face Masks and Staying Home as Virus Spreads," *PBS NewsHour*, April 28, 2020.
368 "Americans wanted": D. Birx, *Silent Invasion: The Untold Story of the Trump Administration, Covid-19, and Preventing the Next Pandemic Before It's Too Late* (New York: HarperCollins, 2022), 181.
369 "equates wearing a mask": Q. Forgey, "'Fauci's a Disaster': Trump Attacks Health Officials in Fiery Campaign Call," *Politico*, October 19, 2020.
369 "I just don't want": D. Victor et al., "In His Own Words, Trump on the Coronavirus and Masks," *New York Times*, October 2, 2020.
369 "Based on the trend": L. Morawska and J. Cao, "Airborne Transmission of SARS-CoV-2: The World Should Face the Reality," *Environ Int* 139 (2020): 105730.
370 "The immensity of such": "An Incalculable Loss," *New York Times*, May 24, 2020.
370 "For society to resume": K. Prather et al., "Reducing Transmission of SARS-CoV-2," *Science* 368 (2020): 1422–24.
372 "Transmission of SARS-CoV-2": S. Miller et al., "Transmission of SARS-CoV-2 by Inhalation of Respiratory Aerosol in the Skagit Valley Chorale Superspreading Event," *Indoor Air* 31 (2021): 314–23.
373 "We are 100 percent": R. Read, "Scientists Challenge WHO on Viral Spread," *Los Angeles Times*, July 4, 2020.
373 "Especially in the last couple": A. Mandavilli, "Infected but Feeling Fine: The Unwitting Coronavirus Spreaders," *New York Times*, March 31, 2020.
373 "Short-range aerosol": World Health Organization, "Transmission of Sars-Cov-2: Implications for Infection Prevention Precautions," 2020.
373 "opinion pieces": J. Conly et al., "Use of Medical Face Masks Versus Particulate Respirators as a Component of Personal Protective Equipment for Health Care Workers in the Context of the COVID-19 Pandemic," *Antimicrob Resist Infect Control* 9 (2020): 126.
374 "This has resulted": Z. Chagla et al., "Re: It Is Time to Address Airborne Transmission of COVID-19," *Clin Infect Dis* 73 (2021): e3981–82.
376 "We have limited time": J. Allen and R. Corsi, "We Can—and Must—Reopen Schools. Here's How," *Washington Post*, July 27, 2020.
379 "I thought, this is insane": Fauci, *On Call*, 396.
379 "the most amazing": Z. Budryk, "Infectious Disease Expert Calls White House Advisers Herd Immunity Claims 'Pseudoscience,'" *The Hill*, October 18, 2020.
379 "Nobody knows": "Transcript: Donald Trump at the GOP Convention," *New York Times*, July 22, 2016.
379 According to Mark Meadows: M. Pengelly, "Trump Tested Positive for Covid Few

Days Before Biden Debate, Chief of Staff Says in New Book," *The Guardian*, December 1, 2021.

380 **"felt like fucking shit":** Fauci, *On Call*, 403.

380 **"the now-famous Skagit":** "Harvard Medical School Grand Rounds—Featuring Dr. Anthony S. Fauci," aired September 10, 2020.

381 **"We don't know yet":** "Skagit Valley Chorale 'Heralding Christmas' 2020 Online Concert," SVC Vocal Course, YouTube, December 2020.

382 **"Are you ready?":** K. Sullivan, "Biden Receives First Dose of Covid-19 Vaccine on Live Television," CNN, December 21, 2020.

383 **"Calling the virus":** Testimony of Linsey C. Marr, Subcommittee on Workforce Protections, Committee on Education and Labor, United States House of Representatives, March 11, 2021.

383 **"A rigid new standard":** American Hospital Association, Statement of the American Hospital Association for the Workforce Protections Subcommittee on the Committee on Education and Labor of the US House of Representatives, March 11, 2021.

384 **"low quality":** C. Heneghan et al., "SARS-CoV-2 and the Role of Airborne Transmission: A Systematic Review," F1000Research 10 (2022): 232.

385 **"If you are fully vaccinated":** E. Cohen and J. Bonifield, "People Vaccinated Against Covid-19 Can Go Without Masks Indoors and Outdoors, CDC Says," CNN, May 13, 2021.

386 **"America is headed":** The White House, "Remarks by President Biden on the Covid-19 Response and the Vaccination Program," August 23, 2021.

387 **"We choose":** B. Stone, "Singing Again: Skagit Valley Chorale Resumes In-Person Practices," *Skagit Valley Herald*, October 17, 2021.

387 **"Current evidence":** D. Lewis, "Why the WHO Took Two Years to Say COVID Is Airborne," *Nature* 604 (2022): 26–31.

388 **"Two years into this crisis":** S. Thrasher, "The Biden Administration Has Failed Its Covid Test," *The Guardian*, January 15, 2022.

CHAPTER 18: A MARK ON THE AIR

390 **"SARS-CoV-2 is an airborne virus":** J. Lazarus et al., "A Multinational Delphi Consensus to End the COVID-19 Public Health Threat," *Nature* 611 (2022): 332–45. A subsequent review concluded that masks generally reduce transmission of Covid. It was based on studies carried out before Omicron. L. Boulos et al., "Effectiveness of Face Masks for Reducing Transmission of SARS-CoV-2: A Rapid Systematic Review," *Philos Trans A Math Phys Eng Sci* 381 (2023): 20230133.

391 **"The trend of the times":** D. Yang, "China's Zero-COVID Campaign and the Body Politic," *Curr Hist* 121 (2022): 203–10.

391 **"There's been so many":** D. Kang, "Packed ICUs, Crowded Crematoriums: COVID Roils Chinese Towns," Associated Press, December 26, 2022.

395 **"I had never experienced":** T. Kawasaki and S. Naoe, "History of Kawasaki Disease," *Clin Exp Nephrol* 18 (2014): 301–4.

403 **"Nobody's seen an epidemic":** M. Lacey, "New Strain of Wheat Rust Appears in Africa," *New York Times*, September 9, 2005.

EPILOGUE: HAPPY BIRTHDAY, CHITA RIVERA

409 **"This is not a sprint":** A. Jarmanning and S. Brown, "Baker Bans Dining in Bars and Restaurants, Closes Schools, Forbids Gatherings of More Than 25," WBUR, March 15, 2020.

409 **"Should not air disinfection":** E. Nardell and R. Nathavitharana, "Airborne Spread of SARS-CoV-2 and a Potential Role for Air Disinfection," *JAMA* 324 (2020): 141–42.

selected sources

PROLOGUE: THAT'S WHERE IT IS

Covid Crisis Group. 2023. *Lessons from the Covid War: An Investigative Report*. New York: PublicAffairs.

Forbes, K. M., T. Sironen, and A. Plyusnin. 2018. "Hantavirus Maintenance and Transmission in Reservoir Host Populations." *Curr Opin Virol* 28: 1–6.

Stokholm, I., W. Puryear, K. Sawatzki, et al. 2021. "Emergence and Radiation of Distemper Viruses in Terrestrial and Marine Mammals." *Proc Biol Sci* 288: 20211969.

Van Bressem, M.-F., P. J. Duignan, A. Banyard, et al. 2014. "Cetacean Morbillivirus: Current Knowledge and Future Directions." *Viruses* 6: 5145–81.

CHAPTER I: THE FLOATING GERMS

Ackerknecht, E. H. 2009. "Anticontagionism Between 1821 and 1867." *Int J Epidemiol* 38(1): 7–21.

Ariatti, A., and P. Comtois. 1993. "Louis Pasteur: The First Experimental Aerobiologist." *Aerobiologia* 9(1): 5–14.

Beretta, M. 2003. "The Revival of Lucretian Atomism and Contagious Diseases During the Renaissance." *Med Secoli* 15(2): 129–54.

DeLacy, M. 2017. *Contagionism Catches On: Medical Ideology in Britain, 1730–1800*. New York: Palgrave Macmillan.

Egerton, F. N. 2012. "History of Ecological Sciences, Part 44: Phytopathology During the 1800s." *Bull Ecol Soc Am* 93(4): 303–39.

Geison, G. L. 1995. *The Private Science of Louis Pasteur*. Princeton, NJ: Princeton University Press.

Henderson, J. 2019. *Florence Under Siege: Surviving Plague in an Early Modern City*. New Haven, CT: Yale University Press.

Lecerf, J.-M. 2023. "*Louis Pasteur à Lille: De la Chimie à la Microbiologie*." *Médecine des Maladies Métaboliques* 17(3): 295–302.

Manchester, K. L. 2007. "Louis Pasteur, Fermentation, and a Rival." *S Afr J Sci* 103(9): 377–80.

Matta, C. 2010. "Spontaneous Generation and Disease Causation: Anton de Bary's Experiments with *Phytophthora infestans* and Late Blight of Potato." *J Hist Biol* 43: 459–91.

Nutton, V. 1990. "The Reception of Fracastoro's Theory of Contagion: The Seed That Fell Among Thorns?" *Osiris* 6: 196–234.

Robertson, L. A. 2022. "The Vanishing Link Between Animalcules and Disease Before the 19th Century." *FEMS Microbiol Lett* 369(1): fnac022.

Roll-Hansen, N. 1979. "Experimental Method and Spontaneous Generation: The Controversy Between Pasteur and Pouchet, 1859–64." *J Hist Med Allied Sci* 34: 273–92.

Ruisinger, M. M. 2020. "*Die Pestarztmaske im Deutschen Medizinhistorischen Museum Ingolstadt.*" *NTM* 28(2): 235–52.

Swanson, M. T. 2010. "Qusṭā ibn Lūqā." In *Christian-Muslim Relations 600–1500*. Edited by David Thomas. Boston: Brill.

Temkin, O. 1977. *The Double Face of Janus and Other Essays in the History of Medicine*. Baltimore, MD: Johns Hopkins University Press.
Weisberg, R. E., and B. Hansen. 2015. "Collaboration of Art and Science in Albert Edelfelt's Portrait of Louis Pasteur." *Bull Hist Med* 89(1): 59–91.
Williams, D. M., and R. Huxley. 1998. "Christian Gottfried Ehrenberg (1795–1876): The Man and His Legacy. An Introduction." *Linnean* 1: 1–13.
Zadoks, J. C. 1985. "Cereal Rusts, Dogs and Stars in Antiquity." *Cereal Rusts Bull* 13(1): 1–10.

CHAPTER 2: THE SANITARIANS

Baldwin, P. 1999. *Contagion and the State in Europe, 1830–1930*. Cambridge: Cambridge University Press.
Blevins, S. M., and M. S. Bronze. 2010. "Robert Koch and the 'Golden Age' of Bacteriology." *Int J Infect Dis* 14(9): e744–51.
Cassedy, J. H. 1962. *Charles V. Chapin and the Public Health Movement*. Cambridge, MA: Harvard University Press.
Duffy, J. 1990. *The Sanitarians: A History of American Public Health*. Urbana: University of Illinois Press.
Fairchild, A. L., D. Rosner, J. Colgrove, et al. 2010. "The EXODUS of Public Health: What History Can Tell Us About the Future." *Am J Public Health* 100(1): 54–63.
Fisher, R. B. 1977. *Joseph Lister, 1827–1912*. London: Macdonald and Jane's.
Howard-Jones, N. 1975. *The Scientific Background of the International Sanitary Conferences, 1851–1938*. Geneva: World Health Organization.
Hume, E. E. 1927. *Max von Pettenkofer*. New York: P. B. Hoeber.
Jones, S. D. 2010. *Death in a Small Package: A Short History of Anthrax*. Baltimore, MD: Johns Hopkins University Press.
Locher, W. G. 2007. "Max von Pettenkofer (1818–1901) as a Pioneer of Modern Hygiene and Preventive Medicine." *Environ Health Prev Med* 12(6): 238–45.
Morabia, A. 2024. *The Public Health Approach: Population Thinking from the Black Death to COVID-19* (Baltimore, MD: Johns Hopkins University Press).
Richmond, P. A. 1954. "American Attitudes Toward the Germ Theory of Disease (1860–1880)." *J Hist Med Allied Sci* 9(4): 428–54.
Rosen, G. 1963. Review of *Charles V. Chapin and the Public Health Movement*, by J. H. Cassedy. *American Journal of Public Health and the Nation's Health* 53(5): 844–45.
Schlich, T., and B. J. Strasser. 2022. "Making the Medical Mask: Surgery, Bacteriology, and the Control of Infection (1870s–1920s)." *Med Hist* 66(2): 116–34.

CHAPTER 3: A WATERMELON DOCTOR

Comtois, P. 1997. "Pierre Miquel: The First Professional Aerobiologist." *Aerobiologia* 13: 75–82.
MacPhail, T. 2023. *Allergic: Our Irritated Bodies in a Changing World*. New York: Random House.
Meier, F. C. 1924. *Extension Work in Plant Pathology, 1923*. USDA.
——. 1930. *The Plant Disease Situation*. USDA.
——. 1931. *Dividends from Barberry Eradication*. USDA.
——. 1936a. "Collecting Microorganisms from Winds Above the Caribbean Sea." *Phytopathology* 26: 102.
——. 1936b. "Effects of Conditions in the Stratosphere on Spores of Fungi." *Natl Geogr Soc Stratosph Ser* 2: 152–53.
——, J. A. Stevenson, and V. K. Charles. 1933. "Spores in the Upper Air." *Phytopathology* 23: 23.
Stakman, E. C., and C. M. Christensen. 1946. "Aerobiology in Relation to Plant Disease." *Bot Rev* 12(4): 205–53.

——, A. W. Henry, G. C. Curran, and W. N. Christopher. 1923. "Spores in the Upper Air." *J Agric Res* 24: 599–606.

Taylor, G., and J. Walker. 1973. "Charles Harrison Blackley, 1820–1900." *Clin Allergy* 3: 103–8.

Trenholm, S. 2012. "Food Conservation During WWI: 'Food Will Win the War.'" Gilder Lehrman Institute of American History.

CHAPTER 4: ETHEREAL SPACE

DasSarma, P., A. Antunes, M. F. Simões, et al. 2020. "Earth's Stratosphere and Microbial Life." *Curr. Issues Mo. Biol* 38: 197–244.

McAdie, A. 1934. "The Discovery of the Stratosphere." *Bull. Am. Meteorol. Soc* 15(6): 174–77.

Rogers, L.A., and F. C. Meier. 1936. "An Apparatus for Collecting Bacteria in the Stratosphere." *J Bacteriol* 31: 27.

The National Geographic Society–U.S. Army Air Corps Stratosphere Flight of 1934 in the Balloon "Explorer." 1935. *Natl Geogr Soc Stratosph Ser* 1.

Von Ehrenfried, M. 2013. *Stratonauts: Pioneers Venturing into the Stratosphere*. New York: Springer.

CHAPTER 5: A PERFECT CYCLE

Buchbinder, L., M. Solowey, and M. Solotorovsky. 1938. "Alpha Hemolytic Streptococci of Air: Their Variant Forms, Origin and Numbers per Cubic Foot of Air in Several Types of Locations." *Am J Public Health Nation's Health* 28(1): 61–71.

Couch, C. 2015. "The Typhoid Buster." *MIT Technology Review*, December 22, 2015.

Fitzgerald, G. J. 2003. "From Prevention to Infection: Intramural Aerobiology, Biomedical Technology, and the Origins of Biological Warfare Research in the United States, 1910–1955." PhD diss., Carnegie Mellon University.

Flohr, C. 1996. "The Plague Fighter: Wu Lien-Teh and the Beginning of the Chinese Public Health System." *Ann Sci* 53(4): 361–80.

Honigsbaum, M. 2019. *The Pandemic Century*. New York: W. W. Norton & Co.

Markel, H., H. B. Lipman, J. A. Navarro, et al. 2007. "Nonpharmaceutical Interventions Implemented by US Cities During the 1918–1919 Influenza Pandemic." *JAMA* 298(6): 644–54.

Mason, C. F. 1917. *A Complete Handbook for the Sanitary Troops of the U.S. Army and Navy and National Guard and Naval Militia*. New York: William Wood & Co.

Pope, V. 1924. "Culture of Oysters Is Aimed at Saving an Old Industry." *New York Times*, July 27, 1924.

Riley, E., and Riley, R. n.d. "Biography of the Riley Family." RRP, Box 4.

Riley, R. 1937. "A Comparative Study of *Streptococcus viridans* Isolated from Room Air and from the Normal Nasopharynx." RRP, Box 1.

Sedgwick, W. T. 1891. "An Epidemic of Typhoid Fever in Lowell, Mass." *Boston Med Surg J* 124(18): 426–30.

——. 1902. *Principles of Sanitary Science and the Public Health*. New York: Macmillan.

Setzekorn, E. 2022. "Disease and Dissent: Progressives, Congress, and the WWI Army Training Camp Crisis." *J Gilded Age Progress Era* 21(2): 93–110.

Selected papers by William and Mildred Wells through 1937

Wells, W. F. 1913. "Use of Alum by Washington Water Works." *Municipal J* 35(5): 130–31.

——. 1916. "Artificial Purification of Oysters." *Public Health Rep* 31(28): 1848–52.

——. 1918. "An Improved Fermentation Tube Battery." *Am J Public Health* 8: 904–5.

——. 1920a. "Purification of Oysters as a Conservation Measure." *Am J Public Health* 10: 342–44.

——. 1920b. "Decline of the Northern Oyster Industry." *Conservationist*, February 1920.

——. 1922. *Studies in Oyster Culture*. New York: J. B. Lyon Co.
——. 1933. "Apparatus for Study of the Bacterial Behavior of Air." *Am J Public Health* 23: 58–59.
——. 1934a. "On Air-Borne Infection: Study II. Droplets and Droplet Nuclei." *Am J Epidemiol* 20(3): 611–18.
——. 1934b. "Viability of Bacteria in Air." *J Wash Acad Sci* 24: 276–77.
——. 1935. "Air-Borne Infection and Sanitary Air Control." *J Ind Hyg Toxicol* 17: 253–77.
——, and H. W. Brown. 1936. "Recovery of Influenza Virus Suspended in Air and Its Destruction by Ultraviolet Radiation." *Am J Epidemiol* 24(2): 407–13.
——, and G. M. Fair. 1935. "Viability of B. coli Exposed to Ultra-Violet Radiation in Air." *Science* 82(2125): 280–81.
——, and E. C. Riley. 1937. "An Investigation of the Bacterial Contamination of the Air of Textile Mills." *J Ind Hyg Toxicol* 19: 513–61.
——, and W. R. Stone. 1934. "On Air-Borne Infection: Study III. Viability of Droplet Nuclei Infection." *Am J Epidemiol* 20(3): 619–27.
——, and M. W. Wells. 1936. "Air-Borne Infection." *JAMA* 107(21): 1698–1703 and 107(22): 1805–9.

CHAPTER 6: THE SCATTERED WORKERS

Gregory, P. H. 1961. *The Microbiology of the Atmosphere*. London: Leonard Hill.
Moulton, F., ed. 1942. *Aerobiology*. Washington, DC: American Association for the Advancement of Science.
Warburton, C. W. 1938. "Fred C. Meier Lost on Hawaii Clipper." *Ext Serv Rev* 9(10): 157.

CHAPTER 7: WAR AT HOME

Edgerton, H. E., and J. R. Killian Jr. 1979. *Moments of Vision: The Stroboscopic Revolution in Photography*. Cambridge, MA: MIT Press.
Hitchens, A. P. 1940. "Experience of the Army and Civilian Conservation Corps in Handling Newly Mobilized Men." *Am J Public Health Nation's Health* 30(11): 1297–1301.
——. 1941. "The Control of Infectious Diseases in Rapidly Mobilized Troops." *Ann Intern Med* 15(2): 172–77.
Jennison, M. W. 1941. "The Dynamics of Sneezing—Studies by High-Speed Photography." *Sci Monthly* 52: 24–33.
Langmuir, A. D., H. S. Ingraham, A. D. Brandt, et al. 1950. "Progress in the Control of Air-Borne Infections." *Am J Public Health Nation's Health* 40(5 Pt 2): 82–88.
Robertson, O. H. 1943. "Air-Borne Infection." *Science* 97(2527): 495–502.
——, M. Hamburger, C. G. Loosli, et al. 1944. "A Study of the Nature and Control of Air-Borne Infection in Army Camps." *JAMA* 126: 993–1000.
Schultz, M. G., and W. Schaffner. 2015. "Alexander Duncan Langmuir." *Emerg Infect Dis* 21(9): 1635–57.
Steinbacher, S. 2010. "The Concentration and Extermination Camps of the Nazi Regime." In *The Routledge History of the Holocaust*. Edited by J. C. Friedman. London: Routledge.
Turner, C. E., M. W. Jennison, and H. E. Edgerton. 1941. "Public Health Applications of High-Speed Photography." *Am J Public Health Nation's Health* 31(4): 319–24.
Weindling, P. 2000. *Epidemics and Genocide in Eastern Europe, 1890–1945*. Oxford: Oxford University Press.
Wheeler, S. M., H. S. Ingraham, A. Hollaender, et al. 1945. "Ultra-Violet Light Control of Air-Borne Infections in a Naval Training Center: Preliminary Report." *Am J Public Health* 35(5): 457–68.

Selected papers by William and Mildred Wells, 1938–1947

Wells, M. W. 1944. "The Seasonal Patterns of Measles and Chicken Pox." *Am J Epidemiol* 40(3): 297–317.

——. 1945. "Ventilation in the Spread of Chickenpox and Measles Within School Rooms." *JAMA* 129(3): 197–200.

Wells, W. F. 1940. "An Apparatus for the Study of Experimental Air-Borne Disease." *Science* 91(2355): 172–74.

——. 1942. "Air-Borne Infection." *Pennsylvania Gazette* 60(6).

——. 1943. "Air Disinfection in Day Schools." *Am J Public Health Nation's Health* 33(12): 1436–43.

——, and W. Henle. 1941. "Experimental Air-Borne Disease: Quantitative Inoculation by Inhalation of Influenza Virus." *Proc Soc Exp Biol Med* 48(1): 298–301.

——, and M. B. Lurie. 1941. "Experimental Air-Borne Disease: Quantitative Natural Respiratory Contagion of Tuberculosis." *Am J of Hygiene* 34: 21–40.

——, J. Stokes Jr., M. W. Wells, and T. S. Wilder. 1940. "Experiments in the Environmental Control of Epidemic Respiratory Infection." *Trans Stud Coll Physicians Phila* 4: 342–45.

——, M. W. Wells, and S. Mudd. 1939. "Infection of Air: Bacteriologic and Epidemiologic Factors." *Am J Public Health Nation's Health* 29(8): 863–80.

——, M. W. Wells, and T. S. Wilder. 1942. "The Environmental Control of Epidemic Contagion: I. An Epidemiologic Study of Radiant Disinfection of Air in Day Schools." *Am J Epidemiol* 35(1): 97–121.

CHAPTER 8: WINGS FOR DEATH

Carus, W. S. 2017. *A Short History of Biological Warfare.* Washington, DC: National Defense University Press.

Cochrane, R. C. 1947. *History of the Chemical Warfare Service in World War II: Biological Warfare Research in the United States.* Washington, DC: Office of Chief, Chemical Corps.

Eickhoff, T. C. 1996. "Airborne Disease: Including Chemical and Biological Warfare." *Am J Epidemiol* 144(8 Suppl): S39–46.

Ewin, T. 2020. "Modern Resonances of Imperial Germany's Biological-Warfare Sabotage Campaign, 1915–18." *Nonproliferation Rev* 27(4–6): 277–87.

Guillemin, J. 2006. *Biological Weapons: From the Invention of State-Sponsored Programs to Contemporary Bioterrorism.* New York: Columbia University Press.

Kabat, E. A. 1983. "Getting Started 50 Years Ago—Experiences, Perspectives, and Problems of the First 21 Years." *Ann Rev Immunol* 1: 1–32.

Poupard, J. A., and L. A. Miller. 1992. "History of Biological Warfare: Catapults to Capsomeres." *Ann N Y Acad Sci* 666: 9–20.

Reed, D. S., A. Nalca, and C. J. Roy. 2018. "Aerobiology: History, Development, and Programs." In *Medical Aspects of Biological Warfare.* Edited by Zygmunt F. Dembek, 855–68. Washington, DC: Walter Reed Army Medical Center.

Regis, E. 1999. *The Biology of Doom.* New York: Henry Holt & Co.

Rosebury, T. 1940. "Biological Research After a Century of Dentistry." *Science* 92(2386): 247–52.

——. 1942. "The Fuller Utilization of Scientific Resources for Total War." *Science* 96(2504): 571–75.

——. 1947. *Experimental Air-borne Infection.* Baltimore, MD: Williams & Wilkins Co.

——, and M. Karshan. 1931. "Studies, in the Rat, of Susceptibility to Dental Caries: I. Bacteriological and Nutritional Factors." *J Dent Res* 11(1): 121–35.

——, and L. M. Waugh. 1939. "Dental Caries Among Eskimos of the Kuskokwim Area of Alaska: I. Clinical and Bacteriologic Findings." *Am J Dis Child* 57(4): 871–93.

CHAPTER 9: A FINE FROZEN DAIQUIRI

Baker, N. 2020. *Baseless: My Search for Secrets in the Ruins of the Freedom of Information Act.* New York: Penguin Press.

Barenblatt, D. 2004. *A Plague upon Humanity: The Secret Genocide of Axis Japan's Germ Warfare Operation.* New York: HarperCollins.

Beedham, R. J., and C. H. Davies. 2020. "The UK Biological-Warfare Program: Dual-Use Contributions to the Field of Aerobiology." *Nonproliferation Rev* 27(4-6): 309–22.

Bernstein, B. J. 1987. "The Birth of the U.S. Biological-Warfare Program." *Scientific American* 256(6): 116–21.

——. 1988. "America's Biological Warfare Program in the Second World War." *J Strateg Stud* 11(3): 292–317.

Cole, L. A. 1988. *Clouds of Secrecy: The Army's Germ Warfare Tests over Populated Areas.* Totowa, NJ: Rowman & Littlefield.

Fearnley, L. 2010. "Epidemic Intelligence: Langmuir and the Birth of Disease Surveillance." *Behemoth* 3(3): 36–56.

Fee, E., and T. M. Brown. 2001. "Preemptive Biopreparedness: Can We Learn Anything from History?" *Am J Public Health* 91(5): 721–26.

Langmuir, A. D. 1951. "The Potentialities of Biological Warfare Against Man—An Epidemiological Appraisal." *Public Health Rep* 66: 387–99.

——, and J. M. Andrews. 1952. "Biological Warfare Defense." *Am J Public Health Nation's Health* 42: 235–38.

——, H. S. Ingraham, A. D. Brandt, et al. 1950. "Progress in the Control of Air-Borne Infections." *Am J Public Health Nation's Health* 40(5 Pt 2): 82–88.

Meselson, M., J. Guillemin, M. Hugh-Jones, et al. 1994. "The Sverdlovsk Anthrax Outbreak of 1979." *Science* 266(5188): 1202–8.

Port, K. L. 2014. *Deciphering the History of Japanese War Atrocities.* Durham, NC: Carolina Academic Press.

Rimmington, A. 2018. *Stalin's Secret Weapon.* Oxford: Oxford University Press.

Shalett, S. 1946. "U.S. Was Prepared to Combat Axis in Poison-Germ Warfare." *New York Times,* January 4, 1946.

Tucker, J. B. 2002. "A Farewell to Germs: The U.S. Renunciation of Biological and Toxin Warfare, 1969–70." *Int Secur* 27(1): 107–48.

Vogel, W. F. 2021. "'The Mighty Microbe Can Go to War': Scientists, Secrecy, and American Biological Weapons Research, 1941–1969." PhD diss., University of Minnesota.

Wheelis, M., L. Rózsa, and M. Dando. 2006. *Deadly Cultures: Biological Weapons Since 1945.* Cambridge, MA: Harvard University Press.

Wright, S. 1990. *Preventing a Biological Arms Race.* Cambridge, MA: MIT Press.

CHAPTER 10: LOCH RAVEN

Hare, R. 1943. "Memorandum on Possible Methods for the Prevention of Rheumatic Manifestations in the Armed Forces." *Can Med Assoc J* 48(2): 116–21.

——. 1945. "Aerial Infection." *Br Med J* 1(4393): 383.

Langmuir, A. D. 1964. "Airborne Infection: How Important for Public Health?" *Am J Public Health Nation's Health* 54(10): 1666–68.

Molteni, M. 2021. "The 60-Year-Old Scientific Screwup That Helped Covid Kill." *Wired,* May 13, 2021.

Randall, K., E. T. Ewing, L. C. Marr, et al. 2021. "How Did We Get Here: What Are Droplets and Aerosols and How Far Do They Go? A Historical Perspective on the Transmission of Respiratory Infectious Diseases." *Interface Focus* 11(6): 20210049.

Riley, R. L., C. C. Mills, F. O'Grady, et al. 1962. "Infectiousness of Air from a Tuberculosis Ward. Ultraviolet Irradiation of Infected Air." *Am Rev Respir Dis* 85(4): 511–25.

——, C. C. Mills, W. Nyka, et al. 1959. "Aerial Dissemination of Pulmonary Tuberculosis." *Am J Epidemiol* 70: 185–96.

Selected papers by William and Mildred Wells, 1948–1963

Riley, R. L., W. F. Wells, C. C. Mills, W. Nyka, and R. L. McLean. 1957. "Air Hygiene in Tuberculosis." *Am Rev Tuberc* 75(3): 420–31.
Wells, W. F. 1951. "Control of Respiratory Infection by Radiant Disinfection of Air." *Proceedings of 3rd Annual Meeting. The American Academy of Occupational Medicine*, 15–23.
——. 1953. "968. Airborne Contagium and Sanitary Ventilation." *VI Congresso Internazionale di Microbiologia, Roma 6-1- Septembre 1953. Riassunti delle Comunicazioni* 3: 263–64.
——. 1955. *Airborne Contagion and Air Hygiene: An Ecological Study of Droplet Infections.* Cambridge, MA: Harvard University Press.
——, H. L. Ratcliffe, and C. Crumb. 1948. "On the Mechanics of Droplet Nuclei Infection. II. Quantitative Experimental Air-Borne Tuberculosis in Rabbits." *Am J Hyg* 47(1): 11–28.

CHAPTER 11: THE BROTHERS RILEY

Abdelfatah, R., and R. Arablouei. 2020. "How One Woman Inspired the Design for the N95 Mask." National Public Radio, May 21, 2020.
Chowder, K. 1992. "How TB Survived Its Own Death to Confront Us Again." *Smithsonian* 23(8): 180–94.
Houk, V. N. 1980. "Spread of Tuberculosis via Recirculated Air in a Naval Vessel: The Byrd Study." *Ann NY Acad Sci* 353: 10–24.
Mateus, B. 2023. "The History and Science Behind Airborne Infections and the Use of Ultraviolet Irradiation for Disinfecting Indoor Air." World Socialist Web Site, February 10, 2023.
Mphaphlele, M., A. S. Dharmadhikari, P. A. Jensen, et al. 2015. "Institutional Tuberculosis Transmission. Controlled Trial of Upper Room Ultraviolet Air Disinfection: A Basis for New Dosing Guidelines." *Am J Respir Crit Care Med* 192(4): 477–84.
Nardell, E. A. 2015. "Transmission and Institutional Infection Control of Tuberculosis." *Cold Spring Harb Perspect Med* 6(2): a018192.
——, S. J. Bucher, P. W. Brickner, et al. 2008. "Safety of Upper-Room Ultraviolet Germicidal Air Disinfection for Room Occupants: Results from the Tuberculosis Ultraviolet Shelter Study." *Public Health Rep* 123(1): 52–60.
——, B. McInnis, and R. L. Riley. 1988. "Ultraviolet Air Disinfection to Reduce Tuberculosis Transmission in a Shelter for the Homeless—Rationale, Installation, and Preliminary Results." *Amer Rev Respir Dis* 137(Supplement): 257.
——, B. McInnis, B. Thomas, and S. Weidhaas. 1986. "Exogenous Reinfection with Tuberculosis in a Shelter for the Homeless." *N Engl J Med* 315(25): 1570–75.
Riley, E. n.d. "More biography 6-5-94." RRP, Box 4.
Riley, R. L. 1976. "Rising Tuberculosis Rate in Baltimore City." *Am Rev Respir Dis* 113(5): 577–78.
——. 1977. "Ultraviolet Air Disinfection for Protection Against Influenza." *Johns Hopkins Med J* 140(1): 25–27.
——. 1980. "Airborne Transmission of Chickenpox." *N Engl J Med* 303(5): 281–82.
——, and J. E. Kaufman. 1971. "Air Disinfection in Corridors by Upper Air Irradiation with Ultraviolet." *Arch Environ Health* 22(5): 551–53.
——, M. Knight, and G. Middlebrook. 1976. "Ultraviolet Susceptibility of BCG and Virulent Tubercle Bacilli." *Am J Respir Crit Care Med* 113(4): 413–18.
——, and E. A. Nardell. 1989. "Clearing the Air." *Am Rev Respir Dis* 139(5): 1286–94.

——, and S. Permutt. 1971. "Room Air Disinfection by Ultraviolet Irradiation of Upper Air." *Arch Environ Health* 22(2): 208–19.

Tsai, K., and C. Tsai. 2020. "Our Dad Invented the N95 Mask: Our Taiwanese American Story." Taiwanese American, September 22, 2020.

Wehrle, P. F., J. Posch, K. H. Richter, et al. 1970. "An Airborne Outbreak of Smallpox in a German Hospital and Its Significance with Respect to Other Recent Outbreaks in Europe." *Bull World Health Organ* 43(5): 669–79.

CHAPTER 12: SACKS OF SUGAR, VIALS OF POWDER

Block, S. M. 2001. "The Growing Threat of Biological Weapons." *Am Sci* 89(1): 28–37.

Braut-Hegghammer, M. 2020. "Cheater's Dilemma: Iraq, Weapons of Mass Destruction, and the Path to War." *Int Secur* 45(1): 51–89.

Duelfer, C. 2004. Comprehensive Report of the Special Advisor to the DCI on Iraq's WMD. Washington, DC: Central Intelligence Agency.

Institute of Medicine. 1992. *Emerging Infections*. Washington, DC: National Academies Press.

Jaax, N. K., P. B. Jährling, T. W. Geisbert, et al. 1995. "Transmission of Ebola Virus (Zaire Strain) to Uninfected Control Monkeys in a Biocontainment Laboratory." *Lancet* 346(8991–8992): 1669–71.

Lederberg, J. 1988. "Medical Science, Infectious Disease, and the Unity of Humankind." *JAMA* 260(5): 684–85.

——. 1997. "Infectious Disease and Biological Weapons: Prophylaxis and Mitigation." *JAMA* 278(5): 435–36.

Leitenberg, M. 1999. "Aum Shinrikyo's Efforts to Produce Biological Weapons: A Case Study in the Serial Propagation of Misinformation." *Terror. Political Violence* 11(4): 149–58.

——. 2009. "The Self-Fulfilling Prophecy of Bioterrorism." *Nonproliferation Rev* 16(1): 95–109.

Miller, Judith, S. Engelberg, and W. J. Broad. 2001. *Germs: Biological Weapons and America's Secret War.* New York: Simon & Schuster.

Morse, S. S. 1991. "Emerging Viruses: Defining the Rules for Viral Traffic." *Perspect Biol Med* 34(3): 387–409.

O'Toole, T. and T. V. Inglesby. 2001. "Epidemic Response Scenario: Decision Making in a Time of Plague." *Public Health Rep* 116 (Suppl 2): 92–103.

——, M. Mair, and T. V. Inglesby. 2002. "Shining Light on 'Dark Winter.'" *Clin Infect Dis* 34(7): 972–83.

Sapp, J. 2021. *Genes, Germs and Medicine: The Life of Joshua Lederberg*. Hackensack, NJ: World Scientific.

Smith, R. J. 1984. "New Army Biowarfare Lab Raises Concerns." *Science* 226(4679): 1176–78.

US Department of State. 2005. "Adherence to and Compliance with Arms Control, Nonproliferation, and Disarmament Agreements and Commitments."

Weyer, J., A. Grobbelaar, and L. Blumberg. 2015. "Ebola Virus Disease: History, Epidemiology and Outbreaks." *Curr Infect Dis Rep* 17(5): 480.

Zilinskas, R. A. 1997. "Iraq's Biological Weapons. The Past as Future?" *JAMA* 278(5): 418–24.

CHAPTER 13: SEA, LAND, FIRE, CLOUDS

Alsante, A. N., D. C. O. Thornton, and S. D. Brooks. 2021. "Ocean Aerobiology." *Front Microbiol* 12: 764178.

Amato, P., M. Joly, L. Besaury, et al. 2017. "Active Microorganisms Thrive Among Extremely Diverse Communities in Cloud Water." *PLoS One* 12(8): e0182869.

——, F. Mathonat, L. Nuñez Lopez, et al. 2023. "The Aeromicrobiome: The Selective and Dynamic Outer-Layer of the Earth's Microbiome." *Front Microbiol* 14: 1186847.

——, M. Ménager, M. Sancelme, et al. 2005. "Microbial Population in Cloud Water at the Puy de Dôme: Implications for the Chemistry of Clouds." *Atmos Environ* 39(22): 4143–53.

Deicke, L. 2022. "Mass Transfer at the Ocean-Atmosphere Interface: The Role of Wave Breaking, Droplets and Bubbles." *Annu Rev Fluid Mech* 54(1): 191–224.

Garcia-Pichel, F. 2023. "The Microbiology of Biological Soil Crusts." *Annu Rev Microbiol* 77: 149–71.

Golan, J. J., and A. Pringle. 2017. "Long-Distance Dispersal of Fungi." *Microbiol Spectr* 5(4).

Huang, J., H. Feng, V. A. Drake, et al. 2024. "Massive Seasonal High-Altitude Migrations of Nocturnal Insects Above the Agricultural Plains of East China." *Proc Natl Acad Sci* 121(18): e2317646121.

Khaled, A., M. Zhang, P. Amato, et al. 2021. "Biodegradation by Bacteria in Clouds: An Underestimated Sink for Some Organics in the Atmospheric Multiphase System." *Atmos Chem Phys* 21(4): 3123–41.

Khodadad, C. L., G. M. Wong, L. M. James, et al. 2017. "Stratosphere Conditions Inactivate Bacterial Endospores from a Mars Spacecraft Assembly Facility." *Astrobiology* 17(4): 337–50.

Kobziar, L. N., P. Lampman, A. Tohidi, et al. 2024. "Bacterial Emission Factors: A Foundation for the Terrestrial-Atmospheric Modeling of Bacteria Aerosolized by Wildland Fires." *Environ Sci Technol* 58(5): 2413–22.

——, and G. R. Thompson. 2020. "Wildfire Smoke, a Potential Infectious Agent." *Science* 370(6523): 1408–10.

——, D. Vuono, R. Moore, et al. 2022. "Wildland Fire Smoke Alters the Composition, Diversity, and Potential Atmospheric Function of Microbial Life in the Aerobiome." *ISME Commun* 2: 8.

Lang-Yona, N., J. M. Flores, R. Haviv, et al. 2022. "Terrestrial and Marine Influence on Atmospheric Bacterial Diversity over the North Atlantic and Pacific Oceans." *Commun Earth & Envir* 3(1): 121.

Lappan, R., J. Thakar, L. Molares Moncayo, et al. 2024. "The Atmosphere: A Transport Medium or an Active Microbial Ecosystem?" *ISME J.* 18(1):wrae092.

Michaud, J. M., L. R. Thompson, D. Kaul, et al. 2018. "Taxon-Specific Aerosolization of Bacteria and Viruses in an Experimental Ocean-Atmosphere Mesocosm." *Nat Commun* 9(1): 2017.

Morris, C. E., F. Conen, J. A. Huffman, et al. 2014. "Bioprecipitation: A Feedback Cycle Linking Earth History, Ecosystem Dynamics and Land Use Through Biological Ice Nucleators in the Atmosphere." *Glob Change Biol* 20(2): 341–51.

Prather, K. A., T. H. Bertram, V. H. Grassian, et al. 2013. "Bringing the Ocean into the Laboratory to Probe the Chemical Complexity of Sea Spray Aerosol." *Proc Natl Acad Sci* 110(19): 7550–55.

Probandt, D., T. Eickhorst, A. Ellrott, R. Amann, et al. 2018. "Microbial Life on a Sand Grain: From Bulk Sediment to Single Grains." *ISME J* 12(2): 623–33.

Šantl-Temkiv, T., P. Amato, E. O. Casamayor, et al. 2022. "Microbial Ecology of the Atmosphere." *FEMS Microbiol Rev* 46(4): 1–18.

Schmale, D. G., and S. D. Ross. 2015. "Highways in the Sky: Scales of Atmospheric Transport of Plant Pathogens." *Annu Rev Phytopathol* 53: 591–611.

Seager, S., J. J. Petkowski, P. Gao, et al. 2021. "The Venusian Lower Atmosphere Haze as a Depot for Desiccated Microbial Life." *Astrobiology* 21(10): 1206–23.

Timerman, D., and S. C. Barrett. 2021. "The Biomechanics of Pollen Release: New Perspectives on the Evolution of Wind Pollination in Angiosperms." *Biol Rev Camb Philos Soc* 96(5): 2146–63.

CHAPTER 14: WE'RE ALL GOING TO GET IT

Abraham, T. 2004. *Twenty-First Century Plague: The Story of SARS*. Baltimore, MD: Johns Hopkins University Press.

Baehr, P. 2008. "City Under Siege: Authoritarian Toleration, Mask Culture, and the SARS Crisis in Hong Kong." In *Networked Disease: Emerging Infections in the Global City*. Edited by S. H. Ali and R. Keil, 138–51. Oxford: Wiley-Blackwell.

Bourouiba, L. 2021a. "Fluid Dynamics of Respiratory Infectious Diseases." *Annu Rev Biomed Eng* 23: 547–77.

——. 2021b. "The Fluid Dynamics of Disease Transmission." *Annu Rev Fluid Mech* 53: 473–508.

——, E. Dehandschoewercker, and J. W. M. Bush. 2014. "Violent Expiratory Events: On Coughing and Sneezing." *J Fluid Mech* 745: 537–63.

Greenfeld, K. T. 2006. *China Syndrome: The True Story of the 21st Century's First Great Epidemic*. New York: HarperCollins.

Horii, M. 2014. "Why Do the Japanese Wear Masks? A Short Historical Review." *Electronic Journal of Contemporary Japanese Studies* 14(2): 1–14.

Hui, D. S. C., and A. Zumla. 2019. "Severe Acute Respiratory Syndrome: Historical, Epidemiologic, and Clinical Features." *Infect Dis Clin North Am* 33(4): 869–89.

Johnson, G. R., L. Morawska, Z. D. Ristovski, et al. 2011. "Modality of Human Expired Aerosol Size Distributions." *J Aerosol Sci* 42(12): 839–51.

Kleinman, A., and J. L. Watson, eds. 2005. *SARS in China: Prelude to Pandemic?* Stanford, CA: Stanford University Press.

Li, Y., X. Huang, I. T. S. Yu, et al. 2005. "Role of Air Distribution in SARS Transmission During the Largest Nosocomial Outbreak in Hong Kong." *Indoor Air* 15(2): 83–95.

Morawska, L., G. Buonanno, A. Mikszewski, and L. Stabile. 2022. "The Physics of Respiratory Particle Generation, Fate in the Air, and Inhalation." *Nat Rev Phys* 4: 723–34.

——, G. R. Johnson, Z. D. Ristovski, et al. 2009. "Size Distribution and Sites of Origin of Droplets Expelled from the Human Respiratory Tract During Expiratory Activities." *J Aerosol Sci* 40(3): 256–69.

Quammen, D. 2012. *Spillover*. New York: W. W. Norton & Co.

SARS Expert Committee. 2003. "SARS in Hong Kong: From Experience to Action." October 2, 2003.

Shadbolt, P. 2013. "SARS 10 Years On: How Dogged Detective Work Defeated an Epidemic." CNN, February 21, 2013.

Varia, M., S. Wilson, S. Sarwal, et al. 2003. "Investigation of a Nosocomial Outbreak of Severe Acute Respiratory Syndrome (SARS) in Toronto, Canada." *CMAJ* 169(4): 285–92.

Yu, I. T. S., Y. Li, T. W. Wong, et al. 2004. "Evidence of Airborne Transmission of the Severe Acute Respiratory Syndrome Virus." *N Engl J Med* 350(17): 1731–39.

Yu, I. T. S., T. W. Wong, Y. L. Chiu, et al. 2005. "Temporal-Spatial Analysis of Severe Acute Respiratory Syndrome Among Hospital Inpatients." *Clin Infect Dis* 40(9): 1237–43.

CHAPTER 15: A STATE OF PREPAREDNESS

Aiello, A. E., R. M. Coulborn, T. J. Aragon, et al. 2010. "Research Findings from Nonpharmaceutical Intervention Studies for Pandemic Influenza and Current Gaps in the Research." *Am J Infect Control* 38(4): 251–58.

Centers for Disease Control. 2010a. "Interim Guidance on Infection Control Measures for 2009 H1N1 Influenza in Healthcare Settings, Including Protection of Healthcare Personnel."

——. 2010b. "The 2009 H1N1 Pandemic: Summary Highlights, April 2009–April 2010." June 16, 2010.

Cowling, B. J., R. O. P. Fung, C. K. Y. Cheng, et al. 2008. "Preliminary Findings of a Randomized Trial of Non-Pharmaceutical Interventions to Prevent Influenza Transmission in Households." *PLoS One* 3(5): e2101.

Frontz, A. J. 2023. "The Strategic National Stockpile Was Not Positioned to Respond Effectively to the COVID-19 Pandemic." Report No. A-04-20-02028. US Department of Health and Human Services, Office of Inspector General.

Holmes, E. C., G. Dudas, A. Rambaut, and K. G. Andersen. 2016. "The Evolution of Ebola Virus: Insights from the 2013–2016 Epidemic." *Nature* 538(7624): 193–200.

Johns Hopkins Center for Health Security. 2017. "The Characteristics of Pandemic Pathogens." November 9, 2017.

——. 2019a. "Preparedness for a High-Impact Respiratory Pathogen Pandemic." September 2019.

——. 2019b. "Tabletop Exercise: Event 201." October 18, 2019.

Loeb, M., N. Dafoe, J. Mahony, et al. 2009. "Surgical Mask vs. N95 Respirator for Preventing Influenza Among Health Care Workers: A Randomized Trial." *JAMA* 302(17): 1865–71.

McDevitt, J. J., P. Koutrakis, S. T. Ferguson, et al. 2013. "Development and Performance Evaluation of an Exhaled-Breath Bioaerosol Collector for Influenza Virus." *Aerosol Sci Technol* 47(4): 444–51.

Milton, D. K., M. P. Fabian, B. J. Cowling, et al. 2013. "Influenza Virus Aerosols in Human Exhaled Breath: Particle Size, Culturability, and Effect of Surgical Masks." *PLoS Pathog* 9(3): e1003205.

——, P. M. Glencross, and M. D. Walters. 2000. "Risk of Sick Leave Associated with Outdoor Air Supply Rate, Humidification, and Occupant Complaints." *Indoor Air* 10(4): 212–21.

Moser, M. R., T. R. Bender, H. S. Margolis, et al. 1979. "An Outbreak of Influenza Aboard a Commercial Airliner." *Am J Epidemiol* 110(1): 1–6.

Nguyen-Van-Tam, J. S., B. Killingley, J. Enstone, et al. 2020. "Minimal Transmission in an Influenza A (H3N2) Human Challenge-Transmission Model Within a Controlled Exposure Environment." *PLoS Pathog* 16(7): e1008704.

Nicoll, A., A. Ammon, A. Amato, et al. 2010. "Experience and Lessons from Surveillance and Studies of the 2009 Pandemic in Europe." *Public Health* 124(1): 14–23.

Pendergraft, M. A., P. Belda-Ferre, D. Petras, et al. 2023. "Bacterial and Chemical Evidence of Coastal Water Pollution from the Tijuana River in Sea Spray Aerosol." *Environ Sci Technol* 57(10): 4071–81.

Rosenthal, E. 2021. "Analysis: How the US Invested in the War on Terrorism at the Cost of Public Health." *KFF Health News*, March 29, 2021.

Roy, C. J., and D. K. Milton. 2004. "Airborne Transmission of Communicable Infection—The Elusive Pathway." *N Engl J Med* 350(17): 1710–12.

Smith, T. 2014. "'The Hot Zone' and the Mythos of Ebola." *ScienceBlogs*, October 20, 2014.

Stolberg, S. G. 2002. "Buckets for Bioterrorism, But Less for Catalog of Ills." *New York Times*, February 5, 2002.

US Senate Committee on Homeland Security & Governmental Affairs. 2022. "Historically Unprepared: Examination of the Federal Government's Pandemic Preparedness and Initial COVID-19 Response."

Van Kerkhove, M. D., S. Hirve, A. Koukounari, et al. 2013. "Estimating Age-Specific Cumulative Incidence for the 2009 Influenza Pandemic." *Influenza Other Respir Viruses* 7(5): 872–86.

Weber, L., L. Unger, M. R. Smith, et al. 2020. "Hollowed-Out Public Health System Faces More Cuts Amid Virus." *KFF Health News*, July 1, 2020.

World Health Organization. 2014. "Infection Prevention and Control of Epidemic- and Pandemic-Prone Acute Respiratory Infections in Health Care."
Yang W., S. Elankumaran, and L. C. Marr. 2011. "Concentrations and Size Distributions of Airborne Influenza A Viruses Measured Indoors at a Health Centre, a Day-Care Centre and on Aeroplanes." *J R Soc Interface* 8(61): 1176–84.
——, and L. C. Marr. 2011. "Dynamics of Airborne Influenza A Viruses Indoors and Dependence on Humidity." *PLoS One* 6(6): e21481.

CHAPTER 16: DISEASE X

Blackwell, T. 2021. "'It's Very Volatile': How a Scientific Debate Over COVID Spread Turned into an Online War." *National Post*, May 13, 2021.
Cheng, P., K. Luo, S. Xiao, et al. 2022. "Predominant Airborne Transmission and Insignificant Fomite Transmission of SARS-CoV-2 in a Two-Bus COVID-19 Outbreak Originating from the Same Pre-Symptomatic Index Case." *J Hazard Mater* 425: 128051.
Global Times Staff Reporters. 2020. "GT Investigates: Review of 40-Day Response." *Global Times*, December 16, 2020.
Hamner, L., P. Dubbel, I. Capron, et al. 2020. "High SARS-CoV-2 Attack Rate Following Exposure at a Choir Practice—Skagit County, Washington, March 2020." *Morb Mortal Wkly Rep* 69(19): 606–10.
Jefferson, T., C. J. Heneghan, E. Spencer, et al. 2022. "A Hierarchical Framework for Assessing Transmission Causality of Respiratory Viruses." *Viruses* 14(8): 1605.
Jimenez, J., L. Marr, K. Randall, et al. 2021. "Echoes Through Time: The Historical Origins of the Droplet Dogma and Its Role in the Misidentification of Airborne Respiratory Infection Transmission." SSRN.
——, L. Marr, K. Randall, et al. 2022. "What Were the Historical Reasons for the Resistance to Recognizing Airborne Transmission During the COVID-19 Pandemic?" *Indoor Air* 32(8): e13070.
Lai, J., K. K. Coleman, S. H. S. Tai, et al. 2023. "Exhaled Breath Aerosol Shedding of Highly Transmissible Versus Prior Severe Acute Respiratory Syndrome Coronavirus 2 Variants." *Clin Infect Dis* 76(5): 786–94.
Lewis, D. 2020. "Is the Coronavirus Airborne? Experts Can't Agree." *Nature* 580(7802): 175.
Li, Y., H. Qian, J. Hang, et al. 2021. "Probable Airborne Transmission of SARS-CoV-2 in a Poorly Ventilated Restaurant." *Build Environ* 196: 107788.
Lu, J., J. Gu, K. Li, et al. 2020. "COVID-19 Outbreak Associated with Air Conditioning in a Restaurant, Guangzhou, China, 2020." *Emerg Infect Dis* 26(11): 2791–93.
McKenna, M. 2022. "When Covid Came for Provincetown." *Wired*, June 14, 2022.
Oshitani, H. 2022. "COVID Lessons from Japan: Clear Messaging Is Key." *Nature* 605(7911): 589.
Ou, C., S. Hu, K. Luo, et al. 2022. "Insufficient Ventilation Led to a Probable Long-Range Airborne Transmission of SARS-CoV-2 on Two Buses." *Build Environ* 207: 108414.
Patranobis, S. 2020. "Doctor Who Treated First 7 Coronavirus Patients in Wuhan Now a Hero in China." *Hindustan Times*, February 2, 2020.
Qian, H., T. Miao, L. Liu, et al. 2021. "Indoor Transmission of SARS-CoV-2." *Indoor Air* 31(3): 639–45.
Reichert, F., O. Stier, A. Hartmann, et al. 2022. "Analysis of Two Choir Outbreaks Acting in Concert to Characterize Long-Range Transmission Risks Through SARS-CoV-2, Berlin, Germany, 2020." *PLoS One* 17(11): e0277699.
Santarpia, J. L., D. N. Rivera, V. Herrera, et al. 2020. "Aerosol and Surface Contamination of SARS-CoV-2 Observed in Quarantine and Isolation Care." *Sci Rep* 10(1): 12732.
Shuren, J., and T. Stenzel. 2021. "South Korea's Implementation of a COVID-19 National Testing Strategy." *Health Affairs*, May 25, 2021.

Tang, J. W., W. P. Bahnfleth, P. M. Bluyssen, et al. 2021. "Dismantling Myths on the Airborne Transmission of Severe Acute Respiratory Syndrome Coronavirus-2 (SARS-CoV-2)." *J Hosp Infect* 110: 89–96.

CHAPTER 17: HISTORY SET US UP

Greenhalgh, T., C. R. MacIntyre, M. G. Baker, et al. 2024. "Masks and Respirators for Prevention of Respiratory Infections: A State of the Science Review." *Clin Microbiol Rev*, 37(2):e 00124–23.

Michaels, D., E. A. Spieler, and G. R. Wagner. 2024. "U.S. Workers During the Covid-19 Pandemic: Uneven Risks, Inadequate Protections, and Predictable Consequences." *BMJ* 384: e076623.

Morawska, L., W. Bahnfleth, P. M. Bluyssen, et al. 2023. "Coronavirus Disease 2019 and Airborne Transmission: Science Rejected, Lives Lost. Can Society Do Better?" *Clin Infect Dis* 76(10): 1854–59.

——, and J. Cao. 2020. "Airborne Transmission of SARS-CoV-2: The World Should Face the Reality." *Environ Int* 139: 105730.

——, and D. K. Milton. 2020. "It Is Time to Address Airborne Transmission of Coronavirus Disease 2019 (COVID-19)." *Clin Infect Dis* 71(9): 2311–13.

Select Subcommittee on the Coronavirus Crisis. 2022a. "The Atlas Dogma: The Trump Administration's Embrace of a Dangerous and Discredited Herd Immunity via Mass Infection Strategy." June 21, 2022.

——. 2022b. "'Now to Get Rid of Those Pesky Health Departments!' How the Trump Administration Helped the Meatpacking Industry Block Pandemic Worker Protections." May 12, 2022.

——. 2022c. "'It Was Compromised': The Trump Administration's Unprecedented Campaign to Control CDC and Politicize Public Health During the Coronavirus Crisis." October 17, 2022.

Wright, L. 2021. *The Plague Year: America in the Time of Covid.* New York: Knopf.

Yorio, K. 2022. "Corsi-Rosenthal Boxes Help Clear the Air at Schools Across the Country." *School Library Journal*, February 4, 2022.

Zelner, J., R. Trangucci, R. Naraharisetti, et al. 2021. "Racial Disparities in Coronavirus Disease 2019 (COVID-19) Mortality Are Driven by Unequal Infection Risks." *Clin Infect Dis* 72(5): e88–95.

CHAPTER 18: A MARK ON THE AIR

Behzad, H., K. Mineta, and T. Gojobori. 2018. "Global Ramifications of Dust and Sandstorm Microbiota." *Genome Biol Evol* 10(8): 1970–87.

Björnham, O., R. Sigg, and J. Burman. 2020. "Multilevel Model for Airborne Transmission of Foot-and-Mouth Disease Applied to Swedish Livestock." *PLoS One* 15(5): e0232489.

Burney, J. A., L. L. DeHaan, C. Shimizu, et al. 2021. "Temporal Clustering of Kawasaki Disease Cases Around the World." *Sci Rep* 11(1): 22584.

Cissé, O., L. Ma, and J. Kovacs. 2024. "Retracing the Evolution of *Pneumocystis* Species, with a Focus on the Human Pathogen *Pneumocystis jirovecii*." *Microbiol Mol Biol Rev* 88(2): e00202–22.

Duplessis, S., C. Lorrain, B. Petre, et al. 2021. "Host Adaptation and Virulence in Heteroecious Rust Fungi." *Annu Rev Phytopathol* 59: 403–22.

Klompas, M., and C. Rhee. 2022. "Optimizing and Unifying Infection Control Precautions for Respiratory Viral Infections." *J Infect Dis* 226(2): 191–94.

Kormos, D., K. Lin, A. Pruden, and L. C. Marr. 2022. "Critical Review of Antibiotic Resistance Genes in the Atmosphere." *Environ Sci Process Impacts* 24(6): 870–83.

Liddicoat, C., H. Sydnor, C. Cando-Dumancela, et al. 2020. "Naturally-Diverse Airborne

Environmental Microbial Exposures Modulate the Gut Microbiome and May Provide Anxiolytic Benefits in Mice." *Sci Total Environ* 701: 134684.
Matthews, K., T. Cavagnaro, P. Weinstein, and J. Stanhope. 2024. "Health by Design; Optimising Our Urban Environmental Microbiomes for Human Health." *Environ Res* 257: 119226.
Mead, H. L., D. R. Kollath, M. M. Teixeira, et al. 2022. "Coccidioidomycosis in Northern Arizona: An Investigation of the Host, Pathogen, and Environment Using a Disease Triangle Approach." *mSphere* 7(5): e0035222.
Pepperell, C. S. 2022. "Evolution of Tuberculosis Pathogenesis." *Annu Rev Microbiol* 76: 661–80.
Pérez-Cobas, A. E., J. Rodríguez-Beltrán, F. Baquero, and T. M. Coque. 2023. "Ecology of the Respiratory Tract Microbiome." *Trends Microbiol* 31(9): 972–84.
Robinson, J. M., and M. F. Breed. 2023. "The Aerobiome-Health Axis: A Paradigm Shift in Bioaerosol Thinking." *Trends Microbiol* 31(7): 661–64.
———, M. F. Breed, and R. Beckett. 2024. "Probiotic Cities: Microbiome-Integrated Design for Healthy Urban Ecosystems." *Trends Biotechnol*, February 16, 2024.
Rossi, F., R. Péguilhan, N. Turgeon, et al. 2023. "Quantification of Antibiotic Resistance Genes (ARGs) in Clouds at a Mountain Site (Puy de Dôme, Central France)." *Sci Total Environ* 865: 161264.
Tang, J. W., R. Tellier, and Y. Li. 2022. "Hypothesis: All Respiratory Viruses (Including SARS-CoV-2) Are Aerosol-Transmitted." *Indoor Air* 32(1): e12937.
Xiao, H., Z. Wang, F. Liu, and J. M. Unger. 2023. "Excess All-Cause Mortality in China After Ending the Zero COVID Policy." *JAMA Netw Open* 6(8): e2330877.
Yang, D. L. 2022. "The Shanghai Lockdown and the Politics of Zero-Covid in China." In *Party Watch Annual Report 2022: Seeking Progress While Maintaining Stability*. Edited by M. Henry, 41–57. Washington, DC: Center for Advanced China Research.
Zhang, N., Y. Guo, B. J. Cowling, et al. 2023. "Explosive Household Spread of the SARS-CoV-2 Omicron Variant and Associated Risk Factors in China in Late 2022." SSRN.

EPILOGUE: HAPPY BIRTHDAY, CHITA RIVERA

Bender, E. 2022. "Safety Is in the Air." *Nature* 610(7933): S46–47.
Blatchley, E. R., D. J. Brenner, H. Claus, et al. 2023. "Far UV-C Radiation: An Emerging Tool for Pandemic Control." *Crit Rev Environ Sci Technol* 53(6): 733–53.
Lewis, D. 2023. "Indoor Air Is Full of Flu and COVID Viruses. Will Countries Clean It Up?" *Nature* 615(7951): 206–8.
Mark-Carew, M., G. Kang, S. Pampati, et al. 2023. "Ventilation Improvements Among K–12 Public School Districts—United States, August–December 2022." *Morb Mortal Wkly Rep* 72(14): 372–76.
Morawska, L., J. Allen, W. Bahnfleth, et al. 2024. "Mandating Indoor Air Quality for Public Buildings." *Science* 383 6690: 1418–20.
Nardell, E. A., D. Welch, R. Hashmi, et al. 2023. "Dynamic Breathing Zone Dose Monitoring of Room Occupants: A New Approach to Demonstrating Safe and Effective Germicidal UV (GUV) Air Disinfection." *Am J Respir Crit Care Med* 207: A1834.
Nardell, E. A., C. J. Roy, M. Barer, et al. 2023. "Remote Aerosol SARS-CoV-2 Transmission from Clinical COVID Patients to Rodent Sentinels." *Int J Tuberc Lung Dis* 27(11 Suppl 1): S634.

index

about the author

CARL ZIMMER writes the "Origins" column for the *New York Times* and has frequently contributed to the *Atlantic*, *National Geographic*, *Time*, and *Scientific American*. His journalism has earned numerous awards, including ones from the American Association for the Advancement of Science and the National Academies of Sciences, Engineering, and Medicine. Zimmer is professor adjunct at Yale, where he teaches writing. He is the author of fourteen books about science, including *Life's Edge*.